Fundamentals of Automotive Electronics

ion
yagement

YLISH CAMPUS

Fundamentals of Automotive Electronics

V.A.W. Hillier TEng., FIMI, MIRTE

Principal Lecturer in Automobile Engineering, Croydon College

STANLEY THORNES (PUBLISHERS) LTD

Originally published in 1987 by Hutchinson Education
Reprinted 1988

Reprinted in 1989 by
Stanley Thornes (Publishers) Ltd
Ellenborough House
Wellington Street
CHELTENHAM GL50 1YD

Reprinted 1992
Reprinted 1993 (twice)
Reprinted 1994

British Library Cataloguing in Publication Data

Hillier, V. A. W.
    Fundamentals of automotive electronics
    1. Motor vehicles—Electronics equipment
    I. Title
    629.2'7   TL272.5
    ISBN 0 7487 0261 X

Library of Congress Cataloguing-in-Publication Data

Hillier, V. A. W. (Victor Albert Walter)
    Fundamentals of automotive electronics

    Includes index.
    1. Automobiles—Electronic equipment
    I. Title
TL272.5.H47   1987   629.2'549   87—603
ISBN 0 7487 0261 X

Typeset in 10 on 12pt VIP Times by
DP Media Ltd, Hitchin, Hertfordshire

Printed and bound in Great Britain at
The Bath Press, Avon.

Contents

Preface

In 1982 vehicle manufacturers spent an average of £25 per car on electronic equipment. By the year 1990 this figure is expected to grow to about £175 per car. This 30% growth rate will create considerable problems for the motor vehicle repair industry unless skilled personnel are available in sufficient numbers to enable faults to be diagnosed accurately and repairs to be carried out economically and efficiently.

Other industries have been able to meet high-tech changes by using technicians who specialize solely in electronics. This is not possible in the motor vehicle repair industry because the electronic applications in this field also embrace many mechanical systems. In view of this, personnel engaged in repair activities must have an electro-mechanical background based on a sound foundation knowledge of electrical and mechanical principles. The combination of this book with my other books on mechanical systems is intended to assist the modern high-tech specialist to acquire this knowledge.

One objective of this book is to provide information to support the study of electrical and electronic systems. The two aspects of electro-technology overlap considerably, so fundamentals of both aspects are covered to enable the reader to develop a knowledge sufficient to permit assimilation of electronic developments expected in the near future.

The broad treatment of topics is intended to enable the material to be used by students preparing for various examinations set by the City & Guilds of London Institute, BTEC and other examining bodies. In addition it is intended that the book will provide valuable support material for automotive electronic courses and personal study activities.

Treatment of the subject is varied to suit the requirements of the reader and with this in mind it is anticipated that the appropriate sections will be read to suit the course of study. Where applicable a cross-reference is made in the book to minimize repetition of fundamental principles.

The electronic world has created many mysteries which are submerged in a sea of technical jargon. Interpretation of this modern-day language is included to enable the subject to be simplified within the bounds of technical accuracy. This should enable the reader to lay a foundation on to which current developments can be built. This basic treatment applies to practical activities such as fault diagnosis and system maintenance. Although an outline is given, no claim is made that it covers every model or that it represents the latest developments. Rapid changes in electronics means that vehicle manufacturers must be responsible for this service.

Apart from cases where extra simplicity is needed, the graphical symbols used for diagrams conform to the recommendations made by the British Standards Institution in BS3939. This guide was revised recently to conform to the recommendations made by the International Electrotechnical Commission (IEC).

V.A.W.H.
March 1986

Acknowledgements

During the time that this book was being researched, considerable assistance was given by many friends in various companies. In particular I wish to thank and acknowledge the technical assistance and material supplied by:

AC Delco
Austin-Rover Group
BMW (GB) Ltd
Robert Bosch Ltd
British Standards Institution
Champion Sparking Plug Co Ltd
Chloride Automotive Batteries Ltd
Citroen Cars Ltd
Honda (UK) Ltd
Ford Motor Co Ltd
FKI Crypton Ltd
Lucas Industries
Toyota (GB) Ltd
VAG (United Kingdom) Ltd
Wipac Group

1 Electrical principles

1.1 Circuit fundamentals

Just like any other subject, electrical and electronic systems obey certain basic rules and laws. To avoid constant repetition throughout the book, the common features have been brought together in this section. This is intended to enable the main principles to be established before they are applied to the individual circuits and components.

A reader who has studied basic electrical theory at school should find that this section revises the important facts associated with the world of electricity.

Electric charge
The word 'electric' is derived from a Greek word *ēlektron* which means amber.

As long ago as 600BC it was known that when this yellow translucent fossil resin was rubbed with a silk cloth, the amber was then able to attract to it particles of dust. Today, a similar effect can be produced after passing a plastics comb through your hair – the comb will attract small pieces of paper.

Experiments performed around the year 1600 indicated that other materials could be 'charged with electricity'. Also, observations showed that when small particles, such as two pith balls, were charged from an electrified material, they would react in different ways when placed close together. When they were each charged from a similar source they would repel each other, but when one was charged from a resinous material and the other ball was charged from a vitreous substance (e.g. glass) the two charged particles would then be attracted together.

In 1747 Benjamin Franklin introduced the names *positive* and *negative* to distinguish between the two electrical charges. Vitreous materials were said to acquire a positive charge and resinous substances a negative charge.

He thought the electrical flow passed from a high potential positive to a lower potential negative source. The choice was unfortunate because a later discovery proved that small electrical charges called electrons moved from negative to positive. By this time many electrical rules and laws had been established, so today the *conventional flow* from positive to negative is applied whenever basic laws have to be used. Actual flow from negative to positive is called *electron flow*.

The form of energy produced by rubbing a material is called *electrostatics*. In a motor vehicle this type of charge is generated at places where friction occurs, e.g. clutch and brake. These charges can build-up in a vehicle, especially on a hot dry day, so the occupants can sometimes detect these charges as they conduct the electric charge to the ground.

Today electric energy is obtained normally from a battery or generator and this form of energy is called *current electricity* to distinguish it from electrostatics. Although the energy is obtained from a different source, the behaviour of both forms is similar.

Electrons and protons
Objects around us consist of various formations of atoms bonded together to form molecules. A molecule of water consists of two atoms of hydrogen and one atom of oxygen; this combination is indicated by the chemical symbol H_2O.

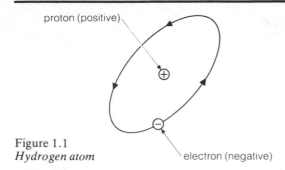

Figure 1.1
Hydrogen atom

Examination of an atom under a powerful microscope shows that it is built up of very minute particles. A hydrogen atom (Figure 1.1) consists of a *proton*, at the nucleus (centre) and one *electron*, which orbits around the proton at high speed. The electron is the lightest particle known at this time; its mass is only about 0.0005 of the mass of the hydrogen atom or stated another way, 9×10^{-28} gram.

Each electron carries a small negative charge and this balances the positive charge held by the proton. The two opposite charges make the atom electrically neutral.

Other materials have different combinations of electrons and protons. A copper atom, Figure 1.2 has 29 electrons; these move in four different orbits around the atom's nucleus. This central region consists of protons and other particles called *neutrons* which have no electrical charge.

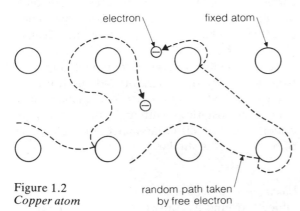

Figure 1.2
Copper atom

Conductors

Some of the electrons in metallic materials are not bonded tightly to their nucleus; instead they drift at random from one atom to another. Mater-ials that have a number of these free electrons make good conductors of electricity because little effort is needed to persuade the electrons to move through the tightly-packed atomic structure.

Copper is a very good conductor so this metal is often used as a material for a cable.

Insulators

Insulation materials have no loosely-bound electrons, so movement of electrons from one atom to the next is very difficult. No material is a perfect insulator because all materials will allow some electron movement if the force is large enough.

Attraction and repulsion

Electrical charges of the same polarity mutually repel one another, whereas charges of opposite polarity attract one another. This behaviour may be likened to that produced by two magnets. When two groups of electrons are brought together, a force exists which attempts to separate the two groups.

The magnitude of the force produced between two electrically charged bodies was studied by the French scientist Coulomb in 1775. To honour his work, the SI unit of electrical charge is called the *coulomb* (C).

This unit is the quantity of charge which passes a section of a conductor in one second when the current flowing is one ampere (A). (The ampere is defined at a later stage on page 5.)

Potential

The potential indicates the amount that a body is electrically charged.

When two bodies are equally charged with electrical energy of the same polarity, e.g. positive and positive, a force is needed to move the two bodies together. The work done in applying this force over a given distance is the *electrical potential*.

Electrical potential is similar to air pressure. Consider as an example a car tyre. To inflate the tyre, work has to be done and this work is stored in the tyre by virtue of its pressure. When you push against the tyre it will deflect and the amount of work needed to do this is calculated by multiplying the force by the distance moved. The amount of work done is an indication of the potential, i.e. a measure of the quantity of energy stored in the tyre.

Electrical potential is based on the 'work per unit of charge'. The *joule* is the unit of work and the coulomb the quantity of charge, so potential is expressed in *joules per coulomb*. To honour the early work of the Italian scientist Volta, the unit of potential is the *volt*.

1 volt = 1 joule per coulomb

Potential difference (p.d.)

The concept of potential difference can be visualized by extending the example of the tyre. Consider two tyres; one at low pressure and one at high pressure. If a hose joins the two tyres together, then air will flow from the high pressure tyre to the low pressure tyre until both pressures become equal. The rate of flow of energy will depend on the pressure difference or as applied to electricity, the potential difference.

P.d. is measured from some reference point and, when this is established, the potentials of the two points can be compared to determine the difference.

Using conventional flow (+ to −), electrical energy is considered to move from a point of high potential to a point of lower potential. If a point A has a higher potential than a point B then work is required to move a positive charge from B to A.

The potential difference between the two points A and B is one volt if the work done in taking one coulomb of positive charge from B to A is one joule.

From this definition it follows that the charge (in coulombs) on an object is proportional to its potential difference (in volts).

Electromotive force (e.m.f.)

A battery and generator are both capable of producing a difference in potential between two points. The electrical force that gives this increase in p.d. at the source is called the electromotive force. The unit of e.m.f. is the volt.

The terminals of a battery and generator are called positive (+) and negative (−) and these relate to the higher potential and lower potential respectively.

Electrical symbols

Diagrams of electrical systems are shown in pictorial or theoretical form; in the latter, graphical symbols are used to indicate the various items that make-up the circuit.

Many separate parts are used in an electrical system of a motor vehicle, so a convention is needed to enable people to understand the graphical symbols. In the past there has been considerable confusion because each country used its own standard. Nowadays many countries have adopted the recommendations made by the International Electrotechnical Commission (IEC).

In the UK the British Standards Institution (BSI) recommend that the symbols shown in BS 3939:1985 should be used. A selection of the main graphical symbols is shown in Table I (overleaf).

Many manufacturers in this country have an overseas parent company so this means that they adopt the standard set for that country. Circuit diagrams of these vehicles contain some strange symbols so care must be exercised when using the diagrams.

Table II (overleaf) shows the main electrical units used in this book.

The electric circuit

The conduction of electricity in metal is due to the drift of free electrons from a lower potential to a higher potential (Figure 1.3). Since many

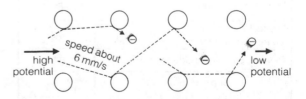

Figure 1.3 *Electron flow from high to low potential*

Description	Symbol	Description	Symbol
Direct current Alternating current		Lamp	⊗
		Fuse	
Positive polarity Negative polarity	+ −	Switch ('make' contact, normally open)	
Current approaching Current receding	⊙ ⊕	Switch ('break' contact, normally closed)	
Battery 12V (Long line is positive)		Switch (manually operated)	
Earth, chassis frame Earth, general		Switch (two-way)	
		Relay (single winding)	
Conductor (permanent) Thickness denotes importance Conductor (temporary)		Relay (thermal)	
Conductors crossing without connecting		Spark gap	
Conductors joining		Generator ac and dc	Ⓖ Ⓖ
Junction, separable Junction, inseparable Plug and socket	○ ●	Motor dc	Ⓜ
Variability; applied to other symbols		Meters; ammeter, voltmeter, galvanometer	Ⓐ Ⓥ ①
Resistor (fixed value)			
Resistor (variable)		Capacitor, general symbol	
General winding (inductor, coil)		Capacitor, polarized	
Winding with core		Amplifier	
Transformer			
Diode, rectifying junction		Junction f.e.t. N-type channel P-type channel	
Light emitting diode			
Diode, breakdown; Zener and avalanche		Photodiode	
Reverse blocking triode thyristor		Thyristor	
Transistor pnp npn			

Table I   Electrical symbols (BS 3939-1985)

Unit	Symbol	Electrical property
Ampere	A	Current
Ampere-hour	Ah	Battery capacity
Coulomb	C	Electrical charge
Farad	F	Capacitance
Hertz	Hz	Frequency (1 Hz = 1 cycle per second)
Volt	V	Potential difference or electromotive force
Watt	W	Power (watt = volt × ampere)

Table II   Electrical units

electrons are involved and the space for the drift is large, then the actual speed of movement of any one electron is very slow; it is only about 6 mm per second.

As the electrons move, they collide with atoms in their path and the resultant impact causes the temperature of the metal conductor to increase.

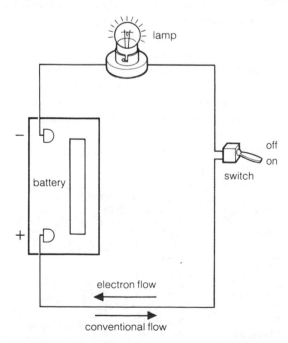

Figure 1.4   *Closed circuit*

A battery is used in Figure 1.4 to start electron movement and provided that a closed circuit (completed path) exists then electrons will flow around the full system.

The electron drift around a circuit is similar to the ball movement in a pipe circuit partially filled with ball bearings (Figure 1.5). When the first ball is struck with a hammer, the force of impact is transmitted to the other balls in turn until all balls are moving around the complete pipework system. To sustain this motion a provision must be made to recharge the balls with energy. The effect of the hammer is performed by a battery in an electrical circuit.

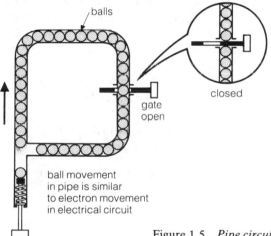

Figure 1.5   *Pipe circuit*

When the movement of the balls has to be stopped, the pipework system is interrupted by a valve. This action is similar to the electrical switch. *Opening the circuit* stops the flow of electrons and *closing the circuit* allows flow to take place.

Electron movement gives an energy flow called *electric current*. Although the existence of electric current flow was known many years ago, the direction was assumed to be from a higher potential (+) to a lower potential (−); an assumption which has since been proved to be incorrect.

Ampere (A)

The ampere is the unit of electric flow and is the rate of electron movement along a conductor. A coulomb is the quantity of electrons so when one coulomb passes a given point in one second then the current is one ampere.

$$1 \text{ ampere} = 1 \text{ coulomb per second}$$

Various standards have been used in the past to define the ampere; nowadays it is defined in terms of the force between conductors. If two parallel conductors are placed a given distance apart, then when current is passed through the conductors, a force is set up which is proportional to the current.

Watt (W)
The watt is a unit of power and applies to all branches of science. It is equivalent to work done at the rate of one joule per second. (One joule is the product of the force, in newtons, and the distance, in metres. $1J = 1Nm$.)

A power of 1W is developed when a current of 1A flows under the 'pressure' or p.d. of 1V.

$$\text{power} = \frac{\text{energy supplied (joules)}}{\text{time (seconds)}}$$

$$= \frac{\text{voltage} \times \text{current} \times \text{time}}{\text{time}}$$

$$\text{watts} = \text{volts} \times \text{amperes}$$

Ohm's Law
In 1826 Ohm discovered that the length of wire in a circuit affected the flow of current. He found that as the length was increased, the current flow decreased and from these findings he concluded that:

'Under constant temperature conditions, the current in a conductor is directly proportional to the p.d. between its ends'.

Today this statement is known as *Ohm's Law*.

Resistance
From Ohm's Law the relationship between p.d. (V) and current (I) is:

$$\frac{V}{I} = R$$

In this case, R is a constant which changes only when the length of the conductor is altered. Evidence shows that the value of the constant is related to the conductor's opposition to current flow, so R is called the resistance and given the unit name of ohm (symbol Ω).

The ohm is the resistance of a conductor through which a current of one ampere flows when a potential difference of one volt is across it.

In practical work the expression $V/I = R$ is often called Ohm's Law. Rearranged it gives:

$$V = IR \quad \text{or} \quad \text{volts} = \text{amperes} \times \text{ohms}$$

If two of these values are known, the third can be calculated, so this expression has a number of practical uses.

When the resistance of a conductor is the main property, it is called a *resistor*.

The resistance of a conductor is affected by its material, temperature and dimensions, namely cross-sectional area and length.

The SI unit of resistivity of a material is the ohm-metre. This is the resistance in ohms of a 1 metre length of material having a cross-section of 1 square metre. Table III shows typical values for some pure metals arranged in the order of resistivity; best conductor is placed first.

Although silver is the leader in respect of electrical conductivity, the lower cost of copper makes this a suitable material for a cable.

Table III Resistivity of some materials used for electrical conductors

Substance	Approximate resistivity (ohm m at 20°)
Silver	1.62×10^{-8} (or 0.000 000 0162)
Copper	1.72×10^{-8}
Aluminium	2.82×10^{-8}
Tungsten	5.5×10^{-8}
Brass	8×10^{-8}
Iron	9.8×10^{-8}
Manganin	44×10^{-8}
Constantin	49×10^{-8}

Since the resistivity is based on the dimension of 1 metre, the effect on cable length and cross-sectional area can be deduced. This shows that a proportional increase in resistivity occurs when either the length is increased or the cross-sectional area is decreased.

Most metals increase their resistivity when the temperature is raised, so these metals are said to have a *positive temperature coefficient*. Conversely if the resistivity decreases with an increase in temperature then the material has a *negative temperature coefficient*; carbon behaves in this manner.

Circuit resistors

Motor vehicle circuits normally consist of a number of resistors controlled by switches and connected to an electrical supply. Resistors can take many forms; they can be a lamp or be a part of some other energy-consuming device.

Basic understanding of circuit behaviour may be helped if the effect of resistors on voltage and current flow is considered.

Resistors may be connected in series, in parallel, or a combination of both.

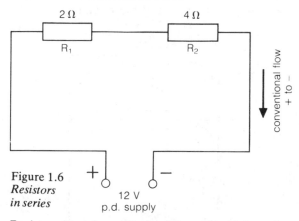

Figure 1.6
Resistors in series
12 V
p.d. supply

Resistors in series　Placing two resistors in series (Figure 1.6) means that the full current must pass through each resistor in turn. When they are connected in this end-to-end manner, the total resistance of the two resistors is the sum of their values, so:

$$R = R_1 + R_2$$

$$R = 2 + 4 = 6\ \Omega$$

Assuming the resistance of the cables is negligible, by applying Ohm's Law the current flow can be calculated.

$$V = IR$$
$$I = \frac{V}{R} = \frac{12}{6} = 2\ \text{ampere}$$

In this case, a current of 2 A will pass through each resistor and around the whole circuit.

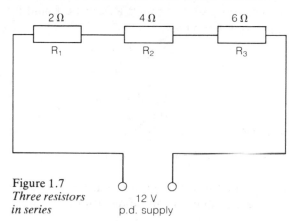

Figure 1.7
Three resistors in series
12 V
p.d. supply

Inserting an additional resistor of 6 Ω in series with the other two (Figure 1.7) will now give a total resistance of:

$$R = R_1 + R_2 + R_3$$

Voltage distribution (resistors in series)　Figure 1.8 shows two resistors in series with an ammeter

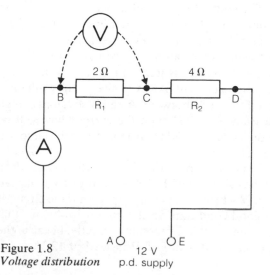

Figure 1.8
Voltage distribution
12 V
p.d. supply

and voltmeter and positioned so as to measure the current and p.d. respectively. It will be seen that the ammeter is fitted in series with the resistors and this means that all current flowing in the circuit must pass through the ammeter, no matter where the meter is inserted in the circuit.

Energy is expended driving the current through a resistor so this causes the potential to drop. The voltage drop (decrease in p.d.) when the current passes through R_1 can be found by applying Ohm's Law:

$$V = IR$$
$$= 2 \times 2 = 4 \text{ volts}$$

A voltmeter, connected as shown, will register the voltage drop. In this case it will register 4 V, so the p.d. applied to R_2 will be:

$$12 - 4 = 8 \text{ V}$$

By moving the voltmeter around the circuit, the voltage distribution can be determined:

Voltmeter position	Potential difference (V)
AE	12
BC	4
CD	8

The voltmeter is a useful meter for locating an unintentional resistance that has developed in a circuit. Figure 1.9 shows a simple lighting circuit consisting of a lamp, of resistance 4 Ω, that is connected to a switch and battery. This lamp requires a current of 3 A to give its full brilliance, but if the cable between A and B develops a 'high resistance' of 2 Ω then the current flow will be reduced to 2 A and the brightness of the lamp will be reduced.

Using a voltmeter to measure the voltage drop of the cable AB will show a reading of 4 V instead of 0 V which is the value expected if the circuit is in good condition. In this case it is obvious that the lamp will not function correctly, because the p.d. across it is only 8 V instead of 12 V.

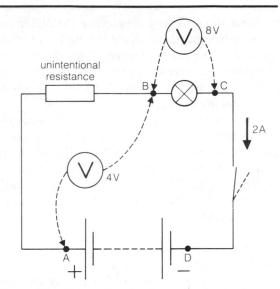

Figure 1.9   *Unintentional resistance in the circuit*

Resistors in parallel   Connecting resistors in parallel (Figure 1.10) ensures that the p.d. applied to each resistor is the same. Current flowing through the ammeter is shared between the two resistors and the amount of current flowing through each resistor will depend on the resistance of that part of the circuit.

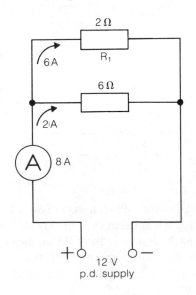

Figure 1.10   *Resistors in parallel*

Applying Ohm's Law to find the current flow in each resistor:

$$I = \frac{V}{R}$$

Current flow through R_1 = $\frac{12}{2}$ = 6 A

Current flow through R_2 = $\frac{12}{6}$ = 2 A

Total current flow through both resistors

$$= 6 + 2 = 8 \text{ A}$$

When calculated in this way, the current through each branch circuit can be found easily. Also it is possible to find the value of a single equivalent resistor (R) which would give the same total current flow as that which passes through both resistors. Applying Ohm's Law:

$$R = \frac{V}{I}$$

$$R = \frac{12}{8} = 1.5 \text{ } \Omega$$

The *equivalent resistance* of a number of resistors R_1, R_2, R_3 can also be found by applying the expression:

$$\frac{1}{R} = \frac{1}{R_1} + \frac{1}{R_2} + \frac{1}{R_3}$$

Figure 1.11 shows a circuit consisting of two lamps in parallel, both controlled by a switch. Arranged in this manner both lamps operate at full brilliance because the battery p.d. of 12V is applied directly to each lamp. Failure of the filament of one lamp has no effect on the other lamp, whereas a break in any part of a series circuit would cause complete failure of all parts of the circuit.

Compound circuit (series–parallel circuit) This circuit uses resistors, or consumer devices, connected so that some parts are in series and other parts are in parallel (Figure 1.12).

When calculating the current flow in these circuits, it is imagined that the parallel resistors are

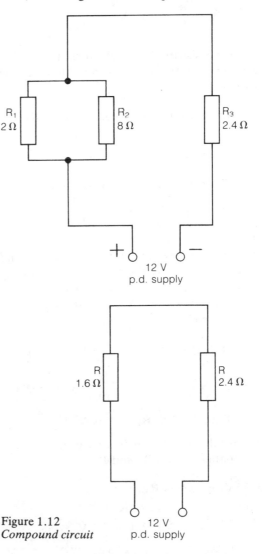

Figure 1.12
Compound circuit

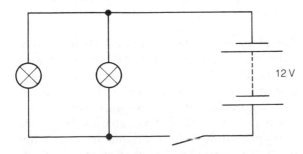

Figure 1.11   *Lamps in parallel*

replaced by a single resistor of equivalent value so as to produce a series circuit.

Equivalent resistance of R_1 and R_2 is found by:

$$\frac{1}{R} = \frac{1}{R_1} + \frac{1}{R_2} \quad \left[\text{or } R = \frac{R_1 R_2}{R_1 + R_2} \right]$$

$$\frac{1}{R} = \frac{1}{2} + \frac{1}{8}$$

$$\frac{1}{R} = \frac{4 + 1}{8} = \frac{5}{8}$$

$$5R = 8$$

$$\therefore R = \frac{8}{5} = 1.6\ \Omega$$

The two resistors R_1 and R_2 working together in parallel give the same current flow as one resistor of 1.6 Ω.

$$\text{Total resistance of circuit} = 1.6 + R_3$$
$$= 1.6 + 2.4$$
$$= 4\ \Omega$$

Current flow is given by applying Ohm's Law:

$$I = \frac{V}{R} = \frac{12}{4} = 3\ A$$

Consideration of the current flow through R_1 and R_2 shows that 3 A is the total current flowing through both resistors. This current divides according to the resistor values – the higher the value, the smaller the current.

$$\text{Current flow through } R_1 = \frac{R_2}{R_1 + R_2} \times I$$
$$= \frac{8}{10} \times 3 = 2.4\ A$$

Current flow through R_2 = 0.6 A

This result may be verified by using the p.d. values applied across R_1 and R_2.

Voltage drop across R_3:
$$V = IR$$
$$V = 3 \times 2.4 = 7.2\ V$$

So p.d. across R_1 and R_2 = 12 − 7.2 = 4.8 V.

Therefore the current flow through R_1:

$$I = \frac{V}{R}$$
$$I = \frac{4.8}{2} = 2.4\ A$$

and current flow through R_2:

$$I = \frac{4.8}{8} = 0.6\ A$$

Circuit terms
Insulated return   The simple circuit shown as Figure 1.13 connects the lamp to the battery and uses a switch to control the 'supply'. To complete the circuit, the cable joining the lamp to the battery acts as a *return* for the current. These terms relate to conventional flow.

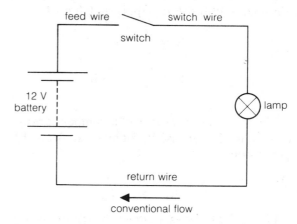

Figure 1.13   *Insulated return circuit*

The term *insulated return* is used whenever an insulated cable provides the return path for the current.

Earth return   In this case the vehicle frame provides the return path for the current (Figure 1.14). Advantages of this system are: reduced cost, lighter weight and simpler circuit layout.

Extra precautions must be taken with this system to prevent chafing of the cable insulation by a sharp part of the vehicle's frame or body. If this occurs, the current will be conducted to earth so, unless the circuit is protected by some form of fuse, a fire can be started.

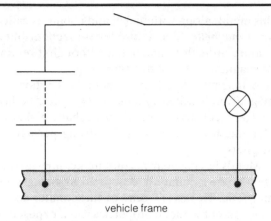

Figure 1.14   *Earth return circuit*

Most vehicles have an earth return system; the exceptions are the special-purpose vehicles such as petrol tankers that have a high fire risk; these use an insulated return layout.

A simpler diagram is obtained when the earth symbol is used (Figure 1.15).

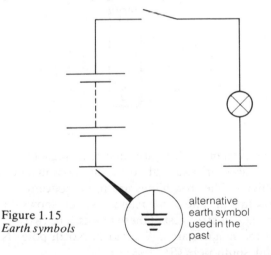

Figure 1.15
Earth symbols

alternative earth symbol used in the past

In North America the term *ground* is used instead of 'earth'.

Earth polarity   The negative terminal of the battery shown in Figure 1.15 is connected to the frame. This arrangement is called *negative earth* and is commonly used on the majority of modern vehicles. Reasons for this choice will be covered at a later stage (see page 124).

In the past, vehicles used *positive earth* whereby the vehicle frame was of positive potential.

Damage to electrical components containing semi-conductor devices will occur if the battery is incorrectly earthed.

Circuit faults
Failure of an electrical system is often caused by a circuit fault. Two common faults are considered at this stage.

Open circuit   A complete circuit is needed if current is to flow around the system. An *open circuit* exists when the circuit is interrupted either intentionally or unintentionally. A switch 'opens' the circuit by breaking the supply wire and a similar effect is produced when either a poor terminal connection or a broken cable stops the current flow (Figure 1.16).

Figure 1.16
Open circuit

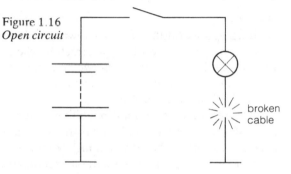

broken cable

Short circuit   When the feed- or switch-wire is damaged and the conductor touches the metal frame, some or all of the current will take this 'easy' path to earth. This alternative path offers the current a short path back to the battery, so the term *short circuit* is used to describe this condition (Figure 1.17).

Figure 1.17
Short circuit

cable rubbing against frame

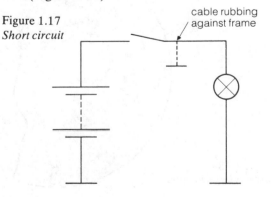

The extent of the short-to-earth, i.e. the resistance of the alternative path, governs the p.d. that is left to act on the lamp in Figure 1.17. As the resistance R is reduced, the p.d. across the lamp is also reduced so the effect of the voltage reduction will be a proportional decrease in the lamp brilliance.

A *dead short* describes a very low resistance path to earth. When this occurs the resistance of the 'new' circuit will be minimal, so the very high current flow that results will soon make the cable glow red-hot. This melts the plastics covering of the cable and often starts a fire. Some circuit-protection device is needed if this danger is to be avoided.

1.2 Magnetism and electromagnetism

Magnetism

As long ago as 600BC it was known that lodestone would always point in one direction when it was suspended. The name *magnet* was derived from the place where this magnetic iron was discovered – Magnesia in Asia.

Because of its directional capabilities, the two ends of a magnet were called North and South. Perhaps the name 'North seeking' would have been more appropriate in view of the fact that the Earth acts as a large magnet (Figure 1.18).

Later discoveries showed that the magnetic effect of lodestone was due to the iron deposits in the stone. Iron can be strongly magnetized, so

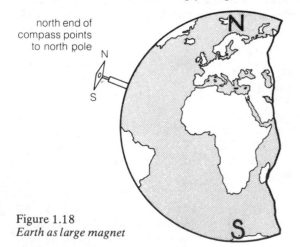

north end of compass points to north pole

Figure 1.18
Earth as large magnet

this metal, along with nickel and cobalt, is called *ferro-magnetic*. Steel is also ferro-magnetic but it is more difficult to magnetize although it retains its magnetism far better than iron.

A magnet made from a metal that retains its magnetism is called a *permanent magnet*. Today, special steel alloys containing cobalt, nickel or aluminium are used to make strong permanent magnets.

Metals having no iron content, i.e. non-ferrous metals, are commonly described as being non-magnetic. However, some of these materials do show slight magnetic properties when exposed to a very powerful magnet.

Magnetic field Iron filings scattered on a sheet of paper placed over a magnet (Figure 1.19) form a pattern due to the presence of a magnetic field.

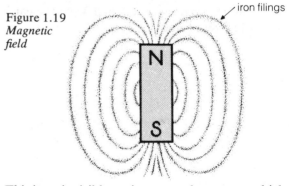

Figure 1.19
Magnetic field

iron filings

This is an invisible region around a magnet which produces an external force on ferro-magnetic objects. The iron filings are more concentrated towards the end of the magnet, so this shows that the field is strongest at these points. The two ends of the magnet are called *poles*: North pole (N) and South pole (S).

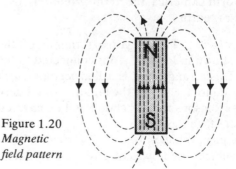

Figure 1.20
Magnetic field pattern

Figure 1.20 shows that the iron filings arrange themselves to form a series of lines which extend from one pole to the other. These lines are called *lines of force*.

Attract and repel action When a magnet is moved towards a suspended second magnet (Figure 1.21) the effect is that:

Like poles repel each other
Unlike poles attract each other

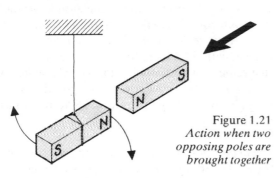

Figure 1.21
Action when two opposing poles are brought together

When this experiment is performed under a sheet of paper containing iron filings, it is seen that the 'unlike poles' of the magnets join together to make a larger magnet and as a result the lines of force pass directly from one magnet to the other. This action produces an external force that makes it difficult to hold the magnets apart.

An opposite effect results when either N and N, or S and S, are brought together. In this case the field of one magnet opposes the field of the other magnet and observation of the filings shows that the lines of force are bent as the magnets come together. The external force pushing the two magnets apart is related to the 'bending' of the lines of force; the greater the field distortion, the larger the magnetic force produced.

The direction of a line of force can be found by using a small compass, which is a magnet pivoted at the centre for rotation purposes. By positioning the compass in various places in the field, the lines of force can be mapped out (Figure 1.22).

Magnetic flux Lines of force passing from N to S indicate a region of magnetic activity around a magnet. This activity is produced by a magnetic

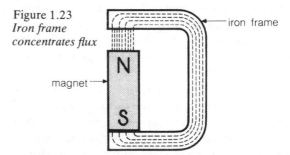

Figure 1.22
Mapping a magnetic field

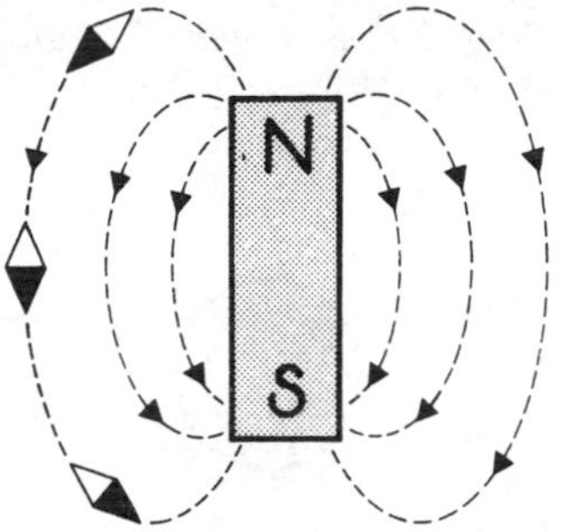

flux. The presence of this magnetic region is detected by the effects it produces.

Sometimes it is said that magnetic flux flows around the magnet from the N pole to the S pole where it then passes internally through the magnet to return to the N pole. This statement suggests movement but, unlike electricity, no flow actually takes place.

Magnetic flux density The density is the strength of a magnetic field at a given point. It is measured by the force that the field exerts on a conductor through which a given current is passed.

Figure 1.23
Iron frame concentrates flux

iron frame

magnet

A ferromagnetic material, placed in a magnetic field (Figure 1.23), provides an easier path for the magnetic flux than through air. An iron frame concentrates the flux where it is required and gives an increased flux density. Iron accepts a magnetic flux easier than air, so iron is said to have a higher *permeability* than air.

Magnetic screening There are cases where a magnetic flux must be either contained, or excluded, from a given region. An iron ring, placed in a strong magnetic field, has a region within the ring that is *screened* or *shielded* from the magnetic flux (Figure 1.24).

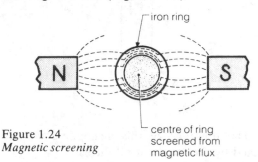

Figure 1.24
Magnetic screening

iron ring

centre of ring screened from magnetic flux

Domain theory of magnetism Various theories relating to the internal changes that take place in a metal when it becomes magnetized have been advanced over the years. The modern view is called the *domain theory*.

Earlier in this book (page 2) the movement of electrons in an atom was described. This showed that electrically charged electrons orbited around a nucleus at a high speed.

The domain theory suggests that when all electrons of one particular atom are moving in the same direction, then magnetism is produced which makes the atom into a tiny atomic magnet. When an atom has an equal number of electrons moving in opposite directions, then it receives no charge from the electrons and no magnetic effect is produced.

A ferromagnetic metal, such as iron and steel, has a composition that consists of a number of domains of 'atomic magnets' arranged so that each group has a common magnetic axis. In Figure 1.25 each arrow represents an atomic magnet; the arrow head indicates the N pole.

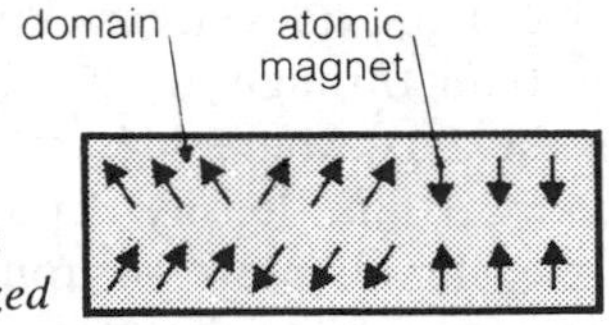

domain   atomic magnet

Figure 1.25
Unmagnetized

When the metal bar is gradually magnetized from some external source, the atomic magnets slowly rearrange themselves. When the external field is very strong, the magnetic axes of all atomic magnets will coincide with the direction of

the external field (Figure 1.26). When this is achieved, the maximum magnetic strength is obtained and no further increase is possible: this condition is called *magnetic saturation*.

Figure 1.26 *Magnetized*

In the case of soft iron, the removal of the external field causes some of the atomic magnets in certain domains to return to their original position. As a result, only a few domains remain aligned with the axis of the bar, so the magnetism that remains is very small. This partial magnetism is called *residual magnetism*; the polarity of the magnetism is set by the external field previously applied.

Demagnetization A permanent magnet will 'lose' its strength if it is exposed to heat or vibration, because these effects allow the atomic magnets to settle back in their preferred positions. To avoid this, a magnet should have a keeper when it has been removed from a component. The keeper becomes an *induced magnet* (Figure 1.27).

Figure 1.27
Keeper

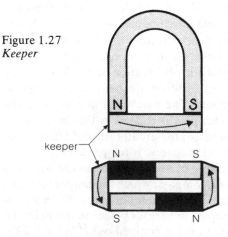

keeper

Reluctance This term is similar to the term 'resistance' as applied to an electrical circuit, except that reluctance refers to a magnetic circuit.

Figure 1.28
Reluctance

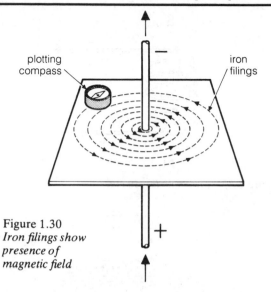

plotting compass   iron filings

Figure 1.30
*Iron filings show
presence of
magnetic field*

Figure 1.28 shows how the reluctance of an air gap is reduced when two poles of a magnet are bridged by a piece of iron.

Electromagnetism
In 1819 Prof. Oersted discovered that a wire carrying an electric current deflected a nearby compass needle. Further investigation showed that the direction of the needle deflection depended on whether the compass was placed under or over the wire and also on the direction of the current (Figure 1.29).

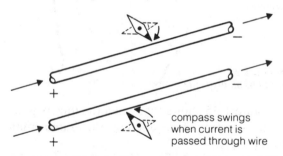

compass swings
when current is
passed through wire

Figure 1.29   *Compass shows presence of magnetic field*

The movement of the compass needle showed the existence of a magnetic flux around the wire. This can be seen from an experiment shown as Figure 1.30. Iron filings, scattered on a sheet of paper, rearrange themselves into a series of concentric rings when current flows in the circuit and the paper is gently tapped. The direction of the flux can be determined by using a plotting compass.

If the current is reversed, a similar pattern is formed but the compass needle points in the opposite direction.

The direction of the flux can be found by imagining a corkscrew placed around the wire.

As current flows along the wire, the direction of the screw indicates the direction of the magnetic flux (Figure 1.31). Known as *Maxwell's Screw Rule*, the method has many practical applications.

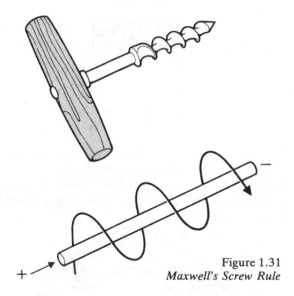

Figure 1.31
Maxwell's Screw Rule

The electromagnet   When current is passed through a wire that is wound to form a coil, a magnetic flux is produced (Figure 1.32). This flux can be concentrated by placing a soft iron core in

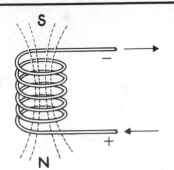

Figure 1.32   *Magnetic flux in coil*

On closing the switch a magnetic flux is set up around each winding. Being as the windings are placed close to each other, the flux blends together to form a common pattern centred around the iron core. This action makes the iron into a magnet; the polarity of the magnet is governed by the direction of the current in the winding.

The relationship between current direction and magnetic polarity can be determined quickly by the *Right-hand Grip Rule*. (This should not be confused with Fleming's rule described on page 20.)

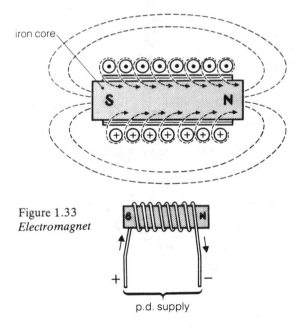

Figure 1.33
Electromagnet

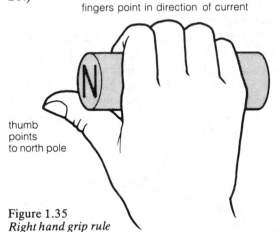

Figure 1.35
Right hand grip rule

the coil. In Figure 1.33 the direction of the current (conventional flow) is shown by the symbols · and +: these represent the point and tail of an arrow that points in the direction of the current (Figure 1.34).

Figure 1.35 shows that when the fingers point to the direction of current flow, the thumb points to the North.

An alternative method is shown as Figure 1.36. The symbols representing N and S indicate the direction of the current in relation to the magnetic polarity.

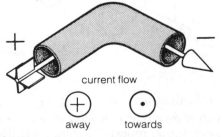

Figure 1.34   *Symbols to represent direction of current*

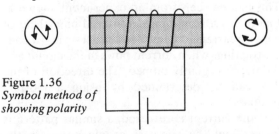

Figure 1.36
Symbol method of showing polarity

The strength of an electromagnet depends on two things; the amount of current that flows through the winding and the number of turns which make the coil. Multiplying these two together gives the unit '*ampere-turns*', so if 5 A flows through a coil winding having 1000 turns, then the magnetic strength will be equivalent to 5000 ampere-turns.

One advantage of an electromagnet is that it can be turned on and off at will. To provide this feature, the core must lose its magnetism quickly, so for this reason iron is preferred to steel for the core material.

The solenoid   This device is used commonly on vehicles to produce movement from an electrical signal, e.g. to control a remote switch or operate a door-locking system.

A solenoid consists of a coil of wire wound around a cylinder into which is fitted a sliding soft iron plunger. A spring holds the plunger away from the coil when the unit is not in use (Figure 1.37).

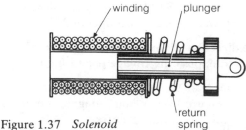

Figure 1.37   *Solenoid*

When current is supplied, the magnetic effect attracts the plunger into the centre of the winding and movement of the plunger operates the appropriate switch or mechanical linkage,

Electromagnetic relay   Some electrical systems require a small current to control the flow of a large current: this duty can be performed by a relay.

One type of relay (Figure 1.38) consists of an L-shaped frame on to which is hinged an armature. A coil, consisting of many turns of very fine enamelled wire, is wound around the soft iron core and a pair of heavy-duty contacts are connected, in series, to a separate circuit which carries a large current.

When a small current is supplied to the coil winding, the core becomes a magnet. At this

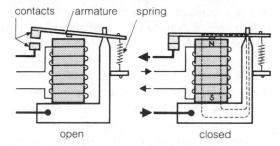

Figure 1.38   *Electromagnetic relay*

point the armature is attracted to the magnet and this causes the main contacts to close.

The voltage at which the contacts close can be altered by varying the strength of the spring.

1.3 Electromagnetic induction

After 1819, when Oersted discovered that magnetism could be produced by an electric current, many scientists searched for a method to establish the reverse effect. It was not until 1831 that Michael Faraday achieved this goal: he showed that electricity could be produced from magnetism. By a series of experiments he demonstrated the principles from which the generator and many other automobile components has been developed.

Faraday's experiments
One of the most important experiments is shown as Figure 1.39. The apparatus he used consisted

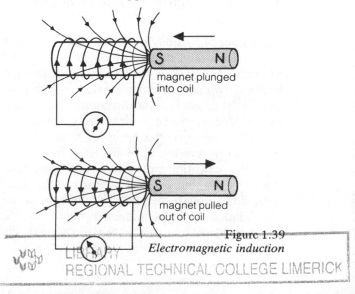

Figure 1.39
Electromagnetic induction

of a coil wound around a paper tube and to the winding he connected a galvanometer (an instrument for detecting the presence and direction of an electric current).

Faraday noticed that when he plunged a magnet into the coil the galvanometer needle moved. Also he saw that as he removed the magnet from the coil the galvanometer needle flicked in the opposite direction. The needle behaviour showed that current was generated only when the magnet was being moved. Furthermore this experiment demonstrated that the direction of the current depended on the direction of movement of the magnet.

Whenever current is generated in this way it is called *electromagnet induction*.

The conclusion made by Faraday was that current is induced into a circuit when a coil winding cuts a magnetic line of force formed around a magnet. Today this conclusion can be expressed as:

An electromotive force is induced whenever there is a change in the magnetic flux linked with the coil.

Faraday discovered that the electromotive force (e.m.f.) induced in the coil winding depended on:

1. Number of turns in the coil
2. Strength of the magnet
3. Speed with which the magnet cuts the coil windings.

Another of Faraday's experiments is shown as Figure 1.40. The apparatus was a coil, wound around a soft iron core, which was positioned between two permanent bar magnets. When he moved the magnets in the direction of the arrows, the galvanometer needle deflected one way, and when he brought the magnets together, the needle moved in the opposite direction.

In this case the induced currents were produced by the changes in the density of the magnetic flux through the coil. The strength of the e.m.f. induced in this manner is proportional to the rate of change of the flux linked with the circuit.

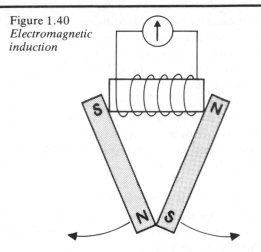

Figure 1.40
Electromagnetic induction

It is found that when the iron core is removed, the reduction in the magnetic flux in the region of the coil is decreased, so this results in a reduction in the e.m.f.

Lenz's Law   In 1834 Lenz stated a law which is related to electromagnetic induction. This law is:

The direction of the induced current is always such as to oppose the change producing it.

This law can be explained by the apparatus shown as Figure 1.41. When a magnet is plunged into a coil, an induced current is generated.

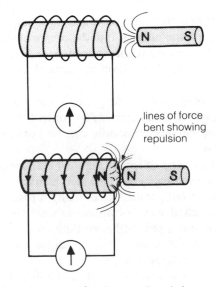

lines of force bent showing repulsion

Figure 1.41   *Apparatus for showing Lenz's law*

According to Lenz's Law this current will always have a magnetic polarity which will repel the magnet, so in this case the current in the coil will form a N pole at the end nearest the magnet.

Mutual induction Faraday's iron ring experiment (Figure 1.42) showed that a coil could be used instead of a magnet to induce a current into an independent coil.

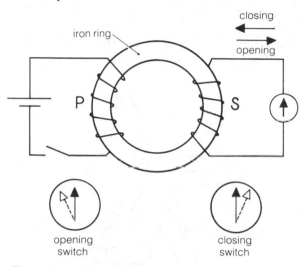

Figure 1.42 *Mutual induction*

He used an iron ring and on opposite sides of this ring he wound two coils: a primary and a secondary. He connected the primary coil to a battery and switch and the secondary coil he joined to a galvanometer.

On closing the switch, the build-up of magnetic flux in the iron ring induced a 'momentary' current into the secondary and this caused the galvanometer needle to kick over. Opening the switch gave a similar effect except the needle kicked in the opposite direction to show a current flow in the reverse direction. He concluded that the breaking of the primary circuit causes the magnetic flux to decay and the change of flux which results induces a current into the secondary coil.

Faraday varied this experiment by using two coils, wound on a piece of wood, placed side by side. The result was similar except that a lower current was induced. The decrease in current was due to the reduced flux concentration brought about by the loss of the iron core.

The term *mutual induction* is used when one coil or conductor induces an e.m.f. into a separate coil or conductor.

Lenz's law also applies to mutual induction so in Figure 1.43 the direction of current in the secondary depends on whether the switch is opened or closed. Closing the switch causes the magnetic flux to build-up in a clockwise direction so, by applying Lenz's Law, the induced current in the secondary gives an anti-clockwise magnetic flux. Once this is established, the Right-hand Grip Rule indicates the current direction in the secondary.

Opening the switch causes the magnetic flux to decay so the induced secondary current will set up a flux in a clockwise direction. This produces a current in the secondary circuit which flows in the opposite way to that obtained when the switch is closed.

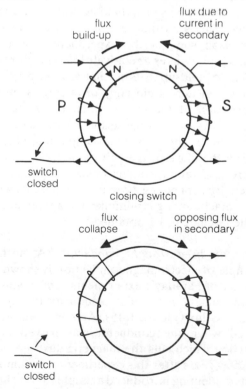

Figure 1.43 *Direction of current in secondary*

Self-induction On opening the switch to break the primary circuit (Figure 1.43) it is seen that a spark jumps across the switch contacts just as the contacts separate. The spark shows that a current of high voltage in excess of about 350 V is induced back into the primary by the decay of the magnetic flux. This effect is called *self-induction* and the e.m.f. causing the effect is termed *back-e.m.f.*

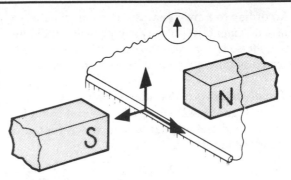

Figure 1.45   *Induction in a straight conductor*

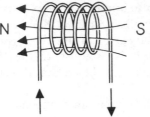

Figure 1.44   *Magnetic field set up by current in coil*

Whenever a current flows through a coil (Figure 1.44) a magnetic field is set up. If the current is either increased or decreased then the back-e.m.f., caused by self-induction, will oppose the change. This is another case of Lenz's Law and is the reason why, in the experiment shown as Figure 1.43, it takes a relatively long time for the current to build-up to its maximum in the primary after the switch is closed. Conversely, a quick interruption of the circuit when the switch is opened will give a large secondary current.

In vehicle ignition systems a sparking plug is used in the place of the galvanometer shown in Figure 1.43, and a contact breaker replaces the switch. By connecting a capacitor in parallel with the contacts, arcing is minimized so a rapid break in the circuit can be obtained.

Induction in a straight conductor A further example of electromagnet induction is shown in another of Faraday's experiments. This involves the movement of a straight conductor through a magnetic flux (Figure 1.45). An e.m.f. was generated when the conductor was moved in a direction which cuts the magnetic flux.

A few years after this experiment was demonstrated, Fleming introduced a simple rule to show the relationship between the directions of the field, current and conductor.

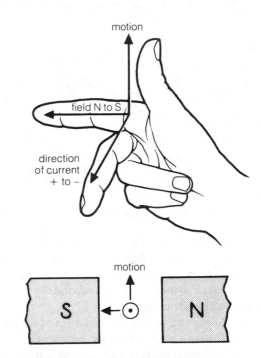

Figure 1.46   *Fleming's Right-hand rule*

Fleming's Right-hand Rule (Dynamo rule) This is applied to electromagnet induction and is shown by Figure 1.46. This rule may be expressed as:

When the thumb and first two fingers of the right-hand are all set at right angles to one another, then the forefinger points in the direction of the field, the thumb to the direction of the motion and the second finger points to the direction of the current.

This can be summarized as:

thu**M**b	**M**otion
fore**F**inger	**F**ield
se**C**ond finger	**C**urrent (conventional flow)

Another rule is credited to Fleming which is used with motors. This will be applied at a later stage in this book (page 25).

The alternating current generator

A simple dynamo is shown as Figure 1.47. This consists of two magnetic poles of a field magnet and a conductor bent to make a loop. Each end of the loop is joined to a slip ring which makes contact with a carbon brush.

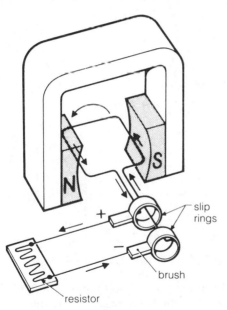

Figure 1.47　*Simple dynamo*

When the conductor loop is rotated, an e.m.f. is generated and this drives a current around the circuit (Figure 1.48).

The direction of the current, as found by Fleming's Right-hand Rule, is shown by the arrows and symbols · and +. Although the current induced in side A flows in the opposite direction to side B, the loop of the conductor ensures that the flow in the circuit at this instant in unidirectional.

As the conductor coil moves away from the

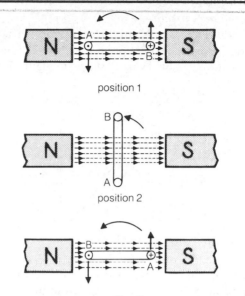

position 1

position 2

position 3

Figure 1.48　*Coil position*

dense magnetic flux region, shown as position 1, the output gradually falls. At position 2 both conductors are moving in the direction of the flux so neither conductor is cutting any of the imaginery lines of force. Being as the flux is uncut, no current will be generated at this point.

In position 3 the coil is situated in a dense flux region so maximum output will be obtained, but at this point the direction of the current in each conductor will be opposite to the flow indicated in position 1.

The e.m.f. output of a generator is shown by the graph (Figure 1.49). It is seen that the e.m.f. will cause the current to flow in one direction and then reverse to flow in the opposite direction. This type of current is called *alternating current* (a.c.).

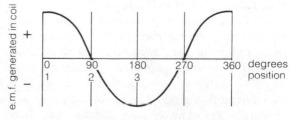

Figure 1.49　*E.m.f. generated in coil*

Alternating current

Waveform terms   When the conductor starts its motion from a vertical point (position 2, Figure 1.48), the e.m.f. will vary as shown in Figure 1.50.

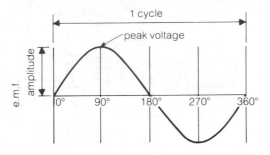

Figure 1.50   *Sine wave*

This shape of curve is called a *sine wave* and when something moves or varies its output in this way the term *sinusoidal* is used.

One complete turn (360°) of the conductor loop gives one complete wave and this is called a *cycle*. The time required to complete one cycle is termed its *period* (or *periodic time*).

A study of the e.m.f. output at a given instant (the *instantaneous voltage*) shows that it builds up to a *peak voltage*, then decreases to zero and reverses to build up to give a peak voltage of opposite polarity. The vertical distance between the two peaks is known as the *peak-to-peak voltage*. This should not be confused with the *amplitude*, which is the vertical distance from the horizontal axis of the graph to the peak.

When the conductor is rotated beyond 360° the e.m.f. cycle is repeated. The number of complete cycles that occur in one second is called the *frequency*; the SI unit for frequency is the hertz (1 cycle/second = 1Hz). A generator frequency depends on its speed of rotation: the faster the speed, the higher the frequency.

For battery-charging purposes, the alternating current must be rectified so that the current is made to flow in one direction only. This undirectional flow, or *direct current*, is achieved when the negative half-wave of the sinusoidal wave is transferred to the positive side of the graph axis (Figure 1.51). This change is called *half-wave rectification*.

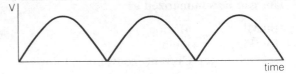

Figure 1.51   *Full-wave rectification*

The e.m.f. from an a.c. generator is not sinusoidal when the poles are shaped to provide a uniformly distributed flux (Figure 1.52). In this case the conductor moves in a dense flux for a large angle of movement so this gives a non-sinusoidal waveform (Figure 1.53).

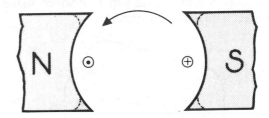

Figure 1.52   *Uniformly distributed flux*

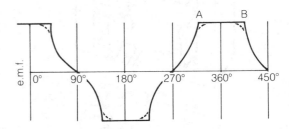

Figure 1.53   *Non-sinusoidal waveform*

By rounding-off the pole shoes as shown by the broken line, the sharpness of the corners at A and B in Figure 1.53 can be reduced.

Current flow   The instantaneous e.m.f. gives a proportional current flow through an external circuit of resistance R. The power (volts × amperes) or heating effect produced by this output is given by:

$$\text{power (W)} = VI$$

where V = volt and I = current (ampere)

From Ohm's Law, V = IR so:

$$\text{power} = (IR) \times I = I^2R$$

This shows that the power is proportional to the square of the current, i.e. when the current is doubled, the power is increased four times.

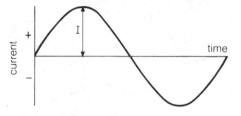

Figure 1.54   *A.C. waveform*

Consider the current shown by the waveform in Figure 1.54. To obtain the heating effect of this current a graph of I^2R is plotted (Figure 1.55).

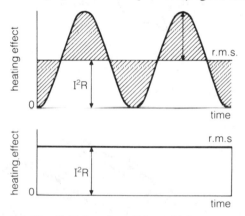

Figure 1.55   *R.M.S. value of sinusoidal current*

This curve shows that the heating effect is unaffected by the direction of the current. By rearranging the various parts of the curve it is possible to make a rectangular pattern. The height of this rectangle represents the *effective* value of the a.c. current, i.e. the value of the direct current which would give the same heating effect in the same resistance.

Another name for the effective value of an alternating current is the *root-mean-square* (r.m.s.) value of the current. Alternating current test meters measure the r.m.s. value of either the current or the voltage.

Sometimes the term *average current* is used. As the name suggests this is the mean current that flows during the cycle and should not be confused with the r.m.s. value.

For sinusoidal waves:

average value   = 0.637 maximum value
r.m.s. value   = 0.707 maximum value

Waveforms can be examined by using a cathode ray oscilloscope (CRO).

Impedance   A back e.m.f. caused by self-induction, develops when the direction of current is reversed. This effect prevents an a.c. current building up to that which would be obtained from a constant d.c. supply.

Impedance of a circuit is the 'opposition' it gives to the flow of alternating current. It is expressed in ohms and calculated by:

$$\text{Impedance} = \text{voltage/current}$$

The frequency of an a.c. current affects the impedance of the circuit.

Multi-phase a.c.   The current delivered by the conductor coil of a simple generator is a single-phase a.c. This means that the output e.m.f. and current corresponds to a single wave as shown in Figure 1.54.

If three conductor coils, A, B and C in Figure 1.56 were equally spaced and connected as shown to four terminals, then the coils will reach their peak e.m.f. at different times.

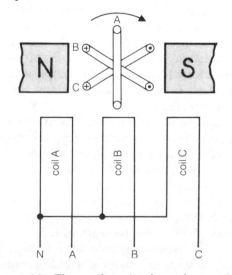

Figure 1.56   *Three coils to give three-phase output*

Output of each coil, measured between the common neutral (N) and the appropriate terminal A, B or C, will give a sinusoidal wave but the waves will be 60° out-of-phase. Figure 1.57 shows a three-phase output obtained from this layout and in this graph the cycle of each phase is emphasized.

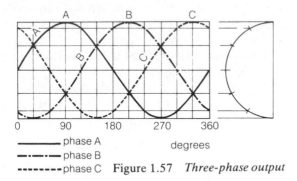

phase A
—·—·— phase B   degrees
------- phase C   Figure 1.57   *Three-phase output*

When a three-phase output is rectified, a smoother current is achieved than that obtained from a single phase.

Eddy currents

In a dynamo, the conductor coils are normally wound on a soft-iron former called an *armature*. The iron concentrates the magnetic flux where it is needed and forms a part of the magnetic 'circuit'.

If the cylindrical armature were made in one piece, rotation of the armature would cause an eddy current to be generated within its iron core. The direction of flow of the current can be obtained from Fleming's Right-hand Rule and in Figure 1.58 it shows how the current makes a path around the armature.

In addition to the problem of heat created by the dissipation of electrical energy, the magnetic effect of the eddy current makes it more difficult to revolve the armature. This is because the eddy current sets up its own magnetic flux as described in Lenz's law. Since the N pole of the flux from the eddy current is moving towards the N pole of the main field, a repulsion action is felt.

Electric machines should have a high efficiency, so steps are taken to avoid the internal energy loss due to eddy currents. This is achieved

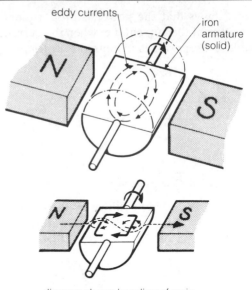

diagram shows bending of main flux by the eddy currents — this causes resistance to motion

Figure 1.58   *Eddy current*

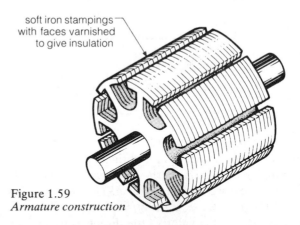

soft iron stampings with faces varnished to give insulation

Figure 1.59
Armature construction

by making the armature, or similar core devices, from thin iron stampings. Each face is coated with varnish for insulation purposes (Figure 1.59).

Eddy currents can be usefully employed in a number of ways. They can be used to damp the needle movement of a test meter and as a brake retarder to resist the rotation of a disc connected to the road wheels (Figure 1.60).

Drag on the disc of a retarder is produced when the main field is energized; the stronger the field, the greater the drag on the disc.

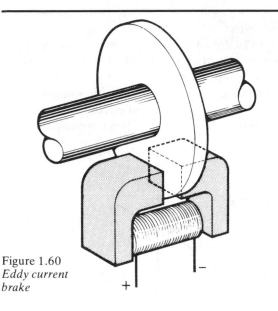

Figure 1.60
*Eddy current
brake*

The transformer

A transformer is a device for stepping-up or stepping-down the voltage.

It consists of two windings, a primary and a secondary, which are wound around a laminated iron core as shown diagrammatically in Figure 1.61. If a higher voltage is required from the secondary, the secondary winding must have more turns than the primary. The relationship between turns and voltage is:

$$\frac{\text{secondary voltage}}{\text{primary voltage}} = \frac{\text{secondary turns}}{\text{primary turns}}$$

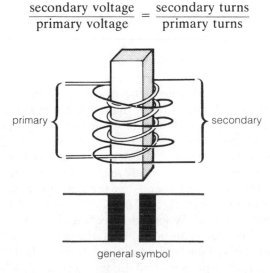

general symbol

Figure 1.61   *Transformer*

It has been stated that induction must be accompanied by a change in magnetic flux, so providing the primary receives an alternating current the output from the secondary will be a.c. at a voltage governed by the *turns-ratio*.

A transformer does not give 'something for nothing': an increase in voltage is balanced by a proportional decrease in current.

The electric motor

When current is supplied by a battery to a conductor placed in a magnetic flux, a force is produced which will move the conductor (Figure 1.62).

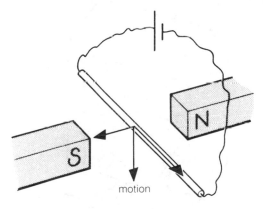

motion

Figure 1.62   *Force on conductor*

Fleming's Left-hand Rule (motor rule) gives the relationship between the field, current and motion. Similar fingers to those used for the Right-hand-Rule (page 21) are employed to give the various directions as follows:

fore**F**inger	**F**ield
thu**M**b	**M**otion
se**C**ond finger	**C**urrent

The cause of the turning motion can be seen when the lines of magnetic force are mapped (Figure 1.63, overleaf). This shows a current being passed through a conductor and the formation of a magnetic field around the conductor. This field causes the main field to be bent and the repulsion of the two opposing fields produces a force that gives the motion.

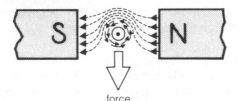

force

Figure 1.63 *Bending of main field*

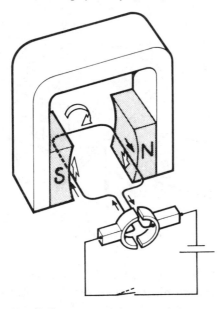

Figure 1.64 *D.C. motor*

Figure 1.64 shows the construction of a simple direct-current motor. The conductor is looped to form a coil and the ends of this coil are connected to a *commutator* (a device to reverse the current in the coil as it rotates). Carbon brushes rub on the commutator to supply the coil with current.

Being as the two sides of the coil in this construction act as two conductors, a greater turning moment (torque) is achieved. In Figure 1.63 the field distortion is shown, and if the lines of force in this diagram are considered to be rubber bands, then the principle of the motor can be seen.

Hall effect

In 1879 Edward Hall discovered that when a magnet was placed perpendicularly to the face of a flat current-carrying conductor, a difference in potential appeared across the other edges of the conductor. This event is called *Hall effect* and the p.d. produced across the edges is termed the *Hall voltage*.

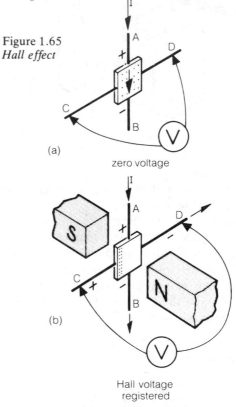

Figure 1.65
Hall effect

(a)

zero voltage

(b)

Hall voltage
registered

Figure 1.65 shows the principle of the Hall effect. In Figure 1.65(a) the vertical edges of the plate have equal potential so the voltmeter registers zero. But when the plate is placed in a magnetic field (Figure 1.65(b)), a p.d. across the edges is shown as a steady reading on the meter.

Hall voltage depends on the current flowing through the plate and on the strength of the magnetic field. When the current is constant, the Hall voltage is proportional to the field strength. Similarly, if the field strength is constant, the Hall voltage is proportional to the current flowing through the plate.

With common metals such as copper, the Hall voltage is very low, but when a semiconductor is used a much higher voltage is achieved. The polarity of the plate edges depends on the type of semiconductor; a p-type has the opposite Hall voltage p.d. to an n-type (see page 38).

The magnetic field which gives the Hall effect does not act as a generator of energy; instead it acts in the form of a switch.

Figure 1.66   *Principle of Hall effect sensor*

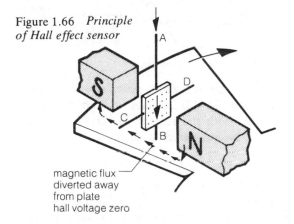

magnetic flux diverted away from plate hall voltage zero

In motor-vehicle work the switching feature is utilized as a circuit breaker and trigger in one type of electronic ignition system. Switching is achieved by using a slotted metal chopper plate to interrupt the magnetic field (Figure 1.66).

Another application of the Hall effect is for measuring the strength of a magnetic field.

1.4 Measuring and test instruments

There is a large range of electrical test equipment available to the electrician. This equipment may be divided into two main categories: basic instruments used by all electricians and special test sets to measure or check the performance of specific items of vehicle equipment. At this stage the former category is considered.

Basic test instruments
Moving-coil milliammeter   This is a galvanometer with a scale graduated to show a current in milliamperes (1 mA = 0.001 A).

Figure 1.67 shows the construction of this type of meter which resembles, in basic principle, the

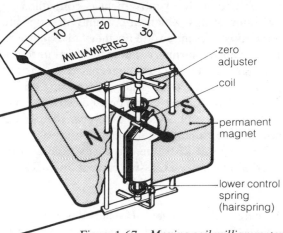

zero adjuster

coil

permanent magnet

lower control spring (hairspring)

Figure 1.67   *Moving coil milliammeter*

layout of a motor. There is a permanent magnet with two shaped pole pieces and between these is placed a fixed iron cylinder. This concentrates the magnetic field and makes the lines of force radiate from the cylinder centre.

The coil is wound on an aluminium former: this is pivoted on jewelled bearings and attached to a pointer which registers on an evenly-spaced scale. The actual number of turns on the coil, and the gauge of the wire, are governed by the purpose for which the instrument is to be used.

By using an aluminium former, the damping effect of the eddy currents allows the pointer to register a reading without oscillating to-and-fro. This 'dead beat' action is obtained by the opposition of the magnetic flux set up by the current induced into the former when it moves in the main field.

When current is passed through the coil (Figure 1.68), the flux distortion causes the coil to

Figure 1.68   *Bending of magnetic flux by current*

force   armature

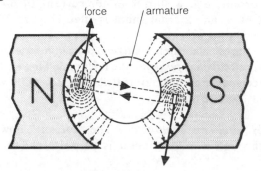

move. The angle of movement is controlled by two hairsprings which are wound in opposite directions to compensate for thermal expansion. These control springs conduct the current to the coil and their strength governs the current required to give a full-scale deflection (f.s.d.).

A moving-coil meter can be used only with d.c. and must be connected so as to give the correct polarity; terminals are marked + and −.

This type of instrument is easily damaged by overload because excessive current damages the hairsprings.

Moving-coil ammeter Having described the milliammeter it would appear that the meter could be scaled-up to read amperes: this would result in a clumsy inefficient meter. Instead the milliammeter is modified to suit the range required by fitting a resistor of low value in parallel with the meter to *shunt* (by-pass) the major part of the current away from the meter (Figure 1.69).

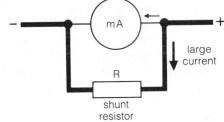

Figure 1.69   *Moving coil ammeter*

By selecting a shunt resistor of suitable value, it is possible to vary the range of the meter to suit the application. The normal ammeter has an internally connected shunt to suit the range but when this meter is to be used to measure a high current, e.g. a starter motor current of about 200 A, an external shunt is fitted (Figure 1.70, facing page).

Since an ammeter is *always placed in series with the circuit components*, the meter must have a low resistance: it must not restrict the current flowing in the circuit.

Moving-coil voltmeter The current flowing through a milliammeter is proportional to the

p.d., so by changing the scale it is possible to use the meter as a millivoltmeter.

When a larger voltage is to be measured, a resistor is fitted in series with the meter to prevent damage from an excessive current. This high-value resistor, called a *multiplier* is fitted internally and the meter is scaled accordingly (Figure 1.71).

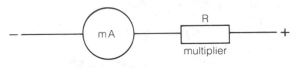

Figure 1.71   *Multiplier resistor*

A voltmeter is *used in parallel* with a circuit component. It should have a high resistance, compared with the resistance across which the p.d. is to be measured, so as to give an accurate reading.

Multi-range meters Since the milliammeter forms the basic meter it is possible to construct a multi-range test meter to cover various ranges.

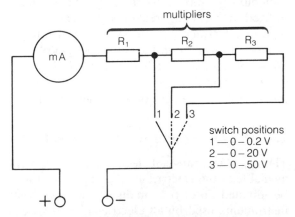

Figure 1.72   *Multi-range voltmeter*

Figure 1.72 shows a circuit of a multi-range voltmeter which incorporates a switch to select the appropriate voltage multipliers. A typical meter of this type has ranges 0–0.2 V, 0–20 V, and 0–50 V.

A multi-range ammeter circuit is shown as Figure 1.73. This uses a universal shunt which has tappings to obtain the various current ranges.

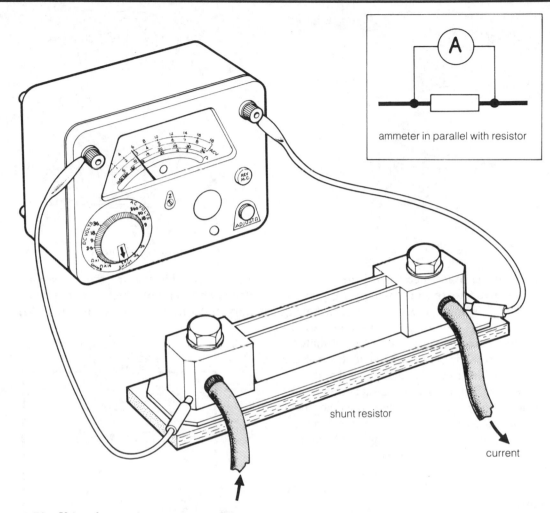

ammeter in parallel with resistor

shunt resistor

current

Figure 1.70 *Using shunt resistor to measure large current*

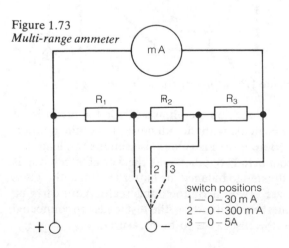

Figure 1.73
Multi-range ammeter

switch positions
1 — 0 – 30 m A
2 — 0 – 300 m A
3 — 0 – 5A

Before using this type of meter, it is wise to initially select a higher range than the maximum expected current so as to avoid overloading the meter.

Moving-iron meters This is a cheaper type of meter which can be used with either d.c. or a.c. currents.

There are two types of moving-iron meter: one operates on magnetic repulsion and the other on magnetic attraction.

Figure 1.74 shows a repulsion-type meter. It consists of a coil which is placed around two pieces of soft iron: one piece is attached to the

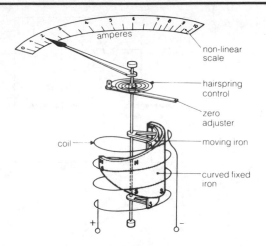

Figure 1.74   *Repulsion-type ammeter*

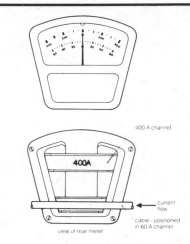

Figure 1.75   *Induction-type meter*

spindle and the other piece cut in the form of a taper and fixed to the meter body. A pointer, secured to the spindle, registers on a scale having unequal spacings. A hairspring provides the control force required for the range of the meter.

When current is passed through the coil the two pieces of iron become magnetized and similar poles are set adjacent to each other. The repulsion of the 'like' poles moves the inner piece of iron towards the narrow end of the taper.

On this type, needle damping is achieved by some form of air vane or dashpot.

Moving-iron meters can be used either as a voltmeter or ammeter by adding suitable shunts or multipliers.

The induction ammeter   This relatively cheap meter is clipped on to the outside of a cable and it registers the current flowing in the circuit (Figure 1.75). It has the advantage that it does not require cable disconnection so it can be quickly fitted, but its accuracy is poor compared with a normal meter.

The instrument measures the magnetic flux around a cable which is set up when current is passing through it.

The ohmmeter   Many multimeters incorporate a circuit for measuring the resistance of a component.

To provide this feature the milliammeter is switched into a circuit which incorporates a small battery and variable resistor (Figure 1.76).

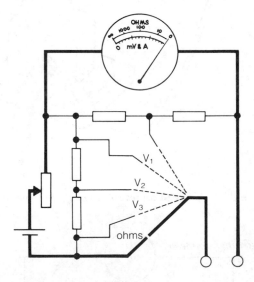

Figure 1.76   *Circuit of multi-meter*

The zero on the ohms scale on the meter is set to coincide with the full deflection of the pointer.

Before the instrument is used the test leads are connected together and the variable resistor is adjusted to balance the battery voltage: this gives a zero reading on the ohms scale. After carrying out the initial setting, the test leads are connected to the resistance to be measured.

The principle of the ohmmeter is based on Ohm's law: R = V/I. For a given voltage it is seen that the resistance is inversely proportional to the current, i.e. as the current increases, the resistance decreases. Scaling the meter in the opposite direction to normal practice, and calibrating the scale by using known resistance values, it is possible to provide a meter which measures resistance in ohms.

The range of an ohmmeter is increased by using a multi-point switch which brings into the circuit various multiplier resistors.

Digital meters The meters previously described show the reading by means of a pointer and a calibrated scale, i.e. an *analogue display*.

Many modern meters make use of semiconductor devices. These meters present the information to the tester by a *digital display*.

Use of test meters
Intelligent use of the three basic meters: voltmeter, ammeter and ohmmeter generally allows the cause of a fault to be quickly diagnosed. The method of performing these tests is considered at this stage.

Voltage checks The p.d. at various parts of a circuit can be determined by connecting a voltmeter between 'earth' and the main points of the circuit as shown in Figure 1.77. A 'good' circuit

gives meter readings (1 to 4) similar to battery p.d. and reading (5) will be zero.

Normally tests should start at the source of power and follow the circuit through to earth. In motor-vehicle work this is not always possible because some points are inaccessible. Instead the initial reading is taken at the source and then this is compared with the p.d. across the consumer component: in the case of the circuit shown in Figure 1.77, this will be the lamp. If the voltage differs by more than the recommended amount, further readings should be taken at the exposed connection points nearer the source to locate the fault.

Two faults identified by this check are: open circuit and high resistance. In Figure 1.77, an open circuit at the switch will give a zero reading at (3) whereas a high resistance at this point will show a p.d. at (3) less than that indicated at (2). For example, if the reading at (2) is 12 V and at (3) the p.d. is 8 V, then this shows the presence of a resistance having a voltage drop of 4 V across it.

Volt-drop test This is carried out by mounting a voltmeter in parallel with a part of the circuit: any reading shown on the meter indicates the p.d. across that part. In the previous example a volt-drop across points (2) and (3) will show 4 V (Figure 1.78). By volt-drop testing all sections of a complete circuit, the sum of all volt-drop readings will equal the battery p.d.

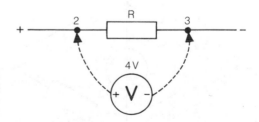

Figure 1.78 *Volt drop test*

Measurement of resistance The ammeter/voltmeter method of measuring resistance can be used if an ohmmeter is not available.

If the value of the resistance in Figure 1.79 is to be found, then the meters should be connected as shown in the diagram. Assuming the volt-drop

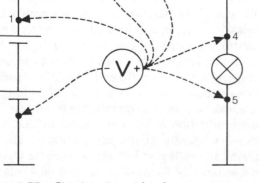

Figure 1.77 *Circuit testing with voltmeter*

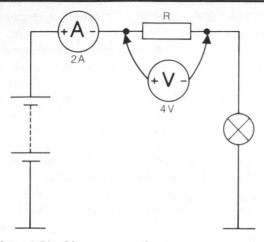

Figure 1.79 *Measurement of resistance*

tery cell and a centre-zero galvanometer, connected across the bridge, shows the difference in potential between points (a) and (b).

Closing the switch causes current to pass across the bridge by taking the two paths as indicated by I_1 and I_2. The current passing along each path is governed by the total resistance of that section of the circuit compared to the total resistance of the alternative path.

If all resistors are of equal value, then the voltage drop across R_1 is similar to R_3. In this case the p.d. at (ab) is zero and a *null deflection* of the galvanometer shows that the bridge is in *balance*.

Other combinations of resistors can be used. When the resistor values produce a balanced bridge, then it indicates that:

$$\frac{R_1}{R_2} = \frac{R_3}{R_4}$$

The bridge method may be used to find the value of an unknown resistor. The bridge is formed by placing resistors of known values at R_2, R_3 and R_4 and the unknown resistor at R_1. The value of R_2 is then varied until the galvanometer shows a zero reading. The unknown resistor's value is given by:

$$R_1 = R_2 \times \frac{R_3}{R_4}$$

A Wheatstone Bridge circuit can also be used to demonstrate the effect of temperature on a resistor. If heat is applied to one of the resistors of the bridge, then the galvanometer will show that the bridge becomes out-of-balance.

This feature can be used to compensate for changes of temperature in a bridge network of resistors. If R_2 in a bridge circuit was an active resistor in a circuit, e.g. a resistor in the air-flow meter of a fuel injection system, then a change in temperature of this resistor would produce a measurement error. This is avoided by fitting, at R_4 a resistor similar to that used at R_2 and arranging its situation in the component so that both resistors operate at the same temperature. In operation this layout will ensure that the change of resistance of R_2 and R_4 will keep the bridge in balance.

across the resistance is 4 V and the current is 2 A, then by applying Ohm's law ($R = V/I$) the resistance is 2 Ω.

Resistance measurement by using a Wheatstone Bridge The bridge arrangement was devised by Charles Wheatstone in 1843 for finding the value of an unknown resistor.

Today the bridge circuit is used for this task and for detecting a resistance change in a part of an electronic circuit.

Figure 1.80 shows a Wheatstone Bridge; it consists of four resistors connected to form four arms of the bridge. Current is supplied by a bat-

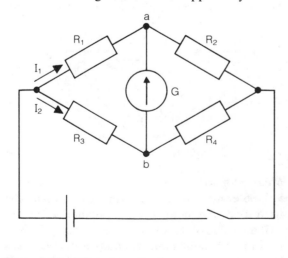

Figure 1.80 *Wheatstone Bridge*

Potentiometer The resistance of a conductor of uniform cross-section is directly proportional to its length.

This can be demonstrated by the layout shown in Figure 1.81. In this case the resistance, as indicated by the potential difference, is measured by a voltmeter. As the voltmeter contact is slid along the wire from 1 to 5, the p.d. registered on the meter decreases proportionally to the distance l.

Figure 1.81
Potentiometer

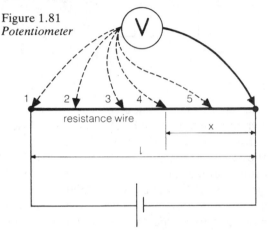

The p.d. across length x is:

$$\frac{x}{l} \times (\text{p.d. across length } l)$$

This is the principle used in a *slide-wire poten-tiometer*.

In science activities, a potentiometer is often used as an instrument to measure e.m.f., whereas in the automotive world a potentiometer is commonly used as a sensor to indicate the position of a moving part of a system.

Figure 1.82(a) shows a simple application of a potentiometer used to indicate a fuel level. In this case, the resistance wire is wound around an insulated former, to increase the length of the wire, and a sliding contact, attached to a float, is used to pass the current to earth. The meter in the circuit is an ammeter which is calibrated to indicate the fuel level.

As the fuel level rises, the contact slides over the wire to decrease proportionally the resistance and increase the current flow.

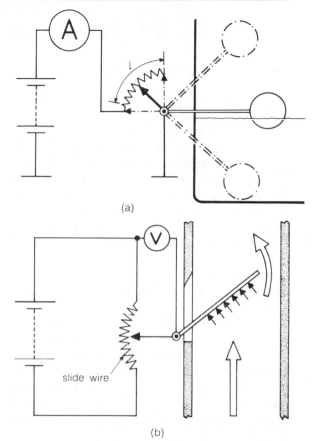

Figure 1.82   *Use of potentiometers*

An alternative circuit, shown in simplified form in Figure 1.82(b) can be used to indicate the position of a part such as a flap or throttle valve in an air intake system. This circuit uses a constant current and utilizes the change in p.d. to indicate the position of the contact on the slide wire. The change in p.d. can be registered on a voltmeter or can be used as a signal voltage to indicate, to an electronic control unit, the flap or valve position.

1.5 Capacitors

The capacitor
A capacitor, or condenser as it was called in the past, is a device for storing electricity for a limited time.

The common type of capacitor consists of two plates separated by an insulator called a *dielectric*

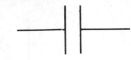

Figure 1.83 *Capacitor symbol*

and represented by the symbol shown in Figure 1.83.

The ability of a capacitor to hold a charge can be demonstrated by using a large capacitor. After charging the capacitor for a few seconds by connecting it to a high voltage d.c. supply, it is possible to obtain a spark when the terminals are connected together. In the past, this storage ability made capacitors attractive for building up high-voltage electrical energy.

Action of a capacitor The action of a capacitor is similar to the water 'capacitor' shown in Figure 1.84(a).

When the pump operates, water is taken from chamber B and is delivered under pressure to chamber A. This causes the piston to move against the spring until a point is reached where the maximum pump pressure equals the spring pressure. At this stage the capacitor is fully charged so if both taps are turned off, energy is stored for future use.

To discharge the 'capacitor', the pump is disconnected and the two sides of the 'capacitor' are inconnected by a pipe (Figure 1.84(b)). On opening the taps, the energy is released and the 'capacitor' is returned to its natural state.

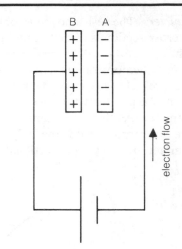

Figure 1.85 *Capacitor action*

The operation of an electrical capacitor is shown in Figure 1.85. When the switch is closed, electrons flow from the negative battery terminal to the capacitor plate A. As this flow takes place, a similar number of electrons move from plate B to the battery. Gradually the plates become charged and the p.d. across the plates is increased. This charge process continues until the capacitor p.d. is equal to the battery p.d. At this point the negative charge at plate A opposes any further flow of electrons.

When the battery is disconnected and the two plates are bridged with a wire, the electrons flow out of plate A and back into plate B. This flow continues until the two charges on the plates becomes equal.

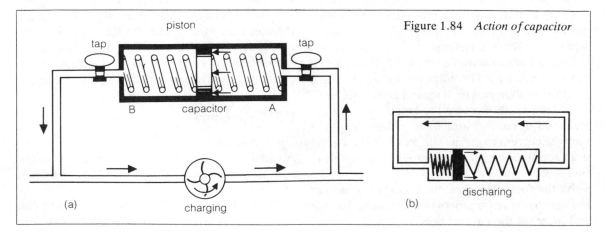

Figure 1.84 *Action of capacitor*

Capacitance Earlier in this book (page 6) it was stated that the quantity of electricity is proportional to its potential difference. Experiments with capacitors show that the ratio between the quantity of charge accepted by the plates, and the p.d. produced across the plates, is the capacitance of the part. This is stated as:

$$C = \frac{Q}{V}$$

where C = capacitance, farads (F)
Q = quantity of charge, coulombs (C)
V = potential difference, volts (V).

The unit of one farad is very large, so small capacitors used on motor vehicles are rated in microfarads (μF) – a millionth part of a farad.

Capacitance depends on a number of factors which include the dielectric material and the plate area.

Dielectric Capacitance can be greatly increased when certain insulation materials are placed between the plates. The electronic structure in these materials is activated by the charge on the plate and in consequence the dielectric encourages the plates to accept a greater charge.

Materials commonly used for a dielectric are mica, waxed paper and, in the case of an electrolytic capacitor, a paste of ammonium borate. Solid materials are ruined if a spark passes through the dielectric whereas liquid materials recover after the p.d. is reduced. This means that the type of capacitor selected for a given task is governed by the maximum voltage of the system in which it is to be used.

Capacitors normally used in motor-vehicle work have a dielectric which completely fills the space between the plates and is made very thin to give a large capacitance. Since the dielectric is an insulator, the safe working voltage may be increased as the thickness is increased.

Area of plates The larger the plate area, the greater the capacitance. So as to save space, the two plates are either rolled to form a cylindrical shape or stacked to give a number of rectangular

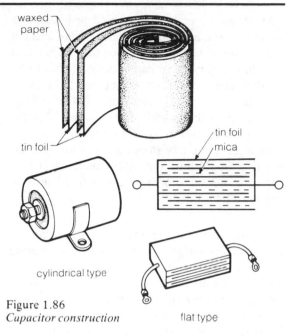

Figure 1.86
Capacitor construction

plates with each alternate plate connected in parallel with the others (Figure 1.86).

Electrolytic capacitor This type consists of two aluminium cylinders with a paste of ammonium borate between them (Figure 1.87).

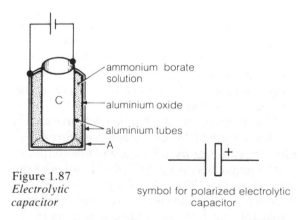

Figure 1.87
Electrolytic capacitor

It is called electrolytic because when a current is passed from the anode (A) to the cathode (C), a reaction takes place which liberates oxygen. This combines with the aluminium to form an aluminium oxide dielectric over the inner surface of (A).

After the dielectric has formed, it becomes a capacitor – the cylinder (A) as one plate and the paste as the other 'plate'. Since the dielectric is very thin, a very large capacitance is obtained.

Two precautions must be taken with this type: the oxide layer will break down if the voltage exceeds the value stated on the case. Also the polarity of the p.d. must not be reversed.

An alternative design of electrolytic capacitor uses two aluminium foils interleaved with paper soaked in ammonium borate and fitted into an aluminium container.

R-C time constant   It takes a given time to charge or discharge a capacitor, so by altering either the supply voltage or capacitance it is possible to obtain any given time interval. The supply voltage and hence the current flow to a capacitor can be controlled by fitting a resistor in series with the capacitor. In this case the time interval for the charge or discharge cycle to take place is called the *R-C time constant*:

$$time = resistance \times capacitance$$

or

$$T = R\text{-}C$$

The product of R and C represents one R-C time constant. In one time constant a capacitor will charge to about 63% of its total charge or discharge to about 37% of its total charge. To fully charge or discharge a capacitor takes about five R-C time constants.

The charging and discharging action is shown in Figure 1.88(a). When the switch is moved to A the capacitor begins to charge and this continues until the p.d. across the capacitor is equal to the battery voltage (Figure 1.88(b)).

Moving the switch to B discharges the capacitor at a rate which initially is quite rapid, but after about one R-C time unit, the rate slows down and the discharge is more gradual.

Increasing the value of either the resistance or the capacitance in the circuit increases the time taken to charge, or discharge, the capacitor; i.e. the R-C time constant is increased.

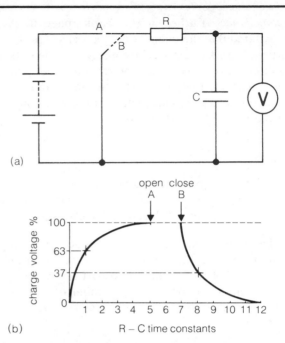

(a)

(b)

Figure 1.88   *Charging and discharging a capacitor*

Use of capacitors   Capacitors are often used in alternating current and radio circuits because they can transmit alternating currents. A direct current cannot be passed through a capacitor. The a.c. application is shown in Figure 1.89.

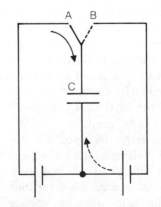

Figure 1.89   *Capacitor subjected to a.c. current*

When the switch is set at position A, the current gives a positive charge to the plate C. Moving the switch to position B causes the plate C to lose its charge and allows plate D to receive a positive charge. Oscillating the switch between A and B gives an alternating current in the capacitor section of the circuit shown.

Capacitors are used for spark quenching, i.e. to prevent arcing at the contacts of a circuit breaker. In an ignition circuit, self-induction causes a high voltage sufficient to produce a spark across the contacts at the instant they break. Fitting a capacitor in parallel with the contact breaker allows the energy to be momentarily absorbed (Figure 1.90). When the capacitor discharges, the contact gap is too large for arcing to occur.

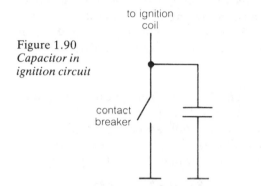

Figure 1.90
*Capacitor in
ignition circuit*

In many ways the quenching action is similar in principle to the way in which a capacitor is used as a smoothing device in a circuit that is subjected to current surges. In this case the capacitor acts as a buffer to absorb the voltage and current 'peaks' which otherwise might cause damage to circuit components or produce radio interference.

One of the many applications of an R-C time constant is an electronic ignition control module (see page 136). The timer feature is used to maintain an ignition output that is constant over a wide speed range. One type of breakerless ignition system (type CD) uses a capacitor to store energy until the spark is required. At this point, the energy is suddenly released and a high voltage discharge through the ignition coil gives a high energy spark at the plug.

1.6 Semiconductors

Semiconductor materials
A semiconductor is a material which has an electrical resistance value lower than an insulator and higher than a conductor.

To be classified as a conductor or insulator the resistivity is lower than 0.000 000 01 Ωm or higher than 10 000 Ωm respectively.

In addition to the resistivity aspect, semiconductor materials, such as silicon and germanium, have an atomic structure which behaves differently from materials that are good conductors. Earlier (p. 3) it was shown that electrical current was produced by the random drift of free electrons. In insulators, and partly in semiconductors, electron drift is limited because there are relatively few free electrons available that are not tightly bound to their atomic 'home'.

Electrons and holes Silicon and germanium have an atomic structure which includes four *valence electrons* in their outermost shell. The name valence is used because these electrons bond with other atoms in the structure in addition to the natural bond to their own nucleus (Figure 1.91). At low temperatures the electrons are

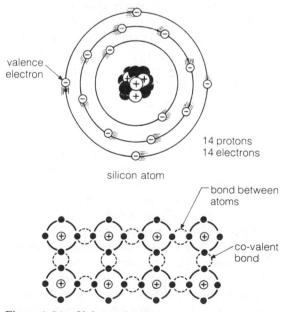

Figure 1.91   *Valence electrons*

firmly bonded to their nuclei, but as the temperature is raised, this bond relaxes and some electrons break free. When this happens the loss of an electron leaves a vacant space called a *hole* (Figure 1.92). The atom is now deficient of a

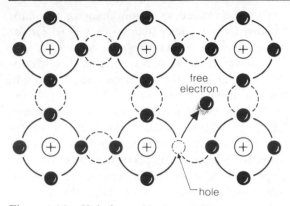

Figure 1.92   *Hole formed by loss of electron*

negatively-charged electron, so the atom is no longer in a neutral charged state; instead the creation of a hole gives the atom a positive charge. This is because the positively-charged nucleus is not balanced by the negatively-charged electrons.

When a hole exists, the positive charge of the atom attracts a valence electron from an adjoining atom and, after transfer has taken place, another hole is formed. This movement of electrons and holes is random: there is no rigid pattern of movement through the semiconductor material at this stage.

Connecting a battery to a semiconductor causes the p.d. to urge all electrons to move in one direction which is towards the positive battery terminal. This electron movement is accompanied by a 'hole drift' in the opposite direction (Figure 1.93)

Whereas the resistivity of good conductors increases with temperature, the electrical resistance of a semiconductor decreases. This is one way in which a semiconductor can be distinguished from a pure metal.

As the temperature of a pure metal is increased, the vibration of the atoms becomes more violent. This action causes the free electrons to collide more frequently with the atoms blocking their path, so in consequence the resistance to current flow is increased and extra heat is generated.

When the temperature of a semiconductor is raised, the increase in thermal energy causes some of the valence electrons to break free from their atomic bonds. These become free electrons and because more current carriers are available, the current flow through the material is increased, i.e. the resistance is decreased.

This negative-temperature-coefficient feature of a semi-conductor is useful for temperature sensing. One application is the sensor unit of a cooling-system temperature gauge. This device is called a *thermistor*.

Semiconductor doping   Doping a semiconductor means the adding of an impurity to a pure semiconductor crystal. This alters the behaviour by changing the number of charge carriers in the material.

N-type semiconductor   This type of semiconductor has a surplus of negatively-charged electrons. It is produced by adding a small trace (about one part in a million) of impurity such as arsenic to a pure silicon or germanium crystal. Arsenic is a

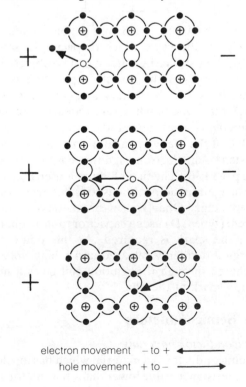

electron movement   – to +   ⟵

hole movement   + to –   ⟶

Figure 1.93   *Electron and hole movement*

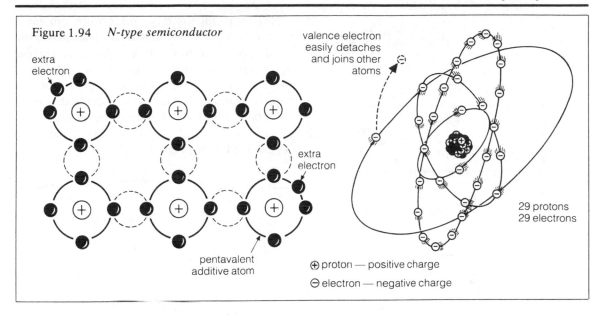

Figure 1.94   *N-type semiconductor*

extra electron

valence electron easily detaches and joins other atoms

extra electron

29 protons
29 electrons

pentavalent additive atom

⊕ proton — positive charge

⊖ electron — negative charge

pentavalent element; this means that it has five valence electrons, so when it is added to a semiconductor, the extra electrons are not able to form bonds with the adjoining atoms. Instead, they remain free to drift at random through the crystal and act as extra charge carriers to those that already exist (Figure 1.94).

The name n-type is short for negative-type. In this case the arsenic is called a donor because the impurity gives extra electrons to the semiconductor.

P-type semiconductor   This type is made by adding a trace of an impurity such as boron or indium to a pure crystal of silicon or germanium. The result is that the crystal becomes short of its full complement of electrons so a positive charge is produced; hence the name p-type.

The impurity used is trivalent, i.e. it has only three valence electrons in the outermost shell of its atom. When this is added to a pure semiconductor crystal, a number of holes are formed in the atom because of the incomplete bond between the impurity and the silicon or germanium (Figure 1.95). This feature explains why the impurities are called *acceptors* – they have the effect of robbing the semiconductor of some of its electrons.

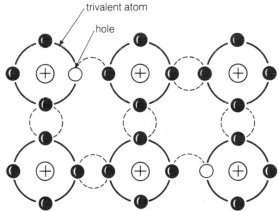

trivalent atom

hole

Figure 1.95   *P-type semiconductor*

A p-type semiconductor in this state may be regarded as a piece of Dutch cheese – full of holes.

P–N junction

When a p- and n-type semiconductor are joined together, the contact region is called a *junction*. At this point the surfaces diffuse together to give a thin region where the electrons and holes penetrate the p- and n-semiconductors respectively.

After the initial electron and hole transfer has taken place, the negative charges on the p-side and the positive charges (holes) on the n-side

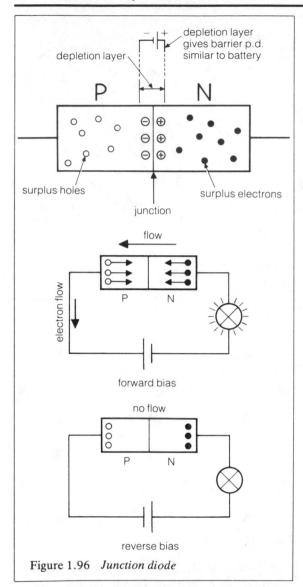

Figure 1.96   *Junction diode*

Reversing the battery connections applies a *reverse-bias* to the p-n junction. Electrons on the p-side are in the minority and since the surplus holes represent positive charges, the conditions will not allow current flow from p to n at this battery voltage. Temperature affects this basic condition, for if the temperature is increased, electrons will be liberated from the p-semiconductor (Figure 1.97).

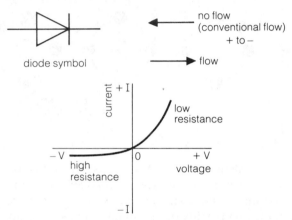

Figure 1.97   *Diode characteristics*

Assuming the voltage and temperature is not high, the diode will act as a 'one-way valve' and will allow current to flow in one direction only. This feature makes it suitable for use as a *rectifier* (a device for converting a.c. to d.c.) and also in situations where a one-way flow is required (Figure 1.98).

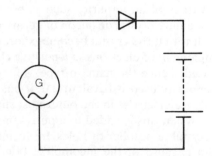
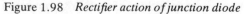

Figure 1.98   *Rectifier action of junction diode*

Direct current is required for charging a battery so the a.c. generated by an alternator must be rectified.

produce a *barrier p.d.* which opposes further diffusion.

Junction diode   Figure 1.96 shows a junction diode placed in a circuit with a battery. With the battery positive connected to the p-semiconductor, the electrons in the n-semiconductor readily flow to fill the holes in the p-semiconductor. The p-n junction is said to be *forward-biased* in this circuit and this will cause the lamp to illuminate.

A single diode in the circuit shown in Figure 1.98 gives *half-wave rectification*, but this is wasteful because it loses half of the available energy (Figure 1.99).

To achieve *full-wave rectification* a bridge circuit having four diodes is needed (Figure 1.100). This circuit ensures that a d.c. flow is obtained irrespective of the direction of current being generated at the source. This type of rectification is used in the alternator.

Zener diode and avalanche diode   Zener discovered that when an increasing reversed-bias p.d. was applied to a junction diode, a point was reached where the diode 'broke down' and

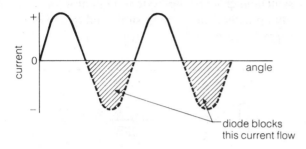

Figure 1.99   *Graph of half-wave rectification*

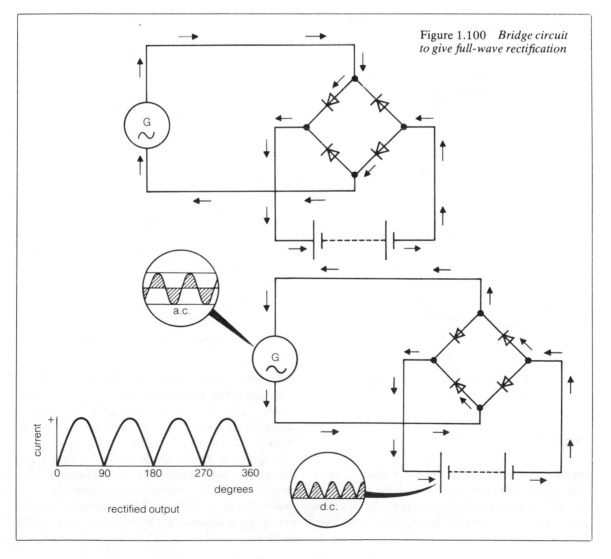

Figure 1.100   *Bridge circuit to give full-wave rectification*

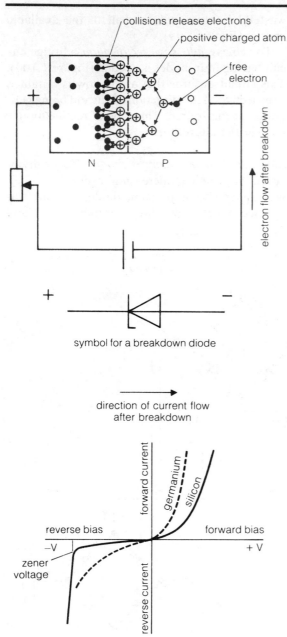

Figure 1.101   *Zener-type diode*

allowed current to flow freely (Figure 1.101). This is called the *Zener effect*. It is due to the high electric field pressure which acts at the junction and causes electrons to break free from their atomic bonds.

This feature makes the Zener-type diode particularly suited for use as a voltage-conscious switch in a charging-system regulator or as a 'dump device' in a circuit subjected to a voltage surge (Figure 1.102).

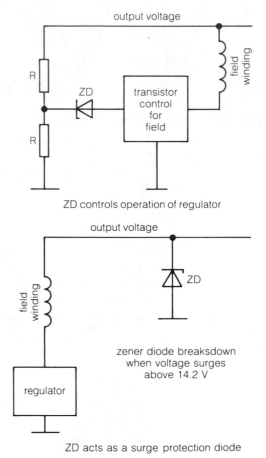

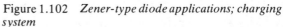

Figure 1.102   *Zener-type diode applications; charging system*

The avalanche effect, when the Zener voltage is reached, has resulted in this type being commonly called an 'avalanche diode' but from a scientific viewpoint the Zener diode is different from a true avalanche diode.

Whereas a Zener diode achieves its voltage reference characteristic by 'tunnelling' of the charge carriers through the junction, the avalanche diode achieves the characteristic by producing a physical bulk breakdown across the junction.

A voltage reference diode with a reverse characteristic below about 4.5V is a Zener diode and a diode having a 'breakdown voltage' above 4.5 V is an avalanche type.

The voltage drop across a diode is constant irrespective of the current it is carrying assuming it is operating within its specified range. This feature also applies to a Zener-type diode so this makes it particularly suitable for applications where a steady voltage is required instead of a varying one. p-n Junctions employed to give steady voltages above about 4.5 V are often called *voltage regulator diodes*.

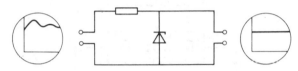

Figure 1.103 *Zener-type diode used as a regulator to stabilize voltage*

Figure 1.103 shows the use of an avalanche diode to stabilize the voltage in a circuit. When the input voltage exceeds the diode's breakdown voltage, the diode conducts and 'absorbs' the excess voltage. During this stage, the output voltage remains constant because it represents the potential difference or voltage drop across the diode. This application is shown on page 135.

Light-emitting diode (LED) In 1954 it was discovered that a diode made of gallium phosphide (GaP) emitted a red light when it was forward-biased.

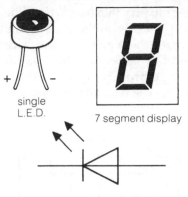

single L.E.D.

7 segment display

general symbol

Figure 1.104 *Light emitting diode*

Since that time, experiments have shown that the colour of the emitted light can be altered by varying the impurities in the material. LEDs are now commonly available in red, orange, yellow and green (Figure 1.104). In addition, a radiation of near infra-red can be obtained: this is used with a phototransistor as a trigger on some optoelectronic ignition systems.

LEDs have a characteristic similar to a common p-n junction.

The transistor
A transistor is formed when two p–n junctions are placed back-to-back. Positioned one way gives a p-n-p type and the other way, an n-p-n type transistor. The three parts are named collector (C), base (B) and emitter (E).

Figure 1.105 shows a diagram of a transistor and the symbols which are commonly used. In

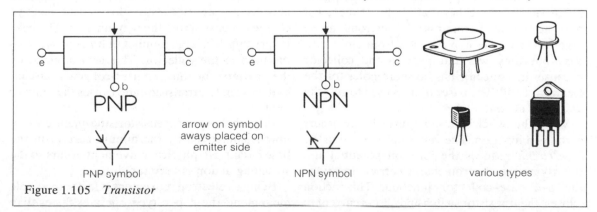

PNP

NPN

arrow on symbol aways placed on emitter side

PNP symbol

NPN symbol

various types

Figure 1.105 *Transistor*

both types, the very thin base is the 'filling' of the sandwich. The arrow in the symbol points either to, or from, the base and is always placed on the emitter side: it shows the direction of conventional flow, i.e. from p to n.

A transistor can be used as a switched device or amplifier.

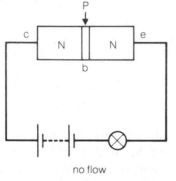

no flow

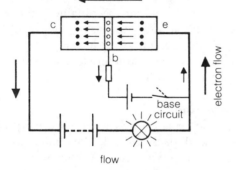

Figure 1.106   *Switching action of transistor*

Switching action   When an n-p-n transistor is connected in the simple circuit shown in Figure 1.106, no current, other than slight leakage, will pass the transistor from p to n. Electrons can be urged to cross the n–p junction from emitter to base but they will not pass to the collector because the collector has no spare holes for the electrons to fill. The base circuit is used to give the switching action.

When the switch is closed, a disturbance occurs at the p–n junction. The very small current in the *base circuit* reduces the junction potential and this allows the electron charge carriers to bridge the p–n base–collector junction. This action 'closes' the transistor switch and allows current to

flow freely in the main circuit. In this diagram, this current operates the lamp.

Very little current is needed to 'switch' the transistor so this feature makes the device suitable for use as an alternative to the electromagnetic relay. The *solid-state* crystal has no moving parts, is resistant to shock and vibration, is small in size and is electrically efficient. As with other semiconductors, it dislikes heat and is instantly damaged if the battery polarity is reversed: this is because a large current flow from collector to base destroys the structure of the base.

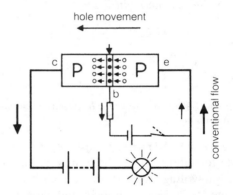

Figure 1.107   *p-n-p Transistor*

The circuit for a p–n–p type differs from a circuit for an n–p–n type transistor: a p–n–p type must have its emitter connected to a positive supply. Current flow through a p–n–p type is obtained by the movement of the positively charged holes (Figure 1.107).

When the base circuit of a p–n–p transistor is closed, the low-voltage battery gives a forward bias to the emitter–base: this causes positive charges to pass across the p–n junction. The base is very thin so a large number of holes pass across the base to the collector. This action, aided by the battery in the emitter-collector circuit, switches-on the transistor and causes the lamp to light.

In both types of transistor, the main current flow is controlled by the minute current in the base circuit. No physical movement occurs so the switching action is very rapid.

Being as electron movement is faster than hole movement, the n–p–n type operates faster than the p–n–p type: this makes the n–p–n type more suited for high-frequency switching operations.

Amplification   In addition to the switching feature, the transistor can also act as a current or voltage amplifier.

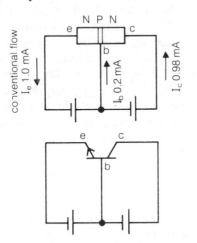

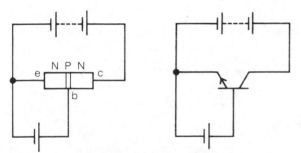

Figure 1.108   *Amplification; C-B circuit*

The amplifying action is shown in Figure 1.108. This shows an n–p–n transistor connected in a *common-base* (C-B) circuit. Typical values for current flow for an a.f. (amplification factor) type transistor are:

$$I_b = 0.02 \text{ mA}, \quad I_e = 1.0 \text{ mA}, \quad I_c = 0.98 \text{ mA}$$

These values show that a base current of 0.02 mA gives a collector current of 0.98 mA and an amplification (magnification) of 0.98/0.02 = 49.

The C-B arrangement is normally unsuitable for current amplification circuits: instead a *common-emitter* (C-E) arrangement is used.

Figure 1.109   *Amplification; C-E circuit*

When the transistor is connected in the C-E mode, the base and emitter are linked together (Figure 1.109).

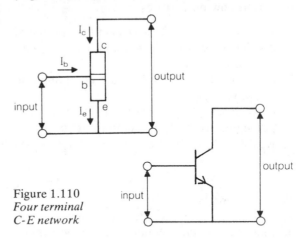

Figure 1.110
*Four terminal
C-E network*

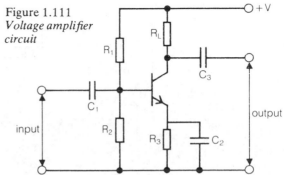

A four-terminal C-E network, shown in Figure 1.110 in simplified form, allows the input and output to be either d.c. or a.c. or a combination of both. The amplification, or forward current gain, shown by the symbol h_{fe} in catalogues, is given by:

$$h_{fe} = \frac{\text{increase in } I_c}{\text{increase in } I_b}$$

The value of h_{fe} depends on the type of transistor and it varies from about 2 to 10 000; a typical value is about 100.

Figure 1.111
*Voltage amplifier
circuit*

When used with suitable resistors and capacitors, the transistor will act as a voltage amplifier (Figure 1.111). A typical *voltage gain* for low frequency a.c. is about 200 when the load resistor R_L for stepping-up the output voltage is about 5kΩ.

The a.f. amplifier circuit shown in Figure 1.111 has the following features:

1. Resistors R_1 and R_2 which act as a potential divider to provide the constant base current, i.e. the base bias.
2. Capacitor C_1 to isolate the d.c. component in the input signal.
3. Capacitor C_2 to prevent feedback of the output signal to the base–emitter circuit.
4. Resistor R_3 to stabilize the circuit for temperature change.

Phototransistor

A phototransistor is a photodiode combined with an amplifier. A *photodiode* is a junction diode that is sensitive to light. Reverse-biasing the diode causes minority carriers to flow in the circuit and this occurs when the diode is shielded from the light. Exposing the diode to light energy produces more electron-hole pairs which pass the junction and increase the current flow.

A phototransistor has the base connected to the emitter via a resistor and when light falls on the emitter side, electron-hole pairs are formed in the base. The base current produced by this action is amplified by the transistor to give a larger collector current.

An optoelectronic transistor as described can be obtained to suit a given light-frequency band. Also they can be supplied to sense electromagnetic radiation waves approaching the infrared (IR) wavelength. A IR phototransistor is used with a LED as a trigger for optoelectronic ignition systems. Another use is for a light-sensing unit in an automatic control system for switching-on the parking lights of a vehicle when it gets dark.

Thyristor

This family of semiconductor devices is a development of the transistor, but whereas the transistor has two p–n junctions, the thyristor has three.

The common type of thyristor is called a *reverse blocking triode* or a *silicon-controlled*

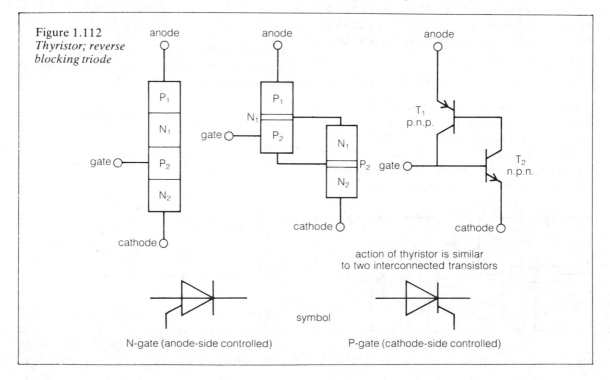

Figure 1.112
Thyristor; reverse blocking triode

rectifier (s.c.r.). The term triode indicates that it has three electrodes or connections; these are anode, cathode and gate.

The main feature of a s.c.r. is its switching action. Applying a small trigger current to the gate switches-on the thyristor and this causes current to flow from the anode to the cathode. Once this main current starts to flow the interruption of the gate current has no effect. Only when the anode-to-cathode voltage is reduced to zero or its polarity is changed is the thyristor switched-off.

This feature makes the thyristor useful in circuits such as a C.D. ignition system where a small trigger current of short duration is all that is needed to start the flow of a large current.

The operation of a silicon p–n–p–n device is similar to two interconnected transistors as shown in Figure 1.112. When voltage is applied to the anode no current will flow to the cathode because both T_1 and T_2 are switched-off. Applying a voltage of similar polarity to the gate will switch-on T_2 and as a result T_1 will also switch-on. Current flow from the collector of T_1 to the base of T_2 will now keep T_2 switched-on even if the gate current is discontinued.

Unipolar transistors

A transistor is called unipolar when it involves only one type of charge carrier such as the majority type. Bipolar transistors, as shown previously, involve the drift of majority and minority carriers through the base region.

Field-effect transistor (f.e.t.)   One type of unipolar transistor recently developed is the field-effect transistor; Figure 1.113 shows an *n-channel junction f.e.t.*

This type of transistor has a p-type silicon region known as a *gate* grown into the sides of a n-type silicon *channel*. The channel is connected to a *drain* (collector) and *source* (emitter). The depletion layer around the p-type gate gives a reverse bias so the gate is negatively biased with respect to the source.

Electrons acting as majority carriers flow from the source to the drain; the rate of flow is limited by the voltage applied to the gate, i.e. as the gate voltage with reference to the source is increased, the electron flow from the source to the drain is reduced. When a sufficiently high gate voltage is applied, the flow between the source and gate ceases altogether.

Control of the main current is achieved by altering the depletion zone around the p-type semiconductor. Raising the gate voltage increases the depletion region and this reduces the width of the channel through which current can flow from the drain to the source (Figure

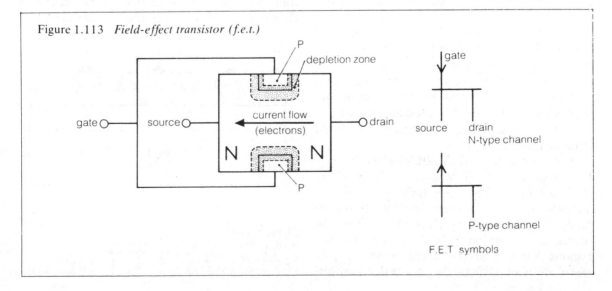

Figure 1.113   *Field-effect transistor (f.e.t.)*

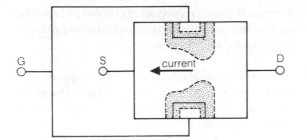

Figure 1.114   *Action of f.e.t. transistor*

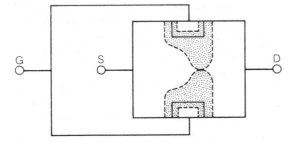

Figure 1.115   *Pinch-off action of f.e.t. transistor*

1.114). When the gate voltage is increased to extend the depletion zone across the channel, a *pinch-off* condition is obtained; at this stage the drain current is cut-off (Figure 1.115).

Compared with a bipolar transistor, an f.e.t. uses a lower current to control the switching and has a much higher input resistance. These advantages make the f.e.t. useful for many applications, e.g. computers used in electronic control units.

Metal oxide semiconductor transistor (m.o.s.t.)
This type has been developed from the f.e.t. It has an input resistance greater than $10^{12}\Omega$ and a gate current that is much lower than the control current for a bipolar transistor. These advantages show why metal oxide semiconductor transistors are now used extensively for circuits involving amplification.

A m.o.s.t. is formed by diffusing two p+ regions into the side of a n-type silicon crystal (Figure 1.116). The surface is then covered with an insulating layer of silicon dioxide. Holes are made in the insulating material to allow the source and drain to be connected to the p+ regions. On top of the silicon dioxide a gate electrode is deposited; this is a thin metal film which is positioned so that it bridges the n-region between the source and drain.

When a negative charge is applied to the gate, positive charges (holes) are attracted to the n-region adjacent to the gate. This builds up an *inversion* layer which forms a p-type channel between the two p+ regions to provide a path for the electrons to flow from the source to the drain. As the gate voltage is increased, the channel gets deeper so this allows a larger current to flow.

This type of m.o.s.t. gives a current flow which increases with the gate voltage so it is said to operate in the *enhancement mode*; types which give a decrease in current for an increase in gate voltage operate in the *depletion mode*.

Both the f.e.t. and m.o.s.t. devices are made in n-channel and p-channel forms.

Microelectronics
Since the introduction of the transistor in the early 1950s, rapid progress has been made to miniaturize components and circuits so as to save space, speed-up operation and reduce costs.

Two basic circuit technologies are used in microelectronics; these are *film* (thick and thin) and *semiconductor integrated circuits* (bipolar and m.o.s.t.). Circuits which combine the two technologies are called *hybrid*.

Capacitors, conductors, resistors, diodes and transistors can all be made in film form on a small chip of silicon. The semiconductor integrated cir-

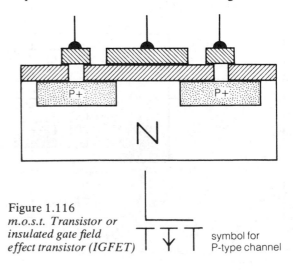

Figure 1.116
m.o.s.t. Transistor or insulated gate field effect transistor (IGFET)

symbol for P-type channel

cuit (I.C.) was introduced in 1956 and since that time improved production techniques have developed the I.C. from *small-scale integration* (s.s.i.) through *large-scale integration* (l.s.i.) to the period today where *very large-scale integration* (v.l.s.i.) is used. The v.l.s.i. type of microcircuitry has over a million devices accommodated on a chip of silicon having an area of a few square millimetres.

1.7 Basic digital principles

Many of the electronic units fitted to motor vehicles operate by means of digital pulses which pass around the system. These pulses transmit data from one part of the system to another by means of a digital code. The alternative method is the analogue system.

Analogue and digital signals
Most measurements are continuous quantities and are called *analogue* quantities. For example, an instrument that uses a needle to sweep across a fixed scale is called an analogue type. Similarly this term can be applied to any sensor or part that provides a continuous signal that can change by an infinitely small amount.

Digital quantities are expressed as whole numbers, so when applied to an instrument, the read-out does not alter until the value has changed by a set amount. A digital watch is an example of this type of measurement. If the unit of time used by the watch is the minute, then only when time has advanced by a full minute does the read-out change. In the case of a watch, smaller digital units can be used to show seconds, 0.1 second, etc.

It is usual to base a digital quantity on two numbers only, e.g. 0 and 1; this is called a *binary* code. A simple switch is a binary device because it has only two positions; on and off.

Electronic circuits using binary input and output signals are relatively cheap to produce. By combining a large number of digital switching circuits it is possible to make a computer. In this unit, digital signals can be stored in a memory bank for use at a later time.

Binary and denary numbers
Numbers in everyday use are based on *denary* notation; this uses the ten digits between 0 and 9. Binary notation is based on the number 2 with each extra digit representing a 'power of 2'. Hence a binary number of 1 1 1 1 is:

$$1 \times 2^3 + 1 \times 2^2 + 1 \times 2^1 + 1 \times 2^0$$

This represents a denary number of
$$8 + 4 + 2 + 1 = 15$$
Table IV (see page 50) enables other values between 0 and 27 to be found. Each binary digit is called a *bit* so 4 bits are needed to represent the denary number 15.

Voltage levels for digital data
Binary notation allows any denary number to be represented by the digits 0 and 1 so circuits which respond to a code based on these two digits are called *logic circuits*. These circuits respond to voltage signals; the voltage needed to produce a pulse to represent the digit 1 is called the *logic level*.

Normally the levels used are:

0–0.8 V	logic 0	low state
2.4–5.0 V	logic 1	high state

Digital electronic control units, used in conjunction with an analogue sensor, incorporate an analogue-to-digital (A/D) converter to change the signal to the pulse form required for the logic circuit. Figure 1.117 shows an analogue signal produced by a sensor and the digital signal given after conversion. When the analogue voltage exceeds 2.4 V, the high-state digital signal is obtained and this is held until the voltage drops below the logic level required to switch it back to the low state (Figure 1.118, p. 51).

Since a digital signal has only two states, it is much easier to transmit, process and store than an analogue signal.

Logical gates
A logic gate is a device or circuit that processes digital signals. It gives a digital output signal in response to one or more input signals.

Table IV   Binary and denary tables

	128	64	32	16	8	4	2	1	Denary (base 10)
	2^7	2^6	2^5	2^4	2^3	2^2	2^1	2^0	Binary (base 2)
Denary									
0	0	0	0	0	0	0	0	0	
1	0	0	0	0	0	0	0	1	
2	0	0	0	0	0	0	1	0	
3	0	0	0	0	0	0	1	1	
4	0	0	0	0	0	1	0	0	
5	0	0	0	0	0	1	0	1	
6	0	0	0	0	0	1	1	0	
7	0	0	0	0	0	1	1	1	
8	0	0	0	0	1	0	0	0	
9	0	0	0	0	1	0	0	1	
10	0	0	0	0	1	0	1	0	
11	0	0	0	0	1	0	1	1	
12	0	0	0	0	1	1	0	0	
13	0	0	0	0	1	1	0	1	
14	0	0	0	0	1	1	1	0	Binary codes
15	0	0	0	0	1	1	1	1	
16	0	0	0	1	0	0	0	0	
17	0	0	0	1	0	0	0	1	
18	0	0	0	1	0	0	1	0	
19	0	0	0	1	0	0	1	1	
20	0	0	0	1	0	1	0	0	
21	0	0	0	1	0	1	0	1	
22	0	0	0	1	0	1	1	0	
23	0	0	0	1	0	1	1	1	
24	0	0	0	1	1	0	0	0	
25	0	0	0	1	1	0	0	1	
26	0	0	0	1	1	0	1	0	
27	0	0	0	1	1	0	1	1	
255	1	1	1	1	1	1	1	1	

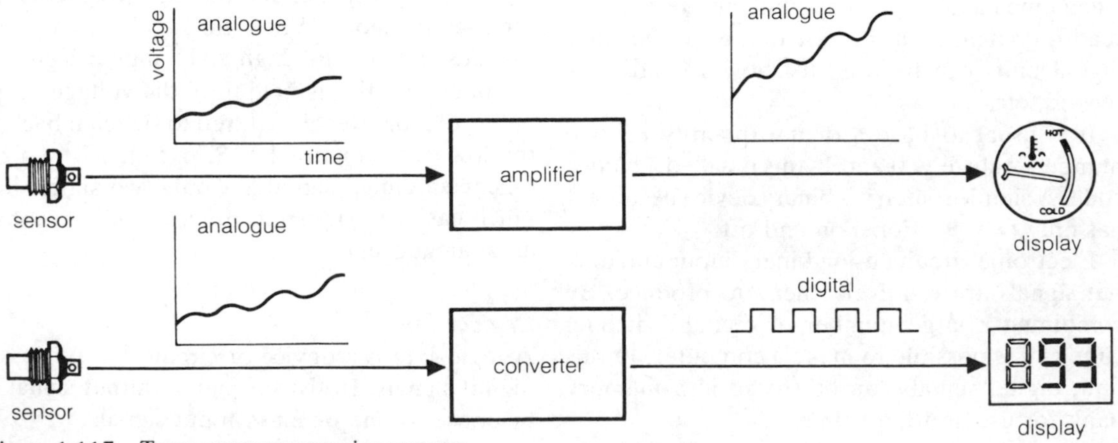

Figure 1.117   *Temperature measuring systems*

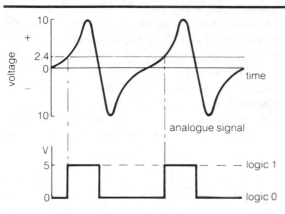

Figure 1.118   *Analogue and digital signals*

There are three basic logic gates; the AND gate, OR gate and NOT gate. From these basic gates can be constructed the NAND gate and NOR gate.

Many logic gates are used in electronic control units that incorporate electronic counters, memory units or a microprocessor; the digital computer has thousands of logic gates in its construction.

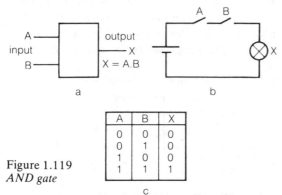

Figure 1.119
AND gate

A	B	X
0	0	0
0	1	0
1	0	0
1	1	1

c

The AND gate   The AND gate shown in Figure 1.119 has two inputs (A and B) and one output (X). It is designed to produce a digital output of logic 1 only when all inputs are set to logic 1. In all other conditions the output will be at the low state, namely logic 0.

Operation of the AND gate is similar to the simple electric circuit as shown in Figure 1.119(b). Representing the inputs by the operation of the two switches, it will be seen that the lamp will light only when both switches are closed.

The logic levels at A, B and X can be shown by a *truth table*; this allows the various input states to be indicated. An alternative method of showing the behaviour of a gate is to use a Boolean algebraic expression; in the case of the AND gate the expression is:

$$X = A \cdot B$$

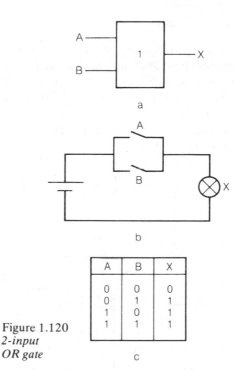

Figure 1.120
*2-input
OR gate*

A	B	X
0	0	0
0	1	1
1	0	1
1	1	1

c

The OR gate   Figure 1.120 shows the symbol for a two-input OR gate. To produce an output of logic 1 from this type of gate requires only one of the inputs to be set at logic 1, i.e. a high-state output occurs when one or both input levels are set to a high state. The logic symbol for a OR gate is:

$$X = A + B$$

The equivalent electrical circuit that produces this action is shown in Figure 120(b). This shows that the lamp will operate whenever one or more of the switches are closed.

The NOT gate   The NOT gate is called an *inverter* because it changes the logic level as the signal pulse passes through the gate. When the input is logic 0, the output is logic 1 and vice versa (Figure 1.121).

The logical symbol $\bar{A}$ is read as 'not A'.

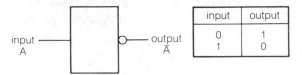

input	output
0	1
1	0

Figure 1.121   *NOT gate*

The NAND and NOR gates   Other logic functions can be obtained by combining the AND and OR gates with a NOT gate; this produces a NAND (NOT–AND) and NOR (NOT–OR) gate respectively.

The symbols and truth tables for these gates are shown in Figure 1.122.

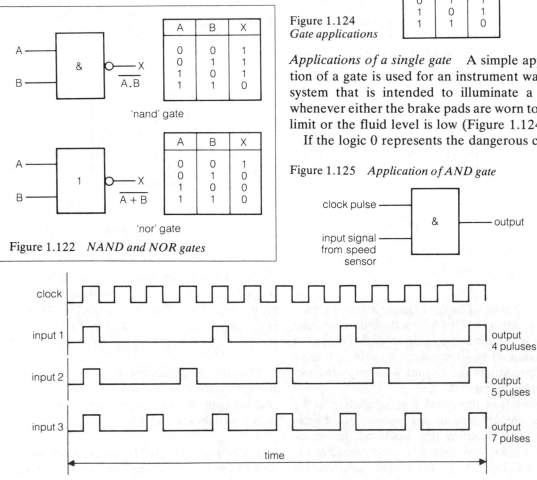

Figure 1.122   *NAND and NOR gates*

In both cases the small circle is the schematic symbol for NOT. The effect of adding the NOT function to the basic AND and OR gates is to invert the output.

Many electronic circuits used for vehicles' components are American-based so circuit diagrams often show symbols that differ from those recommended by BSI. The main variations are shown in Figure 1.123 (see facing page).

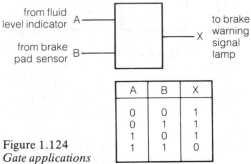

A	B	X
0	0	1
0	1	1
1	0	1
1	1	0

Figure 1.124
Gate applications

Applications of a single gate   A simple application of a gate is used for an instrument warning system that is intended to illuminate a lamp whenever either the brake pads are worn to their limit or the fluid level is low (Figure 1.124).

If the logic 0 represents the dangerous condi-

Figure 1.125   *Application of AND gate*

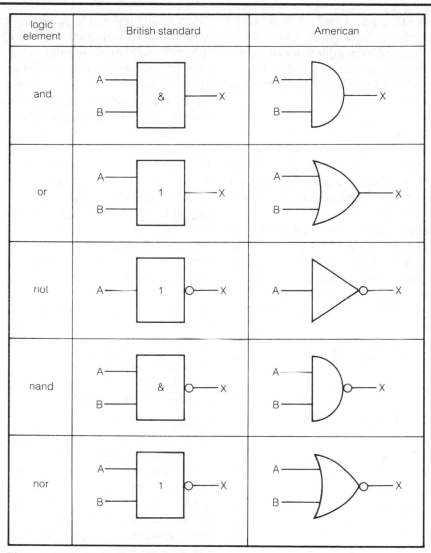

logic element	British standard	American
and	A —— & —— X B ——	A —— X B ——
or	A —— 1 —— X B ——	A —— X B ——
not	A —— 1 o— X	A —— o— X
nand	A —— & o— X B ——	A —— o— X B ——
nor	A —— 1 o— X B ——	A —— o— X B ——

Figure 1.123   *Gate symbols*

tion in each case, and logic 1 is the output signal needed to operate the warning lamp, then by constructing a truth table it will be seen that a NAND gate is the type required for this application.

Another example of a single gate is a case where counting is necessary to obtain the speed of a given component such as a driving shaft.

Figure 1.125 shows this application which uses an AND gate with two inputs A and B. A clock pulse having a constant frequency is applied to A and a square wave pulse, given by a sensor positioned close to the rotating shaft, is applied to B.

Comparing the clock pulse with the pulse form labelled (1) shows that in a given time interval the logic gate will output four pulses. As the shaft speed is increased, the number of output pulses from the gate increases, so this can be used to indicate the shaft speed on an appropriate meter.

In this case the clock pulse is used as a reference signal to enable the varying frequency pulses from the sensor to be counted against a set time interval.

Clock signals are used to control many electronic units involving logic gates. Usually the clock signal is obtained from a *quartz-crystal controlled oscillator*; this produces a stable square wave of frequency 3.2768MHz (3 276 800 oscillations per second).

Combinational logic

In the majority of applications, more than one logic gate is needed to produce a given output signal. When more than one gate is used as a single system the term *combinational logic* is used to describe the system.

As an example of this system, suppose a logic circuit is required to compare two inputs and give an output of logic 1 when the inputs are equal. After constructing a truth table for this set of conditions (Figure 1.126(a)), it will be seen that a circuit similar to that shown in Figure 1.126(b) is needed.

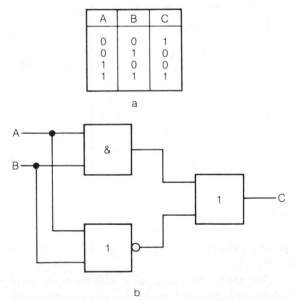

A	B	C
0	0	1
0	1	0
1	0	0
1	1	1

a

b

Figure 1.126   *Combination logic*

This combined gate is used in a computer when it has to compare the data held in its memory unit with a data signal that is transmitted from a sensor. After comparing the two signals, the gate indicates at its output when the two inputs are equal.

Any combinational logic system can be made by using one type of gate for the complete circuit. Either NAND or NOR gates are used; the systems are called *universal NAND logic* and *universal NOR logic systems* respectively. This arrangement gives a cheaper layout and minimizes the risk of incorrect assembly.

Integrated circuit (I.C.)

An integrated circuit consists of a number of gates which are formed upon a single piece of silicon. The first I.C. was made in 1959 when a transistor and resistor were formed on a single *chip* of silicon.

Nowadays these chips are in common use and form 'building bricks' to construct computers and electronic control units.

Logic circuits formed on an I.C. chip use combinations of diodes, resistors and transistors. These make *logic families* which are classified as:

(a) transistor–transistor logic   (TTL)
(b) emitter-coupled logic   (ECL)
(c) complementary metal oxide
      semiconductor logic   (CMOS)

Transistor–transistor logic   TTL circuits are widely used in cheap integrated circuits and cover a large range of logic functions. The power consumed is about 40mW per gate and its speed of switching (propagation delay) is about 9 nanoseconds. It has a good *noise* margin; this means that it resists changing its logic state when small spurious voltages are induced into the transmission line.

Figure 1.127 shows an I.C. having quad 2-input NAND gates.

Emitter-coupled logic   These arrays have a faster switching time than the TTL but they consume more power and are more expensive.

Complementary metal-oxide semiconductor logic   This family is often used in systems having a large number of gates because it consumes very low power (about 0.001mW per gate). The noise margin is high but it has a slow switching speed (about 30 nanoseconds). Also it is susceptible to

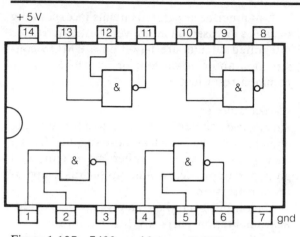

Figure 1.127   *7400 quad 2-input NAND gates*

damage from static charges so special care must be exercised when handling this type of I.C.

MOSFET-type transistors are used for CMOS gates, so this is why the power consumption is low. CMOS units are fitted to electronic-type instrumentation systems.

Bistables

Some integrated circuits use a *bistable* or *flip-flop* to remember a pulse condition that was previously applied to the inputs. The device can be set to remember the two states, 0 and 1, so this makes it suitable to use as a counter or memory store in a computer.

A simple toggle switch is a bistable device (Figure 1.128). When the switch is moved to position A, it remains stable in this position until it is moved to position B; the switch is said to flip-flop between the two states.

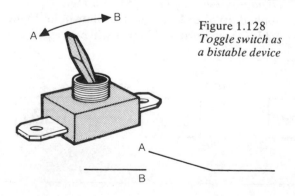

Figure 1.128
Toggle switch as a bistable device

Gates that behave in a manner similar to the toggle switch, i.e. the response to the inputs depends on the previous signals applied to the inputs, give a logic behaviour called *sequential logic*.

R–S flip-flop   The reset–set (R–S) bistable is made by interconnecting two NOR gates or two NAND gates (Figure 1.129).

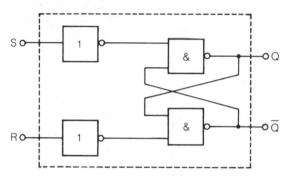

inputs		outputs	
S	R	Q	$\bar{Q}$
0	0	no change	no change
0	1	0	1
1	0	1	0
1	1	*	*

* state is uncertain

Figure 1.129   *R-S flip-flop*

The truth table for the NAND gates shows that when logic 1 is applied to S (set input), the output at Q is logic 1; it will remain in that high state even after S is changed to logic 0. This shows that the high state of S is *latched* into the state of Q.

Only when R is changed to logic 1 and S goes back to logic 0 is Q unlatched; this *resets* the latch and returns the gate to its original condition.

When R and S are set so that they are both at logic 1, the two gates *buck* each other and the final state of the flip-flop is uncertain; this indeterminate state is not permitted so the external circuit is designed to avoid this condition.

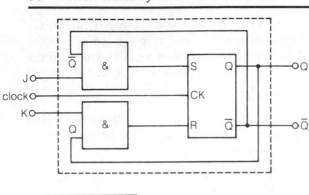

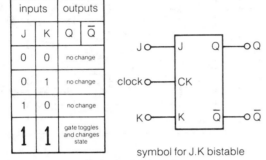

Figure 1.130   *J-K bistable*

A number of integrated circuits incorporating bistables are used on motor vehicles. Often these integrated circuits are based on a CMOS construction and contain four or six latches with a common reset feature.

Electric counter

An electric counter is a logic system for counting digital pulse signals such as those supplied from a transducer for sensing either movement of a given part or physical conditions such as temperature or pressure.

The output from the counter may be used for instrumentation purposes or for controlling the operating mechanism of systems such as ignition timing control and fuel metering.

Often the signal generated by a sensor is of analogue form so this must be converted by an analogue/digital (A/D) converter to a pulse shape required for a logic circuit. A typical pulse shaping device is a *Schmitt trigger* and the layout shown in Figure 1.131 shows the functional role of the trigger when it is used for tachometer operation.

Counters use binary notation to represent the number of pulses measured in a given time. The digital signals given at the outputs from a counter indicate the logic state of the various signal lines. These signals are passed to a decoder which converts the *binary coded decimal* (BCD) to a *seven segment display* (SSD). This decodes the signals by using combinational logic circuits and arranges the appropriate LED segments to illuminate in response to the incoming pulses.

J–K flip-flop   A J–K bistable operates similarly to an R–S type but incorporates two extra AND gates to overcome the indeterminate state produced when both inputs of the R–S are set to logic 1 (Figure 1.130).

When an input is changed so that logic 1 is then applied to both J and K, the flip-flop changes its output to a state opposite to that which existed before the input change.

The J–K bistable shown in Figure 1.130 is based on a clocked R–S flip-flop. This gives a synchronized action whereby the flip-flop can only change its state at a time when a clock pulse is applied to the bistable.

Divide-by-two counter   When a logic 1 pulse is applied to the J–K bistable shown in Figure 1.132(a), an output of logic 1 will be triggered by

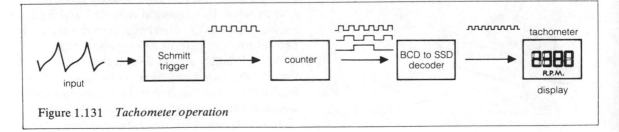

Figure 1.131   *Tachometer operation*

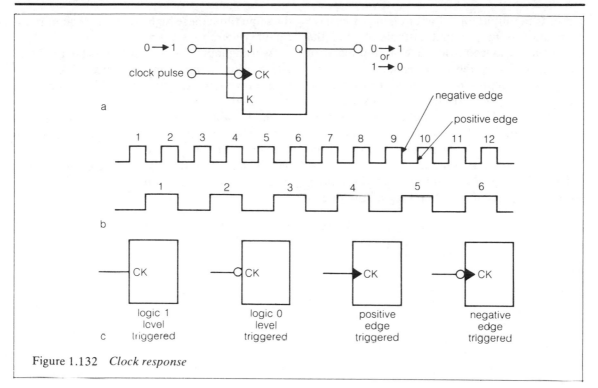

Figure 1.132 *Clock response*

the negative-going edge of the clock pulse (Figure 132(b)). This logic 1 output pulse is held until the next clock signal negative edge toggles the bistable back to its original setting.

In this case the bistable is triggered by the negative edge of the clock pulse but it is also possible to trigger the bistable at other times as shown by the types represented by the symbols in Figure 1.132(c).

Using a bistable as a counter in this way enables the pulse wave frequency to be halved; this is called a *divide-by-two counter*. A single bistable performs this division in a control unit of a fuel injection system (see page 289).

Ripple-through counter Cascading a number of bistables in the form as shown in Figure 1.133 and utilizing the output of one bistable to clock the next bistable makes a multi-stage counter. The three-stage binary counter shown is called a divide-by-eight counter because each stage added increases the division by two; hence three bistables gives $2 \times 2 \times 2 = 8$.

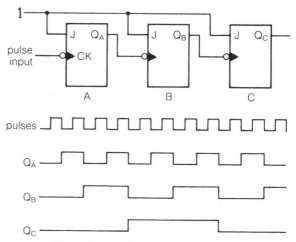

Figure 1.133 *Multi-stage counter*

Applying a pulse of logic 1 to input J of the first bistable A gives an output of logic 1 at Q_A. When the negative edge of the output pulse Q_A is applied to bistable B, this second bistable toggles and outputs logic 1 at Q_B. In a similar way, the output from B acts as a clock pulse for C and this causes C to output logic 1 after eight pulses have been applied to the first bistable A, i.e. the waveform produced will have a frequency of one-eighth that of the input pulse.

The term *ripple-through* or *asynchronous* is used because each bistable operates the next bistable in the chain. In view of the short propagation delay of each bistable there is a short time delay between the original clock signal and the output from the last bistable.

This delay can be overcome by applying a clock pulse to each bistable so that they all operate at the same time; these high-frequency counters are called *synchronous counters*.

By using more bistables it is possible to obtain any division required and by rearranging the circuits, *down counters* and *reversible counters* can be formed.

The bistable layouts as described can be used to store binary codes for use at a later time. When the storage is temporary, the layout is called a *register* but if the data is to be held for a longer time, it is called a *memory*.

2 Vehicle circuits and systems

2.1 Electrical circuits

In bygone days a wiring diagram for a vehicle consisted of a few lines to link the battery with two or three electrical components. Today a diagram for a modern vehicle is a mass of lines that connect with many electrical items scattered all over the page. As time goes by, more and more electrical components are introduced, so the electrical system becomes more complicated by reason of the extra wiring needed to supply and control each new part.

Simplification of the subject is possible by separating the complete electrical system into a number of individual circuits. Not only does this division of the circuits allow study of the behaviour of each part, but it also indicates the method by which an electrician is able to diagnose a fault. The circuit containing the fault is first identified and then the defective part in that circuit is pinpointed.

Vehicle systems
The main parts, systems and circuits are as follows:

Battery   This provides the electrical energy for lighting the vehicle, when the engine is not running, and for cranking the engine for starting purposes. The battery acts as a storage unit for the electrical energy (Figure 2.1).

Charging system   After the battery has discharged some of its energy, more electrical energy must be supplied to restore it to its fully-charged state: the charging system provides this service. In addition, the charging system supplies the complete electrical system with energy at all times when the engine is running.

Starting system   No longer does the driver have to manually crank the engine with a starting handle. Instead the engine 'springs' into life at the touch of a switch: the electrical starting system provides this service.

Ignition system   Engine operation demands a spark in each cylinder at the appropriate time. A voltage much higher than that given by the battery is required, so the ignition system must transform the battery voltage to a value often in excess of 20 kV (20 000 V) to produce the spark.

Lighting system   The law requires that the vehicle must have various lights to show its presence and also to help the driver to see where s/he is going. Bright driving lights must not dazzle the drivers of oncoming vehicles so some arrangement is used to dip the main beam.

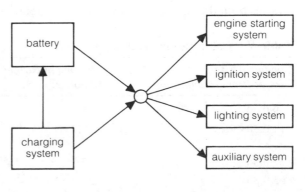

Figure 2.1   *Battery*

Auxiliary equipment   In addition to the main systems, a modern vehicle incorporates many other items which are classified as auxiliaries. This group of components has increased over recent years and will continue to expand to meet the needs of vehicle operation, driver control and comfort of passengers.

The equipment includes: windscreen wipers and washers, horns, direction indicators, heating and ventilation systems, door-locking actuators, electrically-operated windows, and instrumentation.

In-car entertainment   Although this is often classed as an auxiliary, the sophistication and expanse of modern equipment has divorced this section from the general auxiliaries.

Specialists in this field are normally consulted when internal faults are experienced with radios, tape players, etc.

Vehicle circuits
Having identified the main systems it is possible to arrange each system into its appropriate circuit.

Each electrical circuit requires a source of energy, so this means that one electrical feed wire may be incorporated into a number of circuits. Of course, if this feed wire fails, then the fault will show its presence in more than one system.

Electronic systems
Normally the use of the word 'electronic' is applied to any part or system which uses a semiconductor device. Today these devices are widely used in a number of the basic systems in the vehicle. In addition 'electronics' are used for engine management, i.e. controlling the ignition and fuel systems, automatic transmission control and many other specialist duties relating to sensing or controlling the performance of a particular part.

2.2 Cables, terminals and circuit protection

Cables
The various electrical components must be connected to the electrical supply by low-resistance cables. Copper cables, stranded to give good flexibility, are generally used to connect components remotely situated. Where several cables follow a common path they are taped together to form a loom or wiring harness. This construction reduces both the risk of chafing against the metal frame and breakage of the cable due to vibration. To obtain the maximum protection, the cables leave the loom at the point where the component is situated (Figure 2.2).

The quoted size of a cable refers to the wire diameter and the number of strands. If the cable size is too small for its length, or for the current it has to carry, then it will produce a voltage-drop which will affect the performance of the particular item of equipment that it is supplying, e.g. lights will not give their maximum illumination and starter motors will turn slower than normal.

Table V   Cable ratings and applications

Conductor size (No. of strands/diam., mm)	Maximum current rating (ampere)	Application
9/0.30	5.75	Lightly loaded circuits
14/0.30	8.75	Ignition circuits, side and tail lamps, general body wiring
28/0.30	17.5	Headlamps, horns, heated rear windows
65/0.30	35	Ammeter circuit
120/0.30	60	Alternator charging circuit (heavy duty)

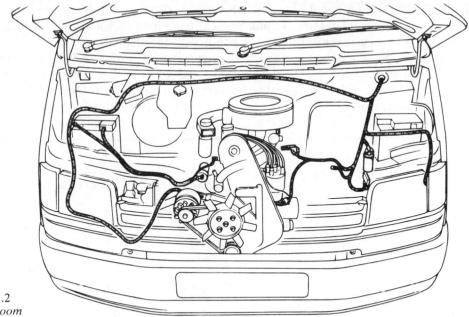

Figure 2.2
Engine loom

Operating temperature of a cable is affected by its resistance and by its ability to radiate its heat to the air. This means that the current density (current per unit of area) of cables bound in a loom is far less than individual cables surrounded by air.

An alternative multi-cable arrangement is obtained by placing the cables side-by-side and welding the cable coverings to a flat strip of plastics material (Figure 2.3). This construction is

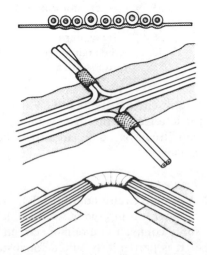

Figure 2.3
Cables placed side-by-side (Rists)

easier to accommodate in narrow spaces and also gives better heat dissipation.

The current to be carried by a cable depends on the circuit to which it is to be fitted: this means that different-sized cables are needed since minimum weight is essential. Most cables, other than thick starter cables, have a wire diameter of 0.30 mm. Table V gives a general guide for cables of average length.

When a defective cable has to be renewed, it should be replaced by a cable of similar size. If a new circuit is to be installed, the maximum current load should be estimated in order to determine the cable size. Assuming the length is not exceptional, a maximum current of 0.5 A per 0.30 mm strand is suggested.

Cable covering In the past, cotton and rubber were used to insulate the conductor, but these materials have been superseded by PVC plastics. The PVC covering does not require the copper to be tinned as was required with rubber if chemical action was to be avoided. PVC has a good resistance to petrol and oil, and although it is non-combustible, it gives off dangerous fumes when heated.

Colour coding   Cables forming part of a loom or a complicated circuit are difficult to trace. To aid the identification of a particular cable, the PVC covering is coloured. Unfortunately, the colour code varies with countries, so the specific code used on a particular vehicle will depend on the home of the parent company.

It is recommended by the British Standards Institution (BSI) that vehicles made in the UK should have a cable colour code which corresponds with the standard AU7.

Being as most wiring diagrams are shown in black and white, a letter code is used on the diagram to identify the colours. Table VI shows some of the main colours used for the principal circuits.

Table VI   Colours for circuits

Circuit	BSI	Letter code British	German
Earth connections	black	B	SW
Ignition circuits	white	W	WS
Main battery feed	brown	N	BR
Sidelamps	red	R	RT
Auxiliaries controlled by ignition switch	green	G	GN
Auxiliaries not controlled by ignition switch	purple	P	VI
Headlamps	blue	U	BL

In addition to the base colour, some cables have a thin tracer line running along the cable. The colour of this tracer identifies the part of the circuit occupied by the cable (Figure 2.4), e.g. a cable shown as UW has a blue base colour and a white tracer.

In view of the different standards used by manufacturers, it is wise to consult the wiring diagram for the vehicle whenever a particular cable or circuit has to be identified.

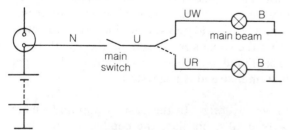

Figure 2.4   *Colour coding*

Circuit numbering   In addition to colour coding, some manufacturers use numbers to identify the circuits as recommended by the German DIN Standard. Table VII shows the main numbers used.

Table VII   Terminal marking according to DIN Standard

Circuit No.	Application
1	Ignition, earth side of coil
4	Ignition, h.t. output
15	Ignition, feed (unfused)
30	Feed from battery
31	Earth
51	Alternator output
54	Ignition, feed (fused)
56	Headlamps
58	Side/tail lamps
75	Accessories

Sub-circuits are identified by adding a number or letter after the main number; e.g. 15-4 is a sub-circuit based on circuit 15, the ignition feed circuit.

Printed circuits
A printed circuit board (PCB), used instead of a number of interconnected cables, provides a more compact and reliable circuit arrangement.

It is particularly suited for instrument panels

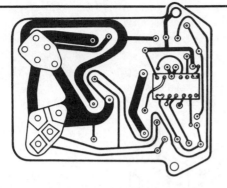

Figure 2.5 *Printed circuit board for a wiper control*

and component sub-assemblies found in electronic control units (Figure 2.5).

The material used for a circuit board has an insulation base on to which is bonded a thin layer of copper. After printing the circuit image on the copper, the board is then immersed in acid. This removes the unwanted copper and leaves a number of thin conductors on to which is soldered the component parts. In the case of a PCB

Figure 2.6 *Single terminals*

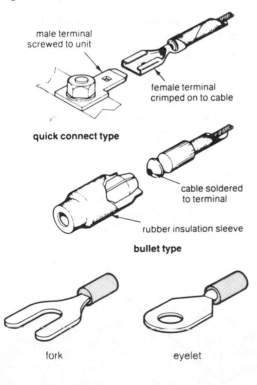

male terminal
screwed to unit

female terminal
crimped on to cable

quick connect type

cable soldered
to terminal

rubber insulation sleeve

bullet type

fork   eyelet

such as an instrument panel, the circuit board is connected to the various cables by means of a multi-pin plug.

The copper forming the printed circuit is very thin, so it must be handled carefully and should not be subjected to a heavy current. Accidental breakage of the copper foil can be repaired by soldering provided the minimum amount of heat is used.

Terminals and connectors
Cables are attached to components by terminals and joined together by connectors.

Terminals   Although the fork and eyelet type terminals are occasionally used, the quick-connect or Lucar type are more common (Figure 2.6).

A suitable crimping tool (Figure 2.7) should be used to fit a terminal: it should be soundly joined to the copper core and must be secured to the insulation covering to resist breakage due to vibration. A terminal should be protected from moisture since this can cause corrosion and give a high resistance.

Figure 2.7 *Crimping tool for fitting terminals*

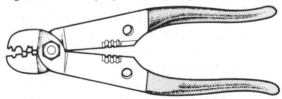

Connectors   Although bullet connectors of the soldered and crimped forms are still used to join two or more cables, the need for greater security and improved protection against the ingress of salt and moisture has demanded the use of more efficient types of connector. These requirements become very important when the connector has to handle low currents.

Figure 2.8 shows a connector manufactured by Rists. The E-type is an environmentally protected connector made in 3, 5, 7 and 9-way forms. This type incorporates sealing features to reduce the risk of electrical breakdown even when it is exposed to roadwash conditions.

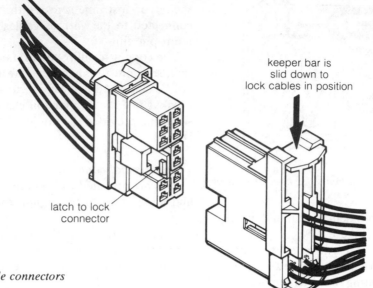

Figure 2.8   *Cable connectors*

The Rists *Total Terminal Security* (TTS) connector is designed to prevent terminal disengagement during assembly of the connector. This disengagement, called *terminal back-out*, is prevented by using keeper bars in the plug and socket. Additional features include a latch, to lock the plug and socket together, and means to prevent the two parts being mated incorrectly.

Electronic control units, as used for engine management systems, require a connector that takes up to 12 cables and a provision to enable the pins to mate progressively when the connector is fitted to the control unit. Figure 2.9 shows a connector suitable for this application. This type

is protected against a hostile environment and includes a latching feature to retain the connector securely in position. *Edge connectors* are used to provide a low-resistance contact with a printed circuit board. Figure 2.10 shows this type of connector.

Cost dictates the quality of a connector, in particular the material and thickness of the contact surfaces. High conductivity sometimes requires the use of precious materials such as silver and gold but rarely can automotive manufacturers afford these expensive materials. The fact that they have to use cheaper connectors can cause problems during fault diagnosis, because

Figure 2.9
Connection to an ECU

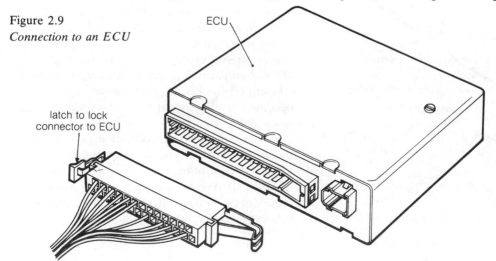

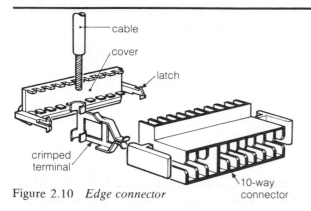

Figure 2.10 *Edge connector*

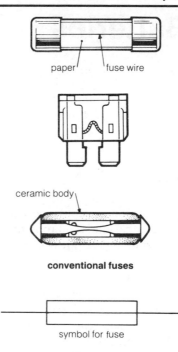

Figure 2.11 *Types of fuses*

some of the low-priced connectors can only be disconnected and rejoined about five times to produce a sound reliable connection. A connector built to an aeronautical specification can be separated many more times but the price of this type is more than ten times as great.

Circuit protection
In the event of a short circuit, a higher current than normal will flow from the battery and this will overload and heat the cable. This can melt the cable insulation and may start a fire. Also the heavy current will soon discharge the battery and immobilize the vehicle. A circuit-protection device, such as a fuse or thermal circuit breaker, reduces the risk of this problem.

Fuses are made in different forms as shown in Figure 2.11; the glass cartridge is the oldest type. This type consists of a short length of tinned wire connected at each end to a metal cap and enclosed in a glass cylinder. A strip of paper, colour coded and marked with the rating, is placed close to the wire. Different ratings are available to suit the various circuits. When the current exceeds the rating, the fuse 'blows', i.e. the wire melts, the paper is scorched and the circuit is broken.

Some fuses, e.g. ceramic type, are rated according to the continuous current that can be carried by the fuse: this is normally half the current required to melt the fuse.

Fuses are either centrally mounted on a fuseboard or placed in a separate fuse holder 'in-line' to protect an auxiliary such as a radio.

Some vehicles have a *fusible link* placed in the main output lead from the battery. This heavy duty 'fuse' melts to reduce the risk of fire if an accident causes the main cable to short to earth.

Thermal circuit breakers These use a bimetallic strip to control a pair of contacts in the main circuit. A current overload heats and bends the strip which opens the contacts and temporarily interrupts the circuit. When this is used in a lighting system, a short circuit will cause the light to go off and on repeatedly, so the driver should be able to bring the vehicle to rest safely. A single fuse of the normal type placed in the lighting system would cause a dangerous situation so separate fuses are used for each headlamp.

Fuse failure When a fuse 'blows', it should be replaced with a fuse of similar rating. If the second fuse fails immediately, then the circuit should be checked to locate the short circuit.

Many faults are caused due to poor cable security or lack of protection at points where the cable passes through holes in metal parts of the vehicle.

3 Batteries

3.1 Battery types

A battery is required to supply electrical energy to meet the electrical load requirements when the engine is not running. It must 'store' electrical energy and then deliver this energy when it is needed at a later time.

A battery fitted to a vehicle fulfils its storage role by an electro-chemical process: the energy delivered by an electrical current produces a chemical change in the battery plates which is reversed when the battery discharges. Current supplied to a battery is called a *charge* whereas the output from a battery is called a *discharge*.

Primary and secondary batteries
Many years ago it was discovered that when two dissimilar metals were placed close together and immersed in an acid solution, an e.m.f. was produced. (Today this effect can be demonstrated by placing two coins made of different metals in a lemon. If a millivoltmeter is connected across the two coins the meter will register a p.d. In this case the acid juice conducts the electrical charges called *ions* from one plate to the other.) A liquid solution which conducts an electrical charge is called an *electrolyte*.

After a short time the surface condition of the plates of a simple battery causes the p.d. to reduce to a low value and in this state the battery is classed as *discharged*. Recharging has no effect on this type of battery so it is discarded. A battery which cannot be 'reversed' is called a *primary battery*: a torch battery is an example of a primary battery.

When a battery allows for the chemical process to be reversed after discharge, i.e. it can be charged and discharged, it is called a *secondary battery* or *accumulator*: this type is used on motor vehicles.

Types of secondary battery
Secondary batteries are classified by the materials used to form the plates and the electrolyte into which the plates are immersed. There are two main types used on motor vehicles:

1. Lead–acid
2. Nickel–alkaline

Lead–acid   This type is used on the majority of vehicles because it is relatively cheap and performs well over a long period of time. Unfortunately all batteries are heavy, so until some new construction is discovered it is expected that the lead–acid type will remain in common use.

For many years the basic construction of a lead–acid type has remained unchanged, but alterations have been introduced recently to minimize the periodic maintenance needed to keep the battery in good condition. The original basic type is called a *conventional* battery; later designs are named *low-maintenance* and *maintenance-free*. The name given to each type indicates the work needed to service the battery.

Nickel–alkaline   As the name suggests this type is a non-acid battery which uses nickel as a plate material. It is more expensive and larger in size than a lead–acid type, but it is more able to withstand heavy discharge currents without damage. Batteries of this type have a very long life so they are used in situations where reliability over a long period of time is a prime consideration.

3.2 Lead–acid batteries

The lead–acid battery is the most popular type used on motor vehicles. It is capable of supplying the large current of 200 A or more demanded by a starter motor, has a fair life of two years or more and is relatively cheap. In the past it suffered the disadvantages of size, weight and the need for periodic attention, but today these drawbacks have been minimized.

The battery consists of a container which houses a number of cells of 2 V nominal voltage that are connected in series by lead bars to give the required voltage (Figure 3.1). Three cells are used in a 6 V battery and six cells for the common 12 V unit. Normally the connecting bars are sunk below the top cover to give a cleaner appearance.

Conventional type
This basic construction is still used on many vehicles.

Plates and separators The cell is made up of two sets of lead plates, positive and negative, which are placed alternately and separated by an insulating, porous material such as PVC plastics, rubber or glass fibre. Each plate consists of a lattice-type grid of *lead–antimony* alloy, into which is pressed the active material. This is a lead oxide paste electrically formed into *lead peroxide* (positive and chocolate in colour) and *spongy lead* (negative and grey in colour).

The surface area of the plates governs the maximum discharge current that can be supplied for a given time, so in order to give maximum output, each cell contains a number of thin plates, each set connected in parallel. Connection within a cell in this manner does not affect the cell voltage.

Container The case is made of semi-transparent polypropylene or of a black hard rubber composition. Recesses are formed in the bottom of each cell to collect active material that falls from the plate grids. This space prevents the material from bridging and short-circuiting the plates.

A moulded cover seals the cells and a removable plug allows the cells to be topped-up with distilled water and also exposes the electrolyte for testing purposes. A small vent hole in the plug

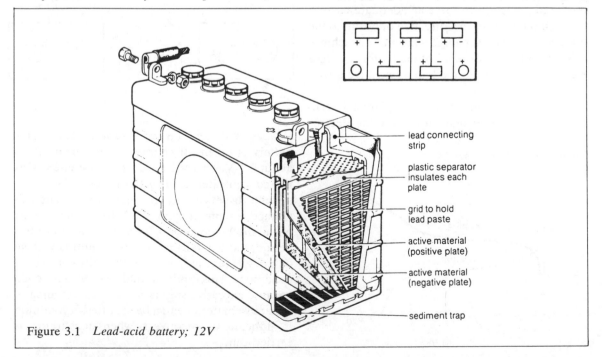

Figure 3.1 *Lead-acid battery; 12V*

lead connecting strip

plastic separator insulates each plate

grid to hold lead paste

active material (positive plate)

active material (negative plate)

sediment trap

allows for the escape of gas during the charging cycle.

Electrolyte Diluted sulphuric acid (H_2SO_4) forms the electrolyte into which the plates are immersed.

Cell action When a battery is fully charged and ready for use the positive plates are lead peroxide (PbO_2) and the negative plates are spongy lead (Pb).

As the battery discharges through an external circuit, the sulphur in the acid combines with some of the lead oxide in the plates and this changes both plate materials to lead sulphate ($PbSO_4$). This substance occupies a larger volume, so if the cell is discharged with an unduly high current, either the plate will buckle or the active material will be dislodged from the grids. The loss of sulphur from the electrolyte to the plates during the discharge process decreases the density of the electrolyte (i.e. reduces the specific gravity), so this feature enables the state of charge to be assessed by using a hydrometer.

To charge a battery requires a d.c. supply at a potential sufficient to force an adequate current through the battery in a direction opposite to the direction of the discharge current. To achieve this, the positive terminal of the battery charger must be connected to the battery positive terminal.

During the charging period the plate materials will return to their original forms and the electrolyte density will increase. When the process is complete, i.e. when the battery is fully charged, the continuance of the charge current will lead to excessive gassing of the cell. The gas consists of hydrogen and oxygen, a *highly explosive* mixture, so a naked flame or electric spark must not be produced in the vicinity of a battery at any time.

The charge and discharge process is shown diagrammatically in Figure 3.2.

Figure 3.2 *Charge and discharge action*

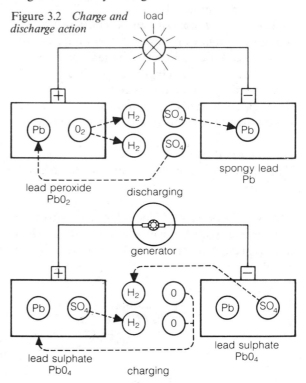

Voltage variation Figure 3.3 shows that when a battery is taken off charge, the terminal p.d. is about 2.1 V. This quickly drops to about 2.0 V as the concentrated acid in the pores of the plate surface disperses into the electrolyte. The cell voltage remains at about 2.0 V for the major part of the discharge period. Towards the end of this period, the p.d. falls more rapidly until a voltage of 1.8 V is reached, which is the fully discharged condition. This limit should not be exceeded because excessive sulphation causes shedding of the active material and a battery in this condition is difficult to reconvert when recharging is carried-out.

Figure 3.3 *Battery voltage*

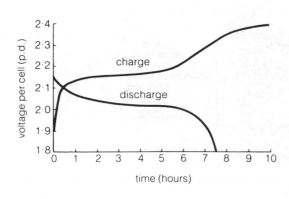

The readings shown in Figure 3.3 represent p.d., so to achieve these results the battery must be supplying a normal discharge current at the time the voltmeter readings are taken.

Terminal p.d. during the charging process rises towards the end of the period from about 2.1 V to over 2.4 V when the cell is fully charged, but the p.d. soon falls to about 2.1 V when the charge current is stopped.

The rise in p.d. when the battery approaches its fully charged state is used to signal the battery condition to the vehicle's charging system. Setting the regulator to limit the maximum generator output p.d. to 14.2 V ensures that the battery cannot be overcharged. When the cell voltage reaches about 2.4 V (i.e. 14.2/6) the p.d. of the generator will equal the p.d. of the battery, so no current will pass to the battery.

Electrolyte density   Automobile starter batteries use an electrolyte of higher density than that used for general-purpose batteries. This greater strength acid is required to allow a high discharge current from a battery of small size. Electrolyte resistance and, in consequence, internal resistance rises outside the relative density limits 1.100–1.300, so these values represent the range in which a battery should operate. Furthermore, a density higher than 1.300 would cause the plates and separators to be attacked by the acid.

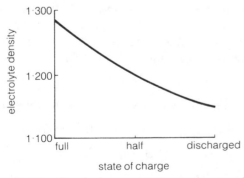

Figure 3.4   *Graph shows variation in density of electrolyte*

Figure 3.4 shows the electrolyte density as the battery state-of-charge varies. Values vary slightly for different makes of battery, but the

following table gives a set of typical values:

Fully charged	1.280
Half charged	1.200
Fully discharged	1.150

These values represent the 'strength' of the electrolyte. This is measured by comparing the mass of a given volume of electrolyte with the mass of an equal volume of pure water. The ratio obtained is termed *specific gravity* (sp. gr.) and is measured by an instrument called a *hyd-*

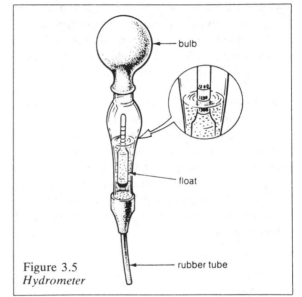

Figure 3.5
Hydrometer

rometer. (Figure 3.5). The reading shown in the diagram is 1.250 and this indicates that the electrolyte is 1¼ times as heavy as pure water. For simplicity, the decimal point is often omitted; in this example the specific gravity is stated as 'twelve fifty'.

The electrolyte expands when the electrolyte temperature is increased, so this must be taken into account if accuracy is required. A temperature of 15°C (60°F) is the standard temperature, so to obtain a true value a correction factor of 0.002 is deducted from the hydrometer reading for every 2°C fall below 15°C and 0.002 is added for every 3°C rise above 15°C.

Freezing of electrolyte   The freezing point of the electrolyte depends on the state of charge, i.e. it depends on the electrolyte density. When the density decreases, the acid strength falls, so the freezing point rises as the electrolyte composition moves towards a pure water state. In consequence, the electrolyte of a discharged battery will freeze at a higher temperature than a fully charged battery.

Capacity   The capacity of a battery is expressed in *ampere-hours* (Ah). This represents the current that a battery will deliver for a given time: it is generally based on a time of 10 hours or in some cases 20 hours; e.g. a battery capacity of 38 Ah, based on a 10-hour rate, should supply a steady current of 3.8 A for 10 hours, at a temperature of 25°C, before the cell voltage becomes 1.8 V (the voltage of a discharged cell).

When the capacity is based on the 20-hour rate, its stated capacity is about 10–20% higher than the capacity given by the 10-hour rate. This increase in capacity is achieved because a lower discharge current is used for the 20-hour rate. The effect of the rate of discharge on the battery voltage is shown in Figure 3.6. This graph shows that the time shortens considerably as the current is increased.

Figure 3.6   *Battery discharge curves*

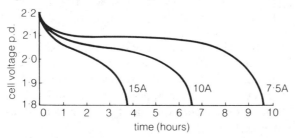

Under engine-starting conditions, the current used is over 200 A so the time that a 38 Ah battery can supply this high current will be considerably less than the calculated time of 0.19 hour.

Capacity of a battery is governed by the plate area, so batteries having many plates of large size have a large capacity. A small-capacity battery may have only five plates (three negative, two positive) per cell, whereas large units may have more than twenty plates per cell.

Capacity reduces as the temperature decreases, so this is an important factor to consider when choosing a battery for low temperature operation.

Reserve capacity   Nowadays the ampere-hour capacity rating has limited appeal; the Reserve Capacity rating has taken its place. This rating indicates the time in minutes that a battery will deliver a current of 25 A at 25°C before the cell voltage drops to 1.75 V.

The standard current of 25 A represents the average discharge on a vehicle if the charging system should fail. The Reserve Capacity indicates the time that a battery will keep the vehicle in operation assuming the electrical load is normal.

A typical value for the Reserve Capacity of a 40 Ah battery is 45 minutes.

Internal resistance   If the voltage could be measured at the source of the energy, i.e. at the plate surface, the voltage obtained would be the electromotive force. This would be the same as the open-circuit voltage measured at the terminals. When current flows from a battery, the resistance of the parts within the battery causes the terminal voltage to fall. In view of this, the p.d. of a battery is less than the e.m.f.

The internal resistance can be found by the method shown in Figure 3.7. A voltmeter of high resistance is connected across the battery and an ammeter is used to measure the current that flows through the external resistor 'R'. The internal resistance 'r' is found by using Ohm's law:

$$V = IR$$

In this case there are two resistors in series so:

$$V = I(R + r)$$

$$R + r = \frac{V}{I}$$

So internal resistance r $= \frac{V}{I} - R.$

The lead–acid battery has a low internal resistance so the comparatively high terminal p.d. makes it attractive for vehicle use. Nevertheless, an internal

Figure 3.7 *Method for finding internal resistance*

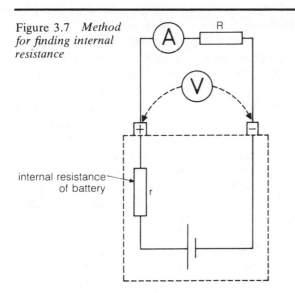

resistance of about 0.005 Ω for a typical battery in good condition causes the battery to become warm and the potential to drop from 12 V when a high discharge current has to be supplied, e.g. a current of 200 A causes a voltage drop of:

$$V = IR$$
$$= 200 \times 0.005$$
$$= 1 \text{ volt}$$

Any increase in the resistance proportionally increases the drop and this has a marked effect on the performance of a battery, especially when it is used to operate a starter on a cold morning.

Internal resistance is the sum of various resistances: these include the following:

● Plates – an increase in plate area decreases the resistance. When a battery is old, a decrease in the active plate area increases the resistance.
● Internal connections – these are large in section to improve the current flow.
● Electrolyte – the resistance increases when the temperature is decreased and also when the acid strength is reduced, i.e. when the battery becomes discharged.

Connecting batteries in series and parallel Occasions arise when two or more batteries are connected together to give a series or parallel arrangement.

Series Figure 3.8 shows two batteries A and B connected in series with the negative terminal of A joined to the positive terminal of battery B. Joined together in this way the nominal p.d. across both batteries (at *x* and *y*) is increased to 24 V but the capacity remains the same at 40 Ah.

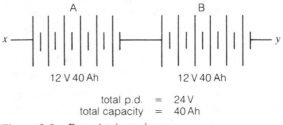

total p.d. = 24 V
total capacity = 40 Ah

Figure 3.8 *Batteries in series*

Parallel Connecting the batteries in parallel (Figure 3.9) allows each battery to give a p.d. of 12 V, so the terminal voltage at *x* and *y* will be similar to that given by one battery. This arrangement doubles the plate area available so the capacity is increased to 80 Ah. No matter how many batteries are connected in parallel, the voltage remains the same. The aim behind this battery arrangement is to increase the capacity of the the system.

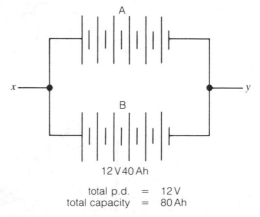

total p.d. = 12 V
total capacity = 80 Ah

Figure 3.9 *Batteries in parallel*

Low-maintenance and maintenance-free batteries
Improved materials and new constructional techniques have either reduced or eliminated the need for a battery to be topped-up periodically with distilled water to replace loss due to gassing. New style batteries which do not need this

maintenance task are attractive to the vehicle owner for obvious reasons.

Use of these batteries has been made possible by the improved control of the charging rate, especially the voltage output, given by an alternator system as compared with a dynamo system.

Gassing has been reduced by changing the grid material from lead–antimony alloy to an alloy of *lead–calcium*.

Low-maintenance This type requires less attention than the conventional battery (Figure 3.10). Under normal-temperature operation and suitable charger conditions, the electrolyte level needs to be checked only once per year, or 80 000 km (50 000 miles).

Other than grid material, the construction of a low-maintenance type is similar to a conventional battery. Since the performance characteristics are based on proven designs, the battery can be used on older vehicles as an alternative to the traditional type.

Maintenance-free This type differs in several respects from a conventional battery: the most significant feature is that the battery is sealed (except for a very small vent hole) and requires no service attention other than to be kept clean.

Figure 3.11 shows a Delco–Remy Freedom battery which first appeared in America in 1971. It is claimed that in addition to being maintenance-free, this battery offers better cold-weather starting power and improved resistance to heat and vibration damage.

The Freedom battery manufacturer has eliminated antimony from the plate grids and this has removed four major causes of early battery failure: overcharge, water usage, thermal runaway and self-discharge. *Thermal runaway* is a condition which occurs in a conventional battery when the battery operating temperature is high or when faulty regulation of the charging system is combined with a rising electrolyte temperature. *Overcharge* is the major cause of gassing in a conventional battery. A Freedom battery uses

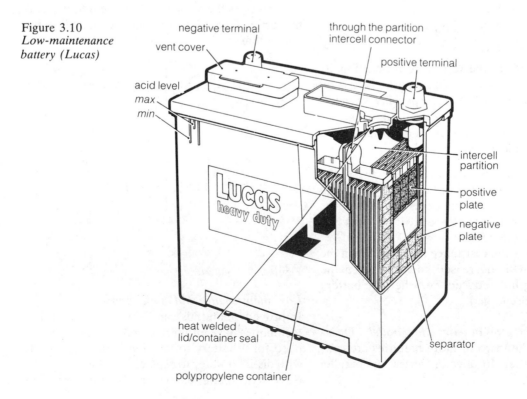

Figure 3.10
*Low-maintenance
battery (Lucas)*

negative terminal

vent cover

acid level
max
min

through the partition
intercell connector

positive terminal

intercell
partition

positive
plate

negative
plate

heat welded
lid/container seal

separator

polypropylene container

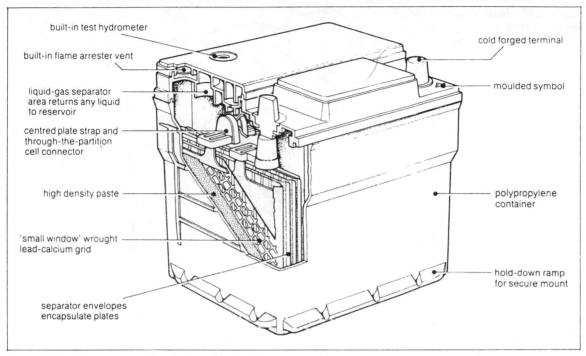

built-in test hydrometer

built-in flame arrester vent

liquid-gas separator
area returns any liquid
to reservoir

centred plate strap and
through-the-partition
cell connector

high density paste

'small window' wrought
lead-calcium grid

separator envelopes
encapsulate plates

cold forged terminal

moulded symbol

polypropylene
container

hold-down ramp
for secure mount

Figure 3.11   *Maintenance-free battery (Delco-Remy Freedom)*

lead calcium (Pb–Ca) for the grid material so with the inherent assistance of the higher e.m.f. given by this construction as it approaches full-charge, it is possible to reduce water consumption during overcharge conditions by over 80%. There is still some gassing, so a gas reservoir is formed in the container to collect the water and return it after cooling to the main electrolyte mass.

This make of battery incorporates a built-in, temperature-compensated hydrometer to indicate the relative density and level of the electrolyte. The indicator displays various colours to show the states of charge. A green-coloured ball shows that the battery is charged and serviceable, whereas a green/black or black signal indicates that recharging is necessary. When a light-yellow signal appears it indicates an internal fault and when this is evident the battery must not be charged or tested. Also when the battery is in this state the engine must not be started with jump-leads. Instead the vehicle should be fitted with a new battery and the alternator should be checked for correct operation.

If the battery is discharged to a point where it cannot crank the engine, and in consequence the engine has to be started by other means, then it will be impossible for the alternator to recharge the battery. When it is in this condition the battery must be removed and bench-charged, because the voltage needed to restore it is higher than the output given by the charging system of the vehicle.

Other design improvements over a conventional battery include; strengthened grid supports, sealed terminal connections and stronger retention supports. These features together with a better efficiency make this battery smaller and lighter in weight than the conventional type.

Figure 3.12 shows a battery made by Chloride called a 'Torque Starter'. This maintenance-free battery uses *Recombination Electrolyte* (R.E.) to reduce the formation of oxygen and hydrogen when the battery is being charged.

In this type each plate is wrapped with a glass micro-fibre separator that absorbs, in its pores, all the liquid electrolyte. There is no free acid in the cell as in conventional batteries.

As the battery approaches its fully-charged

Figure 3.12 *Recombination Electrolyte battery; Chloride Exide Torque Starter*

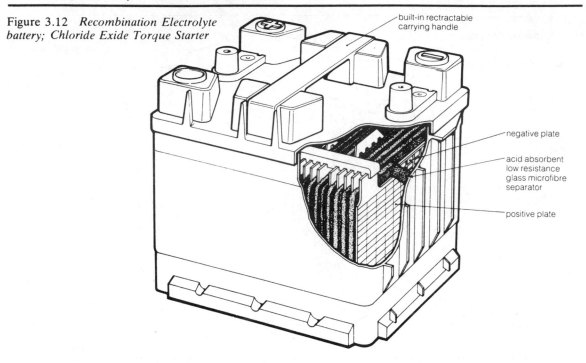

- built-in rectractable carrying handle
- negative plate
- acid absorbent low resistance glass microfibre separator
- positive plate

state, the oxygen liberated at the positive plate passes through the separator pores to the negative plate. After initially reacting to form lead sulphate, the plate then changes to lead after further charging. As a result of this action the negative plate never reaches the right potential for hydrogen to be released, so no water is formed. Since no free oxygen or hydrogen is produced, it is possible to totally seal the battery, other than a small pressure valve set to open if the battery is abused. It is claimed that the R.E. battery can deliver 20% more power than an equivalent conventional battery. Under the Cold Cranking Test the battery can deliver a current of 420 A. These features combined with better resistance to vibration and lightweight construction make the R.E. battery attractive in the replacement and original equipment markets.

3.3 Nickel–alkaline batteries

The nickel–alkaline battery is a strong, long-life battery which withstands greater abuse than a lead–acid type but it is more bulky and expensive.

There are two main types of nickel–alkaline battery. The types are classified by their plate material; nickel–cadmium (Ni–Cd) and nickel–iron (Ni–Fe). The latter type, often called a 'Nife' battery, is less suitable for automobile use so it is not considered in this book.

Nickel–cadmium battery
Figure 3.13 shows a 'cut-open' view of one cell. Both positive and negative plates are made of a

Figure 3.13 *Nickel-cadmium battery*

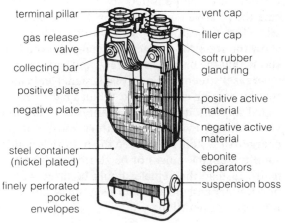

- terminal pillar
- gas release valve
- collecting bar
- positive plate
- negative plate
- steel container (nickel plated)
- finely perforated pocket envelopes
- vent cap
- filler cap
- soft rubber gland ring
- positive active material
- negative active material
- ebonite separators
- suspension boss

nickel-plated steel frame into which are spot-welded a number of flat-section perforated tubes, also made of nickel-plated steel. In the positive plates the tubes are filled with powdered nickel hydroxide and in the negative plates with cadmium oxide.

The plates have lugs by which they are attached to collecting bars to each of which a terminal pillar is also fixed. The plates are assembled into sets in which the negative plates are interleaved between positive plates with ebonite rods between the plates to prevent electrical contact between them.

In what might be called the traditional construction, each cell is enclosed in a nickel-plated steel container having welded seams. The terminal pillars pass through rubber gland rings in the cell lid and are secured by nuts. Each cell has a combined filler cup and vent cap.

An appropriate number of cells is assembled to make up a battery, five cells being used for a 6 V battery, nine for a 12 V and eighteen for a 24 V. Since the steel containers are in electrical contact with the positive plates they must not be allowed to touch one another in the battery crate. Each cell has two suspension bosses welded on opposite sides by which they are located in tough rubber sockets in the wooden crates, gaps being left between adjacent cells.

Electrolyte The electrolyte is a solution of potassium hydroxide (caustic potash, KOH) diluted with distilled water to a specific gravity of about 1.200. The density does not change with the state of charge because it does not chemically combine with the plate material. Instead the electrolyte acts as a conductor for the electrical current and allows oxygen to pass from the negative plates during charge and return during discharge.

Charge and discharge During the charging process the positive plates become oxidized while the negative plates are deoxidized (reduced) from cadmium oxide to spongy cadmium. When the battery is discharged the reverse action takes place. The cell voltage during the charge–discharge cycle varies from about 1.4 V to a minimum of 1.0 V.

Since the active-plate material does not chemically combine with any element in the electrolyte, there is virtually no self-discharge. Therefore the battery can 'stand' for long periods in either the charged or discharged state without causing damage to the battery.

Maintenance of nickel–cadmium–alkaline batteries

Electrolyte Periodically the electrolyte level in the cells should be checked. If the level is below about 40mm (1½ in) above the top of the plates then the cells should be topped-up with *pure distilled water*. Great care must be taken to ensure that no trace of acid is allowed to contaminate the cells, so all equipment used for lead–acid batteries must not be used on nickel–alkaline batteries.

The electrolyte deteriorates with age, so about every four years the electrolyte should be completely changed. When a hydrometer shows that the relative density has fallen to about 1.160 changing is necessary. The ageing process quickens if the electrolyte is exposed to air, therefore the cell vents must be kept closed except when the level is being checked.

In the UK, the electrolyte is supplied in liquid form but for use overseas it is supplied in solid form and must be dissolved in pure distilled water. In both solid and liquid forms the electrolyte must be handled with extreme care and must not be allowed to come into contact with clothing or the skin. It will cause severe burns on the skin which should be covered immediately with boracic powder or washed with a saturated solution of boracic powder. (A supply of boracic powder should always be available whenever electrolyte is being handled.) Prevention being better than cure, it is recommended that goggles and rubber gloves be worn.

Battery tests There is no simple test for the state of charge of a nickel–cadmium–alkaline battery. Neither the cell voltage nor the relative density of the electrolyte give any useful information. In vehicle applications, advantage is taken of the fact that the battery cannot be damaged by overcharging so one should ensure that the charging rate is sufficiently high to provide ample charging. This can be checked by examining the battery from time to time immediately after the vehicle has been running: if the cells are found to be gassing it can be taken as an indication that the state of charge of the battery is being satisfactorily maintained. A further check is the need for topping-up. A reasonable consumption of distilled water is the best indication that the battery is being kept properly charged. Excessive consumption indicates overcharging and a negligible consumption indicates undercharging.

No satisfactory high-rate discharge tester is available for this type of battery, chiefly due to the difficulty of obtaining an adequate area of contact with the steel cell terminals.

General attention The battery should be kept clean and dry and periodically the terminals should be cleaned, fully tightened and lightly smeared with petroleum jelly.

Plastics and wood cell containers should be inspected for damage from fuel oil and hydraulic fluid and the containers should be checked to ensure that no metal objects bridge the metal cells.

The battery should not be discharged below a cell voltage of 1.0 V.

3.4 Charging and maintenance of lead–acid batteries

Battery charging

Charging of a battery on a vehicle is performed by a generator. This provides a d.c. current at a voltage sufficient to overcome both the back-e.m.f. and the internal resistance of the battery.

External battery chargers are used when it is inconvenient to use the vehicle generator. External chargers may be divided into the following categories:

Bench charger Fitted in a well-ventilated, divided-off section of the workshop, this type is generally fixed to the wall above a charging bench. Batteries are connected to the charger either individually or in a balanced series–parallel arrangement. In Figure 3.14 it is seen that the '+' terminals of the batteries are connected to the charger '−'. By setting the output at a nominal 24 V, the total output current from the charger takes three paths; the current flow through any branch of the circuit depends on the state-of-charge of the batteries in that branch.

Figure 3.14
Bench charging

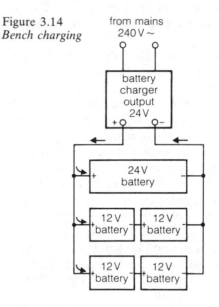

Since the charger is operated from a 240V a.c. mains supply, the charger must incorporate a *transformer* to step-down the voltage to suit the battery, and a *rectifier* to convert a.c. to d.c. By using a circuit to connect the diodes or metal plates in the rectifier, it is possible to obtain full-wave rectification (see page 41). In Figure 3.15 the transformer only has one voltage output, but when additional tappings of the transformer secondary coil are provided, other outputs can be obtained (Figure 3.16). Although the transformer output is marked with the nominal battery voltage, e.g. 12 V, the actual output must be higher than that given by a fully charged battery.

A *constant-voltage* charger gives a voltage output equivalent to the voltage of a fully charged battery, e.g. 14.4 V for a 12 V battery. When a discharged battery is connected to the charger,

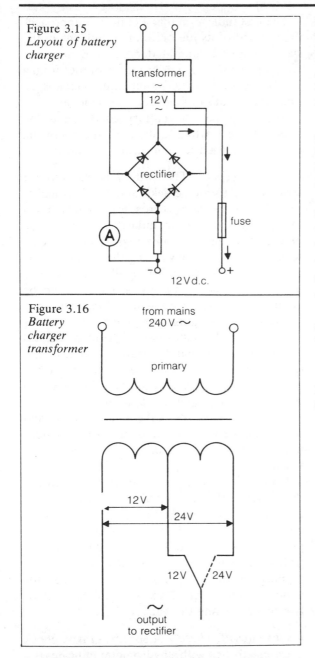

Figure 3.15
Layout of battery charger

transformer
~
12 V
~

rectifier

fuse

Ⓐ

– +
12 V d.c.

Figure 3.16
Battery charger transformer

from mains
240 V ~

primary

12 V

24 V

12 V   24 V

~
output
to rectifier

stant rate. A typical rate recommended for a battery is 1/10 of the ampere-hour capacity, so a 80 Ah battery should be charged at 8 A. Constant-current chargers have the batteries connected in series to ensure that each battery receives the same current.

Fast chargers This transportable type of charger enables a battery to be recharged in about 30–60 minutes to meet workshop requirements. Figure 3.17 shows this type of charger.

Initially the current flow is about 50 A, so special protection devices are incorporated to taper-off the current as it charges to prevent the battery being damaged by overheating. A thermostat is fitted to stop the charge when it senses that the electrolyte temperature exceeds 45°C.

An additional feature of a fast charger is the supply of an extra high current to a discharged battery which enables the engine to be started.

Figure 3.17 *Battery charger (Crypton)*

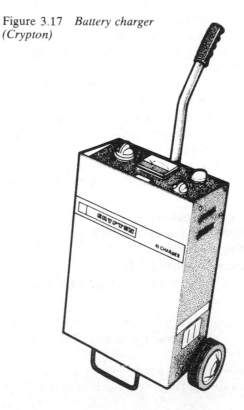

the initial charge current is high, but this gradually falls until it is practically zero after about 6–8 hours. The batteries are normally connected in parallel with this type of charger.

A *constant current* charger has special switching features which allow, by manual or automatic means, for the current to be controlled at a con-

Trickle chargers   This type is intended for use in a car-owner's garage. It provides a small charge current of about 2–4 A to enable the battery to be maintained in a fully charged state in cases where the vehicle is used only infrequently.

Extended trickle charging is detrimental because a low current only activates the surface of the plates.

Battery tests

Personal safety must be observed when batteries are handled or tested. A fully equipped medical kit, including eye-wash facilities, should be available and protective clothing, including eye protection, should be worn. Acid splashes in the eye should be treated immediately with plenty of clean water and medical attention should be sought as soon as possible. Acid on the skin should be washed off with water and neutralized with sodium bicarbonate solution. Acid splashes on clothing must be treated with an alkali, such as ammonia, if holes are to be avoided.

A safety hazard exists during or after battery charging due to the emission of a highly flammable hydrogen gas. Any testing involving production of sparks, e.g. electrical load test, must not be performed until the gas has dispersed from the cell. A similar hazard occurs when a battery is fitted on to a vehicle immediately after the battery has been removed from a charging plant: vehicles fitted with central door-locking systems require special attention in this respect.

To minimize these risks the battery terminals should be removed and fitted in the following order:

Disconnecting a battery –
   earth terminal removed FIRST
Connecting a battery –
   earth terminal fitted LAST

Charging should only be carried-out in a well-ventilated area.

There are three basic checks which are performed on a conventional battery. These are:
* Visual inspection
* Relative density (specific gravity) check
* Electrical load (high rate discharge) test

The first indication that a battery is approaching the end of its life occurs normally when the starter motor is operated on a cold morning. Under these severe conditions the output from a good battery is less than its maximum so this is the time that a battery fault becomes evident.

Assuming that the battery condition is the cause of the starting fault, then the following procedure is used to confirm the diagnosis.

Visual inspection   The battery is checked for terminal corrosion, cracks and for leakage of acid. A white powdery corrosion of metal parts in the vicinity of the battery indicates a past leakage of acid. This should be investigated and the corrosion neutralized by washing the affected parts in ammoniated water. After all traces of acid have been removed the metal parts should be painted, preferably with acid-resistant paint.

Evidence of bulging of the container or cover suggests that the plates have deformed and this generally means that the battery capacity has decreased. Distortion of the container occurs after excessive sulphation or abuse in respect to overcharging or repeated rapid discharging.

Battery life depends on the use of a battery and on the maintenance given to the battery. Under normal conditions a typical life is about 3–4 years.

If the initial checks are satisfactory then the fluid level in the cells should be checked. After ensuring that no naked flame is present, the vent caps are opened and the level checked. If the plates are not covered by at least 10 mm of electrolyte, then top-up with distilled water. After topping-up, the battery should be charged for about 15 minutes at 15–25 A to mix the electrolyte before testing the battery.

Relative density check   A specific gravity check of the electrolyte with a hydrometer indicates the state-of-charge. A fully-charged battery in a serviceable condition should give a reading of:

1.230 with a variation between cells of not more than 50 points (0.050)

This result relates to the sp. gr. at a temperature of 15°C (60°F). When greater accuracy is

Table VIII Results of hydrometer testing

Reading	Variation	Action
1.230	less than 0.050	Battery in good condition; confirm with drop test
1.190	less than 0.050	Discharged battery; recharge for 12 hours at a current equal to about 1/10 of the battery's capacity and retest
Some cells less than 1.200	more than 0.100	Battery should be scrapped

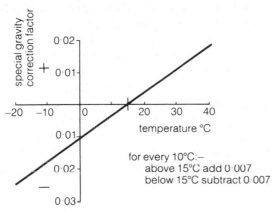

Figure 3.18 *Specific gravity correction*

required, the graph (Figure 3.18) may be used to find the value which is either added or deducted from the reading.

Table VIII shows typical results of a hydrometer test.

Electrical load test   This is also called a *High-rate discharge test* and *Drop test*. It is a severe test and should be performed only on a charged battery, i.e. a battery having a sp. gr. higher than 1.200.

The test simulates the electrical load demanded from a battery during the starting of an engine under cold winter conditions. For this reason, the test should not be extended beyond the time recommended: this is normally about 15 seconds.

The tester (Figure 3.19) consists of a voltmeter and a low-resistance strip which is connected across the battery. The resistance value, and in consequence the current load, can be adjusted to suit the battery being tested. If the load is unknown, then a load equal to three times the

ampere-hour capacity is recommended, e.g. a 100 Ah battery requires a load of 300 A.

The voltmeter indicates the battery p.d. while it is supplying the high current. The voltage given by a good battery varies with the capacity and temperature of the battery. A serviceable battery should give the following:

Capacity (Ah)	Test voltage maintained for 15 seconds
30	9 ± 1
68	10 ± 1
110	11 ± 1

A decrease in temperature reduces these values, e.g. when the temperature changes from about 20°C to −20°C the voltage reduces by 1 V.

Figure 3.19 *High-rate discharge tester*

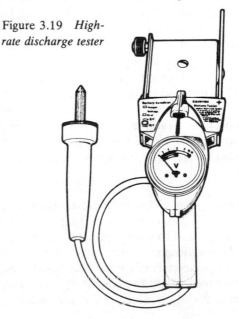

If a charged battery does not maintain the specific voltage for a given time, the battery is unserviceable. In some cases it may be seen that the electrolyte 'boils' in some cells during this test. 'Boiling' is a vigorous reaction which occurs in a faulty cell: this condition should not be confused with 'gassing'.

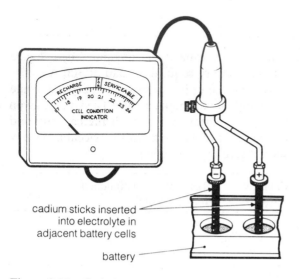

cadium sticks inserted into electrolyte in adjacent battery cells

battery

Figure 3.20   *Cadmium test*

Cadmium test   The tester shown in Figure 3.20 enables the plate condition in each cell to be determined. Cadmium probes are inserted into the electrolyte of adjacent cells and the voltage, indicated on the meter, shows whether the cell is serviceable or requires recharging. Since no chemical action takes place at the cadmium probes, then the result obtained indicates the condition of the negative plates in one cell and the positive plates in the adjoining cell.

A sound battery will give similar results from all cells.

Cold-cranking test   This test indicates the ability of a battery to supply a high current when the battery is exposed to a low temperature. It represents the conditions experienced by the battery during an engine-start in winter. The S.A.E. test determines the current in amperes that a battery can deliver for 30 seconds at a temperature of $-18°C$ ($0°F$) and still maintain a terminal voltage of 1.2 V per cell.

A battery having a rating of 360/60 has a cold cranking current of 360 A and a reserve capacity of 60 minutes.

Self-discharge test   When a battery is left unused for a period of time, a small discharge takes place. This is due to internal chemical action and external leakage caused by a small current flow between the two terminals when moisture and dirt are present on the battery cover.

A normal self-discharge is about 1% of the ampere-hour capacity per day, but if this rate is exceeded it indicates an internal fault.

Maintenance-free batteries have a much smaller self-discharge rate due to the absence of antimony. Since gassing is eliminated the external surface of the battery is kept drier and cleaner. The improved 'stand-time' of these batteries overcomes the need for trickle-charging when the battery is not in use.

Capacity tests   The ampere-hour capacity test and reserve capacity have been described on page 70.

During the life of a battery, active material becomes dislodged from the grids and this falls into the sediment trap at the base of the cell. The reduction in plate area caused by this loss proportionally decreases the capacity.

Another factor affecting capacity is *sulphation*. This is a hard white crystalline substance which forms on the plate surface and acts as a resistance to the passage of charge and discharge currents. Its presence can be detected when the battery is ch ged because the voltage required to overcome the internal resistance will be in excess of 16 V.

Sulphation is caused when the battery is:

• left discharged for a long period of time
• discharged past its normal limit
• topped-up with acid instead of distilled water
• used with a low electrolyte level

Mild sulphation can be overcome by prolonged, repeated charging and discharging at low rates, but it is generally more economical to replace the battery.

Charge rate for conventional batteries It is recommended that the charge current should be 1/10 of the ampere-hour capacity of the battery. Charging at this rate should continue until the sp. gr. values remain constant for three successive hourly readings and all cells are gassing freely.

During charging, the electrolyte should be maintained at the indicated level by topping-up with distilled water.

It takes about 16 hours to recharge a battery at the normal rate from a sp. gr. of 1.190 to its fully-charge state.

Battery replacement
Conventional batteries are supplied as new in the following forms:

- Charged and filled ready for use
- Dry-charged
- Dry-uncharged

Dry-charged batteries This form has the plates in a charged state but the cells contain no electrolyte.

After the battery has attained room temperature it is filled, to the indicated level, with sulphuric acid diluted to give a sp. gr. of 1.260 at 15.5°C (60°F). The temperature and sp. gr. of each cell is then taken and the battery is allowed to stand for 20 minutes. After this time, the temperature and sp. gr. are measured. Assuming the temperature has not risen by more than 6°C (10°F) or the sp. gr. has not dropped by more than 0.010, then the battery is ready for use. If either the temperature or sp. gr. readings are outside the limit, then the battery should be recharged at the normal rate.

Electrolyte preparation Sulphuric acid is normally supplied as a concentrated solution of sp. gr. 1.840 in a large carboy. Special care must be exercised when diluting the acid to the required strength.

Suitable protective clothing and goggles must be worn and the vessel used for the mixing should be either glass or earthenware.

The acid must be added slowly to the water because the opposite way causes a violent reaction, so remember:

always add *A*CID to *W*ATER

(N.B. the 'A' and 'W' are in alphabetical order.)

To obtain a final sp. gr. of 1.260, 1 part of acid (at 1.840 sp. gr.) is added to 3.2 parts of distilled water.

In countries where average temperatures are normally above 26°C, an acid strength of 1.210 is used for filling a battery: this strength is obtained by mixing 1 volume of acid to 4.3 parts of distilled water. The graph (Figure 3.21) shows the quantity of water required to obtain a given sp. gr. value.

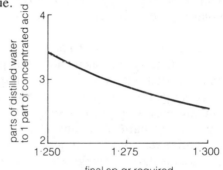

Figure 3.21 *Graph shows water required to dilute concentrated acid*

Dry uncharged batteries Depending on the battery type, filling is carried out in either one or two stages. The main steps for the two-stage are:

1. Half fill each cell with diluted acid.
2. Allow the battery to stand for 6–12 hours.
3. Fill with diluted acid to indicated level.
4. Allow to stand for 2 hours.
5. Charge at initial charge rate recommended for the battary (this is about 2/3 of the normal charge rate). Electrolyte temperature during this phase should not exceed 37°C (or 48°C in climates normally above 26°C, 80°F).
6. After a minimum of 48 hours the charge is discontinued when the voltage and sp. gr. readings show no increase over five successive hourly checks.

7. Adjust electrolyte strength to the recommended value by withdrawing some electrolyte and replacing it with either acid or distilled water. After adjustment, charge for one hour to ensure the acid and water is mixed.

As the name suggests, the one-stage method is to fill the battery to the indicated level in one step; the other operations are similar.

Attention to low-maintenance batteries
Whereas the electrolyte level of a conventional battery should be checked every few weeks, the reduced water-loss of a low-maintenance battery means that this operation only needs to be performed every 12 months.

In other respects this type of battery is treated in a similar way to a conventional battery.

Attention to maintenance-free batteries
Being as the battery is sealed, no topping-up needs to be carried-out. A normal visual inspection is necessary for cracks and corrosion and where an indicator is provided, a sp. gr. check is made. When the battery is fully charged the built-in hydrometer displays a green dot but when the area is dark, recharging is necessary.

On rare occasions the hydrometer may show a light-yellow signal. This indicates an internal fault and to avoid a possible hazard the battery must not be charged or 'jump-started'. Instead the battery should be replaced and the alternator checked for correct operation.

Voltmeter test for state-of-charge   Where no hydrometer indicator is provided, a voltmeter is used to ascertain the state-of-charge.

The method used is:

1. Switch-on headlamps for 30 seconds to remove the 'surface charge'.

2. Switch-off lamps and any external loads on the battery such as courtesy door lights.

3. Measure voltage across the battery as shown in Figure 3.22.

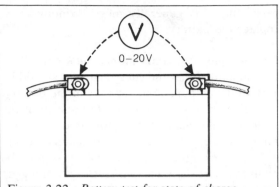

Figure 3.22   *Battery test for state-of-charge*

Typical voltage values are:

Voltmeter reading	Condition of battery
Less than 12.2 V	Deeply discharged
12.2–12.5 V	Partially discharged
Greater than 12.5 V	Fully charged

Charging   A deeply discharged battery cannot be recharged on the vehicle or by a charger which is incapable of giving a terminal voltage of 13.9–14.9 V. On no account should the voltage exceed 15 V because this high voltage will cause the battery to gas and electrolyte will be lost.

When a constant-current charger is used, an initial current of 25–30 A should be set so as to keep the terminal voltage to 14.0–14.4 V. The charge current must be monitored and adjusted to maintain the stated terminal voltage.

The battery is considered as fully-charged when the terminal voltage remains constant over two hours.

Electrical load test   This may be carried-out in a manner similar to that used with conventional batteries. The maker's recommendations should be consulted for the current load and voltage output of a serviceable battery.

General battery faults
Table IX gives some typical faults relating to batteries.

Most battery faults become apparent at a time when winter is approaching. The extra demand

Table IX   Battery faults

Fault	Cause
Undercharging	• Low alternator output, perhaps due to a slipping drive belt • Excessive use of the battery, which may be due to a short circuit • Faulty alternator regulator • Terminal corrosion
Overcharging (excessive gassing)	• Defective cell in battery • Faulty alternator regulator
Low battery capacity	• Internal or external short between cells • Sulphation • Loss of active material from plates • Low electrolyte level • Incorrect electrolyte strength • Terminal corrosion

on the battery results in either sluggish operation of the starter motor or a short period that the motor can be operated at a suitable speed. The remedy is to apply the basic test sequence:

Visual inspection – recharge – hydrometer test – load test

The result of this test sequence will confirm if the battery has reached the end of its useful life.

Terminal corrosion   This creates a high resist-ance which can cause the battery to be suspected of being faulty. If a terminal is coated with a white powder or green-white soft paste then this should be removed by immersing the terminal lug in ammoniated warm water or soda dissolved in water.

After cleaning both contacting surfaces the terminals should be coated with petroleum jelly and tightened securely.

Various types of connector are shown in Figure 3.23.

Figure 3.23   *Battery connectors*

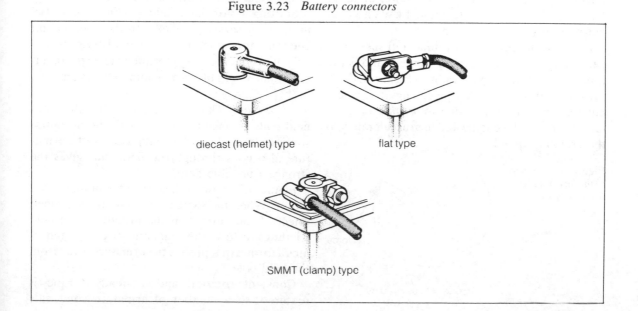

diecast (helmet) type

flat type

SMMT (clamp) type

4 Starting-motor systems

4.1 Light vehicle starting systems

A starting motor converts electrical energy supplied from the battery into mechanical power. The system must supply sufficient power to enable an engine to be cranked, i.e. turned-over, at a speed of about 100 rev/min so as to atomize the fuel and compress the air–fuel mixture sufficient to start the engine. In addition, the speed must be adequate to allow the momentum of the moving parts to 'carry' the engine over from one firing stroke to the next.

Power requirements
The power needed to attain a suitable speed depends on the size and type of engine and on the ambient conditions. Whereas a normal start of a warm 1½ litre engine requires a power of about 1.2 kW, this is increased to about 4 kW on a cold morning in winter.

Power is the product of torque and speed. By definition *torque* is a turning moment and is the force exerted at a given radius (Figure 4.1). A starting motor pinion driving an engine flywheel must exert sufficient torque to *break-away*, i.e. initially move, the engine and then accelerate it to the cranking speed.

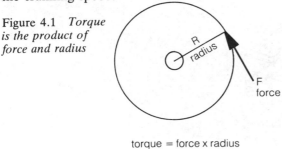

Figure 4.1 *Torque is the product of force and radius*

torque = force x radius
T = F x R

To provide high power, the motor circuit must be of low resistance to enable a current of up to 500 A to flow freely. The cables and switches must withstand the large load and the motor must be capable of converting the energy in an efficient manner. Naturally a starting system will not function properly unless the battery can provide the high current that is demanded. Similarly the battery p.d. should not fall excessively as this will affect the motor speed.

Types of motor
Motor vehicles use a d.c. motor which is based on the principle described on page 25. In the past, motors with electromagnetic field systems have been popular, but improved permanent magnet materials have enabled the construction of lightweight, and more-compact starting motors.

Starting motors have field windings connected in series or series–parallel with the main circuit and armature. Since a series-wound motor is capable of producing a high torque at low speed, it is particularly suitable as an engine-starting motor.

Series motor This type of motor has the thick field coils arranged in series with the armature windings and all current that passes to the armature also goes through the field: this gives the strongest possible field.

Figure 4.2 shows a diagram of a simple series motor. When the switch is closed, the combined effect of the current in the armature and field windings distorts the magnetic flux; this generates a torque that pushes the armature away from the field pole.

Constant rotation and a steady torque is required, so a number of armature conductor

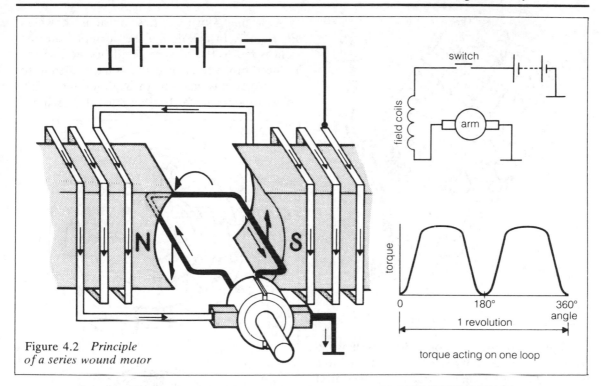

Figure 4.2 *Principle of a series wound motor*

torque acting on one loop

coils are needed: these coils are set in slots around a laminated soft-iron core. The end of each coil is soldered to a copper commutator segment which is insulated from the adjacent segments by mica (Figure 4.3). Armature conductors are made in the form of thick copper strips to provide a high current flow.

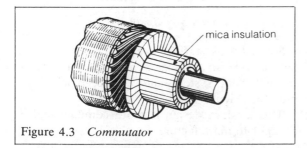

Figure 4.3 *Commutator*

Comparatively hard brushes, often of a composition of carbon and copper, are used and these are pressed against the commutator by springs of a spiral shape (Figure 4.4).

Using a normal armature and commutator allows the brushes to 'feed' the armature conductor that is positioned where the field flux is most

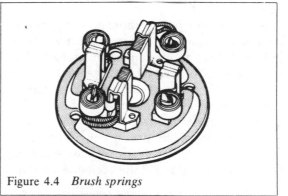

Figure 4.4 *Brush springs*

dense. When this conductor is pushed away, another conductor takes its place. By using a number of conductors, a near-uniform rotation is obtained. A typical armature has about 30 slots for conductors: the larger the number of slots, the smoother the motion.

Figure 4.5 shows the construction of the field coils. These are made of copper or aluminium alloy and are wound in a direction which produces 'N' and 'S' poles. Each coil is bound with tape to provide insulation.

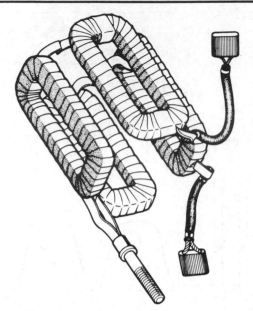

Figure 4.5   *Field coil construction*

In the case of the two-pole field coils shown in Figure 4.5, one end of the coil is connected to a brush and the other end is attached to the starter supply terminal.

By using more poles a more powerful motor is obtained. Figure 4.6 shows a 2-brush, 4-pole motor in which the total magnet strength of the series-wound field is doubled because the current is made to form other field paths. The polarity of the poles is N–S–N–S and the diagram shows how the yoke forms a part of the magnetic circuit.

As the motor is in use for only short periods of time, plain, oil-impregnated sintered bronze brushes are suitable for the armature bearings.

Figure 4.6   *A 4-pole, 2-brush motor*

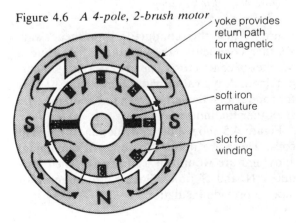

yoke provides return path for magnetic flux

soft iron armature

slot for winding

A 4-pole, 4-brush series motor is shown in Figure 4.7. In this design the current from the field is fed to the two insulated brushes, so the reduced brush resistance allows more current to flow. Current is the same throughout the circuit, so two other insulated brushes must also be fitted.

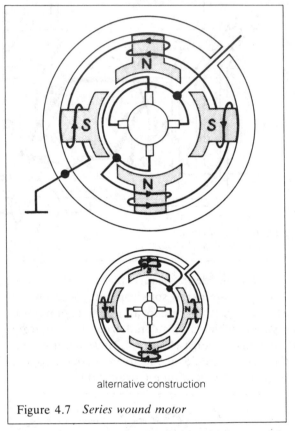

alternative construction

Figure 4.7   *Series wound motor*

Some Lucas starting motors, which use this field-and-brush arrangement, have a 'wave-wound' field system and a 'face-type' commutator (Figure 4.8).

This modern design is more compact and is cheaper to manufacture.

Series-parallel motor Figure 4.9 shows a series-parallel motor. This has the field coils in series with the armature but connects the two pairs of field coils in parallel. Current flowing to the armature divides as it enters the motor; half passes through one pair of field coils and the remainder flows to the other pair.

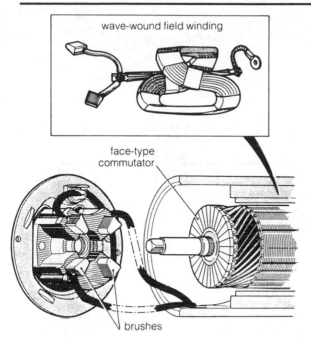

Figure 4.8 *Wave-wound field and face-type commutator*

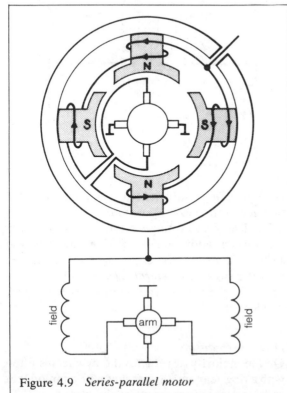

Figure 4.9 *Series-parallel motor*

A lower field resistance is achieved with this arrangement, so the motor can handle more current and give a higher torque output.

Characteristics of a series motor Torque output of a motor is directly proportional to the product of magnetic flux and current. In a series motor these are at a maximum at zero speed so the torque will be at a maximum when the armature is 'locked'.

As armature speed is slowly increased, a *back-e.m.f.* is generated and this causes the current to gradually decrease. Back-e.m.f. is due to the tendency of the motor to act as a generator. As the armature moves through the magnetic flux, an e.m.f. is induced into the conductors. As the polarity of the induced e.m.f. is opposite to the p.d. applied to the motor, the e.m.f. acts against the supply p.d., hence the term back-e.m.f.

The increase in back-e.m.f. with increase in speed, and in consequence the decrease in current, causes the torque output of the motor to gradually fall. This characteristic makes the series motor very suitable for engine starting. A very high torque is required to give engine break-away but a much lower torque is needed to overcome the resisting torque of the engine at cranking speed.

Speed of a motor varies inversely as the field strength, i.e. when the field strength is decreased the armature speed is increased. Figure 4.10 shows the effect of load on a series motor. Under

Figure 4.10 *Effect of load on starting motor*

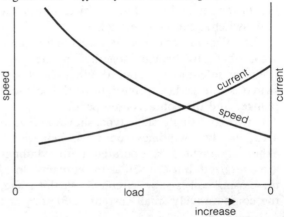

a heavy load the field is saturated and no large variation in speed occurs with slight alteration in load, but when the load is considerably reduced, the rise in the back-e.m.f. causes the magnetic flux to diminish: this results in a rise in the motor speed.

The graph shows that a series motor will overspeed if it is allowed to run free without load: this may cause serious damage to the motor.

The efficiency of a starting motor is between the limits 50–70%.

Starting-motor circuit

Ideally the battery should be situated as close as possible to the motor to minimize voltage drop of the cables. A thick cable such as 37/0.90 is used and a solenoid is fitted to act as a remote-controlled switch to limit the length of the main cable.

Although an insulated-return system is sometimes used, the earth-return arrangement shown in Figure 4.11 is the most common. Since the circuit may have to carry a current of up to 500 A, it is essential that all connections are clean and secure and that the earth-bonding strap joining the engine to the vehicle body is in good condition.

Solenoid   The solenoid shown in Figure 4.11 is a single-coil, single-stage type which can be operated only after the ignition has been switched-on. In the past, an independent starter switch was used but it is now common practice to combine the starter and ignition switches: by turning a key both switches can be operated.

The solenoid has copper contacts of adequate area, which are brought together in one stage when the solenoid is energized. When the starter switch is released, a spring returns the solenoid plunger and the contacts are opened.

A two-coil, single-stage type, shown in Figure 4.12, has two windings connected in parallel. When the switch is operated both windings are energized, but as soon as the contacts close, the closing coil is short-circuited. At this stage, the comparatively small current needed by the

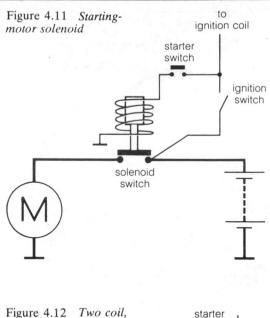

Figure 4.11   *Starting-motor solenoid*

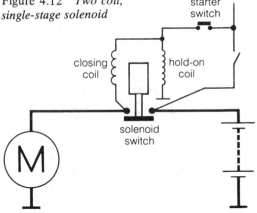

Figure 4.12   *Two coil, single-stage solenoid*

hold-on coil is sufficient to hold the solenoid plunger in the closed position.

Vehicles fitted with an automatic transmission require an additional switch in the solenoid switch circuit. This extra switch, called an *inhibitor switch* (or *neutral safety switch*), is set to open and prevents starter operation when a gear is selected.

Drive mechanisms

On the majority of vehicles the flywheel is fitted with a ring gear and this meshes with a pinion that is driven by the armature of the motor.

Gear ratio The ratio of the flywheel gear and the starter pinion is governed by the characteristics of the motor. Figure 4.13 shows a typical output for a motor and in this case the maximum power is developed when the motor speed is 1000 rev/min. Assuming the cranking speed required is 100 rev/min, then the ratio needed for this motor is given by:

$$\text{ratio} = \frac{\text{motor speed}}{\text{cranking speed}} = \frac{1000}{100} = 10{:}1$$

Normally the pinion has about 9 teeth so the number of teeth on the flywheel is set to give the appropriate ratio; in this case 90 teeth.

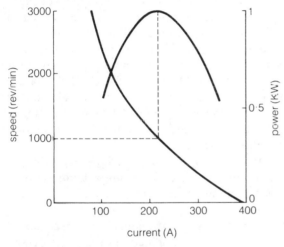

Figure 4.13 *Motor performance*

Pinion engagement The pinion is meshed with the flywheel only when the starting motor is operated. Engagement can be made in one of two ways:
• inertia engagement
• pre-engagement

Inertia engagement Inertia is the natural tendency of a body to resist any change to its velocity. In the case of a starting motor, the inertia of a pinion is utilized to move the pinion along its shaft (i.e. to move it axially) and slide it into mesh with a flywheel gear. When the pinion becomes fully engaged, the rotation of the pinion drives the flywheel.

Figure 4.14 shows the main features of a Lucas 'S' type inertia drive. The pinion is mounted on a helical screwed sleeve that is splined to the armature spindle and retained by a strong compression spring.

When the motor is operated, the combined effect of the sudden rotation of the armature and the inertia of the pinion causes the pinion to move along the helical sleeve in a direction towards the motor. During this movement, the pinion slides into mesh with the flywheel teeth aided by the chamfer on the teeth. The sudden shock as the pinion starts to drive is cushioned by the large compression spring. While the pinion is driving, the reaction to the variable resisting torque of the engine causes the helical sleeve to move axially. This movement is absorbed by the spring so the shocks are damped.

When the engine starts, the speed of the flywheel throws the pinion along the helical sleeve to disengage the drive. This ejection is quite rapid so the large compression spring is again used to cushion the shock as the pinion hits its stop.

The nominal distance between the pinion and the flywheel teeth for the drive shown in Figure 4.14 is only about 3 mm, so a thin wire retaining spring is fitted to prevent the pinion vibrating along the helix and touching the flywheel when the engine is running.

Movement of the pinion for engagement is either towards the motor (inboard type) or away from the motor (outboard). The pinion, fitted to

Figure 4.14 *Lucas 'S' type inertia drive*

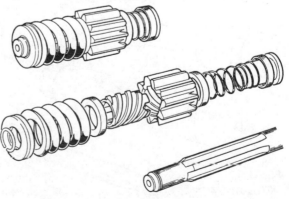

the commonly used inboard arrangement, is close to the motor when it is driving so the bending stress in the shaft is less than that given by the outboard type.

An alternative type of inertia drive is the Lucas 'Eclipse' shown in Figure 4.15. This type is similar to the Bendix drive commonly used on American vehicles.

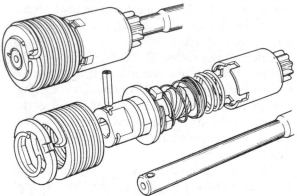

Figure 4.15 *Lucas Eclipse-type drive*

Inertia action is the same as that used in Figure 4.15 but one difference is that the Bendix type uses the main spring to transmit the drive from the armature shaft to the helical sleeve. In addition to the torsional load, the spring also acts as a cushion to absorb the shocks of engagement and disengagement.

An engine that has a small flywheel requires a small pinion to give the required gear ratio. In this case the pinion is mounted on a barrel which has a suitable mass and diameter to give sufficient inertia for engagement. A Lucas 'SB' barrel-type drive is shown in Figure 4.16.

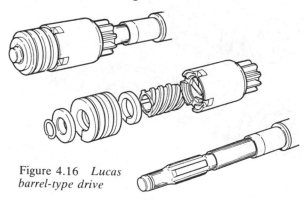

Figure 4.16 *Lucas barrel-type drive*

All inertia-type drives rely on the pinion sliding on the helical sleeve. This is not possible if the sleeve is lubricated with normal engine oil because the wet surface collects dust from within the clutch housing. One recommended lubricant is Molykiron (S.A.E.5), but if this type of lubricant is not available then the helix should be left in a dry state after cleaning.

Pre-engaged drive This type was originally introduced for diesel engines, but nowadays it is used on many petrol engines.

The pinion engagement is performed by an electrical solenoid which is integral with the starting motor. In addition to its mechanical engagement role, the solenoid acts as a relay switch to delay the passage of the full motor current until the pinion has fully meshed with the flywheel. When the engine fires, the pinion does not eject until the driver releases the switch. This feature overcomes the problem of premature ejection of the pinion during isolated firing stroke. Better starting and reduced wear of the flywheel teeth is therefore achieved.

After the engine has started, overspeeding of the motor is avoided by using a unidirectional (over-running) clutch between the pinion and the armature.

Figure 4.17 shows the constructional details of one type of pre-engaged motor. A solenoid plunger is connected to an operating lever, which is pivoted to the casing at its centre and forked at its lower end to engage with a guide ring. This ring acts against the unidirectional roller clutch and pinion. Helical splines, formed on the armature shaft, engage with the driving part of the unidirectional clutch. These splines cause the pinion to rotate slightly when the clutch and pinion are moved axially. A strong return spring in the solenoid holds the lever and pinion in the disengaged position.

When the starter switch is operated, the two-coil solenoid winding becomes energized and the plunger is drawn into the core. This initial action causes the lower end of the operating lever to move the guide ring and pinion assembly towards the flywheel teeth. This movement, aided by the

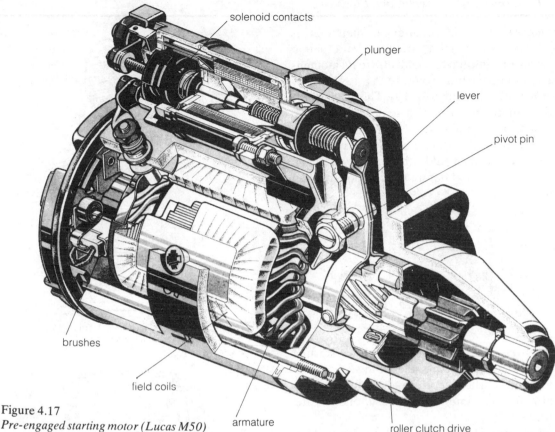

Figure 4.17
Pre-engaged starting motor (Lucas M50)

solenoid contacts

plunger

lever

pivot pin

brushes

field coils

armature

roller clutch drive

slight rotation of the pinion, normally gives full meshing of the gears. After this initial action, extra travel of the solenoid plunger causes the main contacts to close: this connects the battery to the motor.

Drive from the armature is transmitted to the unidirectional clutch and pinion by helical splines.

Sometimes the initial movement causes the pinion teeth to butt against the flywheel teeth and this prevents full engagement. When this occurs, a spring in the linkage flexes and allows the solenoid plunger to operate the main switch. As soon as the armature and pinion start to move, the teeth engage and the meshing spring pushes the pinion to its driving position.

After the engine has fired, the pinion speed will exceed the armature speed. If the motor is still in use, the rollers in the unidirectional clutch will be unlocked and the clutch will slip to protect the motor (Figure 4.18).

Release of the starter switch de-energizes the solenoid and allows the return spring to open the switch contacts. This occurs well before the pinion disengages and so avoids overspeeding of the motor. Further movement of the plunger causes the operating level to fully withdraw the pinion from the flywheel.

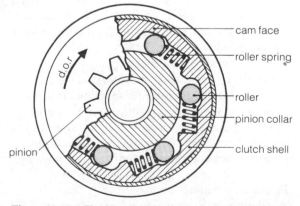

cam face

roller spring

roller

pinion collar

clutch shell

pinion

d.o.r

Figure 4.18 *Unidirectional clutch to protect motor*

Armature braking   After disengagement there is a tendency for a large armature and pinion assembly to continue to rotate due to its momentum. On some starting motors this is prevented by using a disc brake (Figure 4.19). The driving part of the unidirectional clutch is designed to rub against a part of the casing when the pinion has fully returned. This device minimizes noise and tooth wear which would otherwise occur if the starter was operated before the components had come to rest.

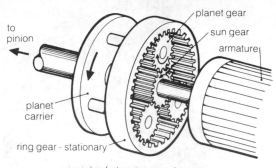

principle of planetary gearing

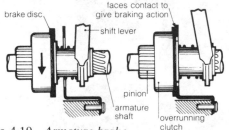

Figure 4.19   *Armature brake*

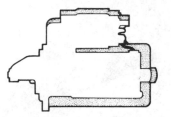

comparison of size — permanent magnet type and conventional type motor

Low-power indexing   As applied to starting motors, indexing means the lining-up of the pinion and flywheel teeth to allow full engagement before maximum power is supplied by the motor.

Low-power indexing is used on Lucas M50 starting motors. These motors have a two-stage solenoid that enables one field winding to be energized before the other field windings come into operation. The low-powered initial rotation of the pinion reduces the problem of tooth abutment.

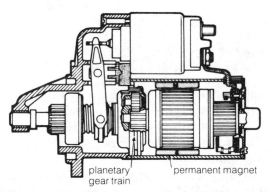

Figure 4.20   *Permanent-magnet starting motor (Bosch type)*

Pinion setting   A pinion of an inertia drive should have the correct out-of-mesh clearance so that it does not contact the flywheel before it starts to revolve. Lucas 'S' type drives should have a clearance of about 3 mm.

On some pre-engaged motors, an adjustable pivot pin for the operating lever allows the pinion to be set in the correct position. After energizing the solenoid from a 6 V supply, the pin is adjusted until the recommended clearance is obtained between the pinion and end housing.

Permanent-magnet motor

A significant improvement in performance-to-

weight ratio is achieved by using a permanent–magnet field system. Used in conjunction with a planetary gear-reduction drive, this type of motor is very compact and is suitable for use as a starting motor for a car engine.

Figure 4.20 shows the basic construction of a permanent–magnet motor. This type uses a planetary gear train made of a new plastics–steel material. The sun gear is attached to the armature and the output is taken from the three planets which revolve around the inside of a fixed ring gear.

4.2 Heavy-vehicle starting systems

An engine having a capacity of more than 3 litres needs a starting motor of considerable power, especially if the engine is a compression-ignition (C.I.) type. This type of engine must be cranked at a speed of at least 100 rev/min to initiate combustion, whereas a petrol engine will usually start if it is rotated at about 50–75 rev/min.

The C.I. engine always draws in a full charge of air and has a very high compression ratio, so the maximum torque required to drive the engine over 'compression' is greater than a petrol engine. Having passed top dead centre (t.d.c.) the high pressure in the cylinder accelerates the piston rapidly, so this causes both the rotational speed and the resisting torque loading on the starting motor to vary considerably.

To obtain high starting power many vehicles use a 24 V system. For a given electrical power requirement, the doubling of the voltage, compared with a light vehicle's 12 V system, reduces the current by half. Without the extra 12 volts, the current load on the battery and starter circuit would be exceptionally high, especially on a cold morning in winter.

Another problem experienced with large engines is the high torque required to overcome breakaway and inertia of the heavy parts. It needs a strong drive system and the pinion of this system must be fully engaged before full power is applied to the motor.

Two types of motor and drive system are used; these are:

- Axial (sliding armature)
- Coaxial (sliding gear)

Axial (sliding armature) starting motor
The main features of this type are its size and robust construction. Engagement of the pinion to the flywheel is obtained by arranging the complete armature assembly to slide axially through the motor casing. Figure 4.21 shows a simplified construction.

The motor is shown in the rest position and in this state the armature is held by a spring so that it is offset to the field poles. When the field is ener-

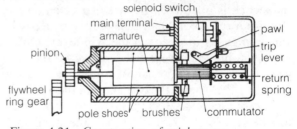

Figure 4.21 *Construction of axial starter*

gized, the armature is pulled to the left and the pinion is slid into engagement with the flywheel.

Figure 4.22 shows the electrical circuit which incorporates three field windings. The main winding is the usual thick-section, low-resistance winding and is connected, in series, to the armature. The auxiliary winding is wound with thinner wire and has a relatively high resistance; it is also connected in series with the armature but in parallel with the main winding. The holding winding is also a high-resistance winding but is connected in parallel with the armature as well as with the other windings.

The starter is operated through a two-stage solenoid switch, mounted on the starter, and is energized by the driver's switch in the cab. When

Figure 4.22 *Circuit of axial motor*

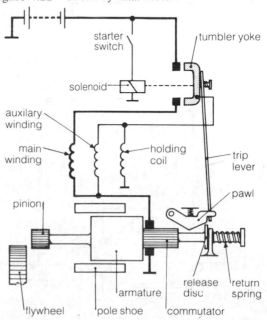

the switch is operated, the first pair of contacts closes but the second pair is held open by a pawl which engages in a slot in the trip lever. Only when the pinion is near fully engaged does the pawl allow the second pair of contacts to meet.

Figure 4.23 shows the operation of this type of motor. In Figure 4.23(a) the first pair of contacts has been closed which energizes the auxiliary windings, holding windings and armature. This action causes the armature to rotate slowly and move axially to a position where it is central to the field poles. At the same time, the pinion is slid into mesh with the flywheel gear.

When the pinion is near to full engagement, the release disc on the armature strikes the pawl and causes the trip lever to close the second pair of contacts (Figure 4.23(b)). Current now flows through the main windings, which allows the motor to develop its full torque.

As cranking speed increases, back-e.m.f. causes the current through the main and auxiliary windings to decrease, especially when the engine fires spasmodically but does not actually start. In this condition, the magnetic strength in the main and auxiliary windings is insufficient to oppose the armature return spring and hold the pinion in full engagement. This is prevented by the holding winding because the current in this winding is not affected by the back-e.m.f. generated by the rotating armature.

After the pinion has de-meshed and the armature has returned, the momentum of the rotating mass tends to keep the armature rotating. This is resisted by the 'generator effect' produced by the interaction of the holding winding and the armature. This electrical reaction to the armature 'brakes' the armature and quickly brings it to rest to enable the driver to re-engage the starter without damage to the gear teeth.

The pinion is connected to the armature shaft through a small multi-plate clutch. This serves two functions:

- It is arranged to slip if the torque applied to it exceeds a pre-determined limiting value, thus safeguarding the starter from damage should the engine backfire.
- It is arranged to disengage when the engine starts and drives the pinion faster than the armature, thus preventing the armature being damaged by excessive speed.

Figure 4.23 *Action of axial motor*

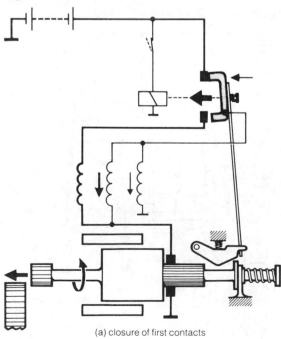

(a) closure of first contacts

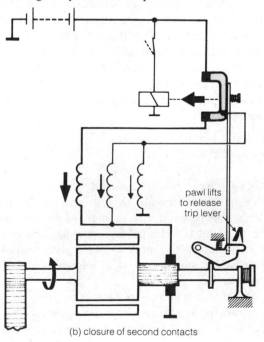

pawl lifts to release trip lever

(b) closure of second contacts

Coaxial (sliding gear) starting motor

Lucas CAV can offer this type of motor in 12 V or 24 V versions (see Figure 4.25) as an alternative to the axial type, whereas Bosch use it for the heavy end of their range.

As with the axial starting motor, this type moves the pinion into engagement under reduced power and only when it is fully meshed is full power applied. The main difference is in the way the pinion is slid into mesh with the flywheel. Instead of the whole armature assembly moving axially, the pinion only is made to slide into mesh by a solenoid mounted in a housing co-axially with the shaft.

Figure 4.24 shows the main details of the circuit of a coaxial starting motor. The main terminal is connected directly to the battery and the

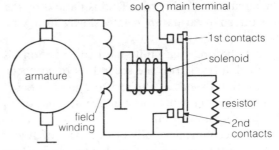

Figure 4.24 *Circuit of coaxial starter*

terminal marked 'sol' is connected to the battery via a starter switch in the cab. When this switch is operated, the two-stage solenoid is energized, which moves the pinion into mesh and at the same time closes the first set of contacts, the second set being kept open by a trip lever. At this

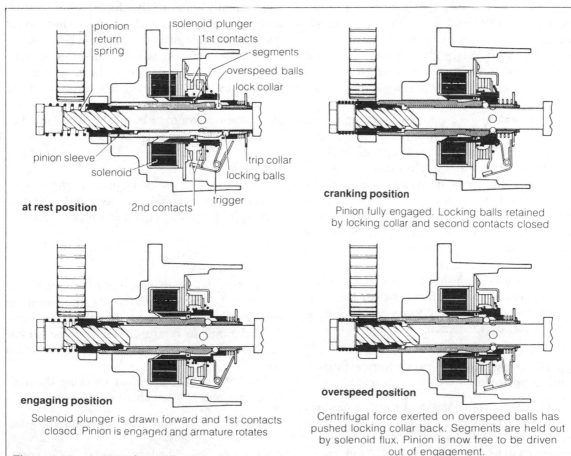

at rest position

cranking position
Pinion fully engaged. Locking balls retained by locking collar and second contacts closed

engaging position
Solenoid plunger is drawn forward and 1st contacts closed. Pinion is engaged and armature rotates

overspeed position
Centrifugal force exerted on overspeed balls has pushed locking collar back. Segments are held out by solenoid flux. Pinion is now free to be driven out of engagement.

Figure 4.25 *Action of coaxial starter*

stage, current to the main field is limited by the resistor, so the armature rotates slowly during the engagement period.

Just before the fully-meshed position is reached, a lever trips the second set of contacts. This action by-passes the resistor, gives full current to the main field and allows the motor to produce its maximum torque.

The mechanical details of one type of Lucas CAV motor are shown in Figure 4.25. This design uses four steel balls to lock the pinion sleeve to the shaft to avoid premature ejection of the pinion when the engine fires spasmodically. When the engine starts normally, overspeeding of the motor is prevented by utilizing the centrifugal effect on a set of steel balls positioned adjacent to the locking balls. When a given speed is reached, the outward force on the balls moves the locking collar and allows the pinion to disengage.

A return spring at the flywheel end of the armature shaft assists pinion disengagement and holds the pinion clear of the flywheel while the engine is running.

Bosch sliding-gear motors have the main solenoid placed at the opposite end of the motor to the pinion. This solenoid keeps the pinion in full engagement until the driver releases the starter switch. Drive from the armature to the pinion is transmitted by a multi-disc clutch. In addition to its torque-limiting duty, this clutch also prevents overspeeding by releasing the plates and slipping when the engine starts and overruns the motor.

Some motors use a shunt field winding to limit the no-load speed and others have a brake winding which comes into action when the driver releases the starter switch.

4.3 Starting-motor maintenance and fault diagnosis

The most important part of a starting system is the battery; if this is not in first-class condition, then the speed will be low and the duration of cranking will be limited. It should be noted that a fault attributed to a battery may be due to other factors such as: a defective charging system, a short to earth in another system, or overload of the battery due to driving or seasonal conditions.

Attention to warning instruments and observation of the engine's starting performance, enables the driver to recognize the initial conditions that soon develop into a major fault.

As a very large current has to be provided by a battery to the starting motor, a drop in the battery p.d. occurs. When this drop is excessive, the ignition system is 'robbed' and the voltage output of the coil may then be insufficient to give a suitable spark at the plugs, although the starting motor is still functioning, albeit at a low speed. This condition can be confirmed when an engine cannot be started with the motor, but can be started easily by 'bump-starting' (rolling the vehicle in top gear and suddenly releasing the clutch). Normally this problem is associated with a battery fault, but on many modern engines the problem is overcome by fitting a cold-start ballast resistor in the ignition supply lead. In addition to the resistor lead, a separate lead is fitted between a terminal on the starter solenoid and the ignition coil; this lead allows the resistor to be by-passed when the starter is operated (see page 117).

Maintenance
Routine attention should be given to the battery, especially the terminals. All terminals and connectors in the starter circuit should be clean and secure.

Fault diagnosis
If the likely defects of each part of the system are considered, together with the possible symptoms given by each defect, then it is possible to reverse the order and, as a result, the electrician should be able to offer a probable cause of a particular fault.

The parts of the system which cause the main problems, and the possible faults are:

- battery   low state of charge or defective.
- terminals   high resistance due to corrosion or slackness.

- cables   broken or partially broken, especially the earth-bonding strap between the engine and frame.
- solenoid   dirty contacts or faulty connection between windings and terminals.
- starter switch   high resistance at contacts or broken cables.
- motor   brushes not bedding or dirty commutator.
- pinion   not meshing or jammed due to a worn flywheel ring gear.

These faults are incorporated in Table X. The second column suggests some initial checks which should be made to enable the actual cause to be diagnosed. Location of the precise cause often requires the use of test equipment.

A starting motor that seizes the engine when the motor is operated indicates that the teeth on the flywheel have worn to the extent that they allow the gears to jam together as they attempt to mesh.

An engine always comes to rest at the start of one of the compression strokes, so the flywheel becomes burred due to pinion entry in these

Table X   Fault diagnosis

Symptom	Result of initial check	Possible cause
Low cranking speed	Lights dim when starter switch is operated	1. Discharged or defective battery 2. Poor connections between battery and solenoid 3. Tight engine
Starter does not operate	No lights or lights go out when starter is operated	1. Discharged or defective battery 2. Poor connections between battery and solenoid or between battery and earth 3. Severe short circuit to earth in starting motor
	Solenoid 'clicks' when starter switch is operated; lights unaffected	1. Poor connection between solenoid and motor 2. Broken or insecure earthstrap 3. Defective solenoid 4. Defective motor – most probably commutator or brushes
	No 'click' from solenoid; lights unaffected	1. Defective solenoid 2. Defective starter switch 3. Poor connections between starter switch and solenoid 4. Defective inhibitor switch (auto. transmission)
	Repeating 'clicking' from solenoid; lights unaffected	Broken holding coil in solenoid
	Repeated 'clicking' from solenoid; lights dim	Discharged or defective battery
	Lights dim when starter is operated and engine has seized	1. Pinion teeth jammed in flywheel 2. Engine has seized due to engine problem
Starter 'whines' but pinion does not engage		1. Dirt on helix (inertia drive) 2. Defective pinion engagement system (pre-engaged)

positions. The number of places that wear occurs around the circumference depends on the number of cylinders of the engine, e.g.:

4-cylinder engine   2 places
6-cylinder engine   3 places
8-cylinder engine   4 places

Whereas in the past a spanner was used to wind-out the pinion after it had jammed, on a modern unit it is necessary to slacken-off the securing bolts of the motor. In emergency, it is sometimes possible to free a pinion by rocking the car backwards and forwards with top gear engaged and the ignition off, but this practice may result in a bent armature spindle.

Most starter-ring gears are held on to the flywheel by an interference fit. When a new gear has to be fitted, the worn gear is removed by drilling a hole in the gear and then splitting the ring with a chisel. The new ring is preheated to the recommended temperature and, while it is still hot, it is tapped into place making sure that the chamfer on the teeth is positioned on the side of pinion entry.

Circuit tests
Prior to carrying-out these checks, the battery condition should be determined by using a hydrometer and heavy-discharge tester.

The tests shown in Figure 4.26 are made with a moving-coil voltmeter (0–20 V range) and all readings are taken *when the engine is being cranked*. To prevent the engine starting, the low-tension ignition circuit should be disconnected or, in the case of a diesel engine, the stop control should be operated.

Although the diagram shows an inertia drive

Figure 4.26　*Starting motor circuit tests*

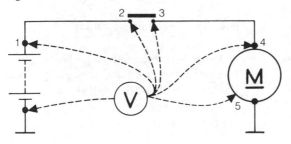

motor, the same method is used on a pre-engaged motor.

Test 1　The meter is connected across the battery and the reading is noted. If the voltage is less than 10 V the battery is suspect.

Tests 2, 3 and 4　Readings taken at the points shown should be similar to test 1. If the reading at point 4 shows a voltage drop greater than 0.5 V, then the cause of the resistance should be investigated.

Test 5　This test measures the voltage drop on the earth side of the circuit. If the reading is greater than 0.25 V then all earth connections should be checked, especially the engine-bonding strap.

A high resistance at any connection point causes a drop in voltage; this can be verified by connecting a voltmeter across the part of the circuit where the resistance is suspected (Figure 4.27). If the voltage applied to the starter is lower than specified, the current will be reduced and this will lower the motor's speed and torque.

Figure 4.27　*Volt drop test to locate resistance*

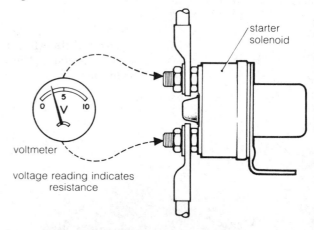

Brushes and commutator
If the battery and external circuit are serviceable but the motor fails to operate, then the motor should be removed, dismantled and inspected. The earth terminal of the battery must be disconnected before attempting to remove the motor.

Commutator　This should be cleaned with a petrol-moistened cloth; any burnt spots can be removed by using fine glass-paper, but not emery cloth.

When a satisfactory surface cannot be obtained, it will be necessary to skim the commutator in a lathe.

Brushgear   All brushes must move freely in their boxes, the springs must exert sufficient force and each brush should not be less than the specified length, e.g. in the case of Lucas motors, the minimum length is 9.5 mm and 8 mm for face type and cylindrical commutators respectively.

Fitting new brushes involves the soldering of the brush leads to the appropriate connector points. In the case of aluminium-alloy field coils, the old brush lead is cut at a given distance from the field and the new lead is soldered to the old lead.

A brush used on a cylindrical commutator must be bedded-in by using glass-paper to cut the brush to the same contour as the commutator.

Bench testing the motor
A number of tests can be applied to the motor; these include the following:

Light-running test   Under free-running conditions on a test stand, the power of the motor can be measured (Figure 4.28). The time taken for this test must be short because the centrifugal

Figure 4.28   *Position of meters for light-running test*

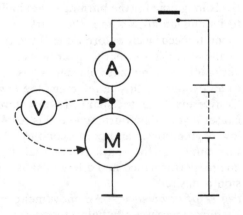

effect at the no-load speed can cause damage to the armature windings.

A Lucas 9M90 motor should 'draw' 840 W from a 12 V supply.

Locked-torque test   A torque arm is clamped to the pinion and a spring balance is used to measure the force exerted by the motor (Figure 4.29). When the armature is locked the motor produces its maximum torque and consumes its maximum current. Any high resistance in the motor reduces current and as a result the torque is lowered.

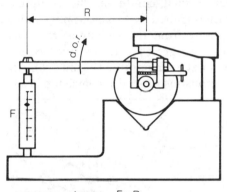

torque = F x R

Figure 4.29   *Locked-torque test*

A Lucas 9M90 motor should produce a maximum torque of 12.88 Nm (at 20°C) and 'draw' 3220 W from a 7 V supply.

Running-torque test   The torque can be measured by using a dummy flywheel on a special test bench which loads the motor and simulates the resisting torque of an engine. By noting the output torque t (Nm) and speed n (rev/s), the power (W) is obtained from:

$$\text{Power} = 2\pi nt$$

Comparing the output power with the input power allows the efficiency of the motor to be calculated:

$$\text{Efficiency (\%)} = \frac{\text{output power}}{\text{input power}} \times 100$$

5 Combustion and ignition

5.1 Combustion process

Normal combustion in a petrol engine
Under normal conditions the compressed-air mixture is ignited by a sparking plug and a flame, originating from the vicinity of the plug electrodes, progresses across the combustion chamber at a regular rate (Figure 5.1). Although the combustion process is completed in a fraction

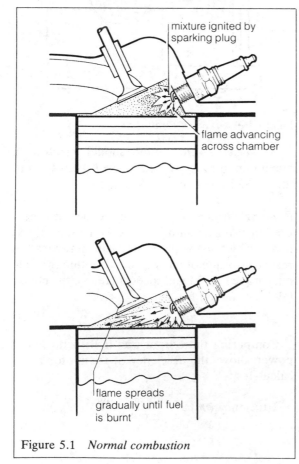

Figure 5.1   *Normal combustion*

of a second, the pressure caused by the heat of the burning gas rises steadily to give a smooth start to the power stroke.

Power output and economy depend on the flame speed; this can be varied by altering the following:

- *compression ratio* – A high flame speed is obtained when the fuel and air particles are closely packed together: this occurs when the compression ratio is high or when the engine inhales a large volume of petrol–air mixture (i.e. when the volumetric efficiency is high).
- *air/fuel ratio* – The highest flame speed and highest power is obtained from a ratio which is slightly richer than the chemically correct figure of 15 parts of air to 1 part of petrol. When the mixture is weakened from the ratio that gives highest power, the flame speed decreases considerably.
- *ignition timing* – The maximum gas pressure developed during combustion should occur about 12° after t.d.c. Since it takes a comparatively long time for the burning gas to build up to its maximum pressure, the spark must be timed to occur well before t.d.c. If the timing of the spark is too early (over-advanced), a very rapid burning takes place which is similar to an explosion; this can damage the engine. Conversely a late spark (over-retarded) causes a very slow burning and this results in low engine power and poor economy. Accurate timing of the spark is essential if good engine performance and a low exhaust emission is desired.
- *degree of turbulence* – Air movement in the chamber increases the flame speed.
- *quantity of exhaust gas present in the chamber –*

Any exhaust gas that is combined with the new mixture slows-down the burning and this lowers the maximum combustion temperature. In high-performance engines it is sometimes necessary to recirculate back some of the exhaust gas in order to reduce pollution from the exhaust gas.

Combustion in a Diesel engine

The compression-ignition (C.I.) engine has no sparking plug to initiate combustion. Instead an injector is used to spray fuel oil into the combustion chamber just before the piston reaches t.d.c. at the end of the compression stroke. Because the compression ratio of a C.I. engine is much higher than that used with a petrol engine, the heat generated by the high compression of the air is sufficient to ignite the fuel oil as it is sprayed into the chamber.

A special provision must be made when a cold C.I. engine is to be started. This is necessary because the low temperature of the cylinders absorbs the heat generated by the compression of the air; the result is that the injected fuel oil will not ignite. One way of overcoming this problem is to fit a heater plug in the combustion chamber

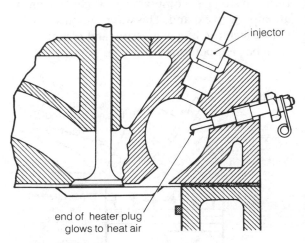

Figure 5.2 *Diesel heater plug*

(Figure 5.2). This device is an electrically-heated bulb or wire filament which glows red-hot when in use. Prior to starting a cold engine, the heater

plugs are switched-on manually or automatically for a few seconds to raise the temperature of the air in the cylinder.

Combustion faults – spark-ignition engines

Two combustion faults associated with a petrol engine are:

- combustion knock (often called detonation)
- pre-ignition

Both faults reduce engine power and if they are allowed to continue for a long period of time, they can seriously damage the engine.

Combustion knock (detonation) To obtain good engine power, a high flame speed is necessary, since the quicker the fuel burns, the higher will be the temperature of the gas. This requirement suggests that the factors controlling flame speed should be set so as to give the highest speed, but when this is attempted various combustion faults develop. To illustrate this problem, consider an engine which has a provision for varying the compression ratio. As the compression pressure is increased, the flame speed is also increased, but when a certain compression is reached, the flame speed suddenly rises to a figure which equals the speed of sound. No longer is the fuel burnt in a progressive manner; instead the petrol–air mixture explodes to give a condition called *detonation*. The compression pressure at which detonation occurs depends on many factors; these include the grade of petrol used and the type of combustion chamber.

Examination of the combustion process prior to the onset of this severe detonation would show that a portion of the petrol–air mixture remote from the sparking plug (i.e. the end gas) would not be performing in the normal way. Instead, this pocket of gas, which has been compressed and heated by the gas already burnt, will spontaneously ignite and cause a rapid build up of pressure. This condition is called *combustion knock* and Figure 5.3 indicates the main features.

Although detonation and combustion knock are two different combustion faults, the general effects are similar. For this reason the two condi-

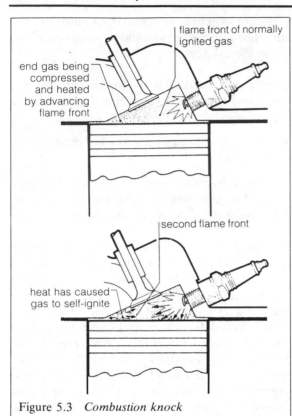

Figure 5.3   *Combustion knock*

tions are often grouped together. In both cases they occur *after the spark*.

Effects of detonation and knock depend on the severity of the condition, but the main results are:

- pinking – A sound produced by the high-pressure waves; the noise described by some people as a 'metallic tapping sound' and by others as the 'sound of fat frying in a pan'.
- shock loading of engine components.
- local overheating – piston crowns can be melted.
- reduced engine power.

These effects show that detonation and knock should be avoided, and if this is to be achieved attention must be given to the items that promote detonation. The main factors are:

- compression pressure and grade of petrol – The compression ratio is linked to the fuel used; engines having high compression ratio generally require a fuel having a high octane number.

- air–fuel ratio – Weak mixtures are prone to detonate.
- combustion chamber type – The degree of turbulence and provision for cooling of the end-gas have a great bearing on the compression ratio which is used.
- ignition timing – An over-advanced ignition promotes detonation.
- engine temperature – Overheated components increase the risk of detonation.

Pre-ignition   This condition applies when the petrol–air charge in the cylinder is fired by a red-hot particle before it is ignited by the sparking plug. The incandescent objects may be a carbon deposit or any protruding component which forms a part of the combustion chamber. Pre-ignition always occurs *before the spark* and is an undesirable condition which can lead to detonation, melting of the piston crown and other forms of damage. Normally pre-ignition gives a considerable reduction in the engine's power output and is often accompanied by the sound of 'pinking'.

Figure 5.4 shows pre-ignition of a charge, and in this case it indicates a situation whereby the rising piston is compressing a gas which is attempting to expand. This will result in a considerable increase of gas pressure and combustion-chamber temperature.

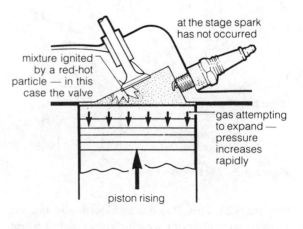

Figure 5.4   *Pre-ignition*

Running-on An engine which continues to run after the ignition is switched-off is caused by the petrol–air charge being ignited by a 'hot spot', such as a sparking plug, valve or carbon deposit, that glows and fires the charge at about t.d.c. Cooling-system faults and excessive carbon deposits are often responsible for this condition.

Weak mixtures used on modern engines to obtain a low exhaust emission of pollutant gases cause the various parts within the combustion chamber to operate at a higher temperature than that used in the past. Running-on in these engines is avoided by fitting a special device to starve the engine of fuel when the ignition is switched-off. One system uses a solenoid valve to close the slow-running petrol jet in the carburettor and another arrangement has an electrically-operated valve which allows air to by-pass the carburettor and enter the manifold.

Combustion and fuel requirements
To obtain good burning, the fuel must be mechanically broken-up (atomized) as it leaves the carburettor and be vaporized to form a gas as it passes through the warm intake manifold. Furthermore the petrol should have a sufficiently high octane rating to avoid detonation and be metered to give a suitable air–fuel ratio to meet the engine conditions. An incorrect ratio causes problems as shown in Table XI.

Exhaust-gas composition
Clean air is made up of nitrogen and oxygen; petrol consists of carbon and hydrogen. When air and petrol are mixed together in the correct chemical proportions and ignited, the resulting combustion liberates heat and changes the chemical structure to form carbon dioxide and water (Figure 5.5).

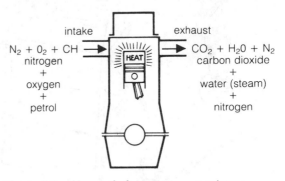

Figure 5.5 *Chemical changes–correct mixture*

A rich mixture as produced when the ratio is less than 15:1, by mass, has insufficient air to give complete combustion, so some of the fuel particles are exhausted before they have been fully burnt. This forms a poisonous gas called *carbon monoxide* (CO) and legislation exists to limit the amount of CO that is exhausted from a modern engine.

Table XI Effects of incorrect mixture strengths

Fault	Effect	Symptom
Rich mixture	Incomplete combustion	1. Black smoke from exhaust 2. High exhaust pollution
	Slow burning	1. Low power output 2. High fuel consumption
	Sparking plugs soon soot-up	1. Misfiring 2. Poor starting
Weak mixture	Slow burning	1. Low power output 2. High fuel consumption 3. Overheating
	Detonation	1. Pinking (excessive knocking) 2. Low power and if condition persists will give blue smoke from exhaust showing piston failure

Equipment as shown in Figure 5.6 is available for measuring the CO content in the exhaust. This enables the air–fuel ratio of the engine to be set so as to give good combustion and a 'clean' exhaust.

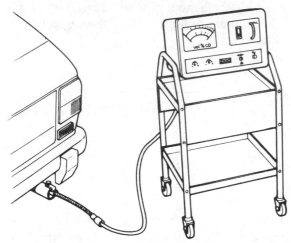

Figure 5.6   *CO test on exhaust gas*

5.2 Basic ignition systems

Engines using petrol as a fuel require a spark to start the combustion process; the timing of the spark is critical if maximum power and economy are required.

History of the internal combustion engine shows that the jump-spark system used today was gradually developed through the stages of hot wire, break spark and trembler coil, with each step showing a definite improvement over its predecessor.

There are two jump-spark ignition generator systems in use today; they are the *battery-coil* and the *magneto*, the latter confined mainly to the small engines used on motor cycles and lawn mowers.

Ignition requirements
The number of sparks required depends on the type of engine; two-stroke engines require one spark per cylinder per revolution of the crank, whereas four-stroke engines need only one spark per cylinder for every other revolution.

The conditions inside the cylinder at the time of ignition govern the voltage required to produce a spark. Whereas it needs only a few hundred volts to make a spark jump across a plug gap of 0.6 mm (0.024 in), the voltage rises to over 8 kV when the pressure is raised to represent cylinder conditions. Today most ignition systems are capable of supplying a voltage in excess of 28 kV.

Although this high voltage can be produced, it does not mean that the system always operates at this voltage. The voltage produced is only that needed to make a spark jump the spark gap. Since engine conditions such as compression pressure and fuel mixture change with speed and engine load, the voltage needed by the plug varies from about 10 kV during cruising to over 20 kV when the engine is being accelerated. This difference in voltage requirement shows up when an ignition fault is present. Although the engine may perform well when it is under light load, the demand for either high power or high speed is answered by a poor response from the engine because the ignition system will not give the high voltage required.

In addition to the voltage aspect, the spark produced at the sparking plug electrodes must also have sufficient energy to produce a spark of very high temperature and enough heat to initiate the burning of the fuel droplets situated between the plug electrodes. A normal mixture in a warm engine needs about 0.1 mJ of energy per spark, but this has to be increased considerably during cold starting or when the engine is operated on a weaker-than-normal mixture as is the case when good economy and low emission are required.

Widening the sparking-plug gap to meet these operating conditions increases both the energy and the voltage outputs, but as the output is raised, an increase in erosion wear at the sparking-plug electrodes and distributor components is caused as well as placing extra stress on the ignition coil.

The duration of the spark in milliseconds is an indication of the energy content of a spark; a typical time is 1 ms (0.001 second).

Coil-ignition system

This battery–inductive ignition system was introduced by C.F. Kettering of Delco in 1908 but it was not until the mid–1920s that it was commercially accepted as a successor to the magneto.

Up to that time very few vehicles needed a battery, so the magneto was used because it was a self-contained ignition generator. With the introduction of electric lighting, a battery then had to be carried. This step, together with the difficult starting characteristics of the magneto–ignited engine, brought about a change to the battery–inductive system which is commonly known as coil ignition.

Conventional coil-ignition circuits

Figure 5.7 shows the main details of a coil-ignition circuit. The heart of the system is the ignition coil; this transforms the low-tension (l.t.) 12 V supply given by the battery to the high-tension (h.t.) voltage needed to produce a spark at the sparking plug.

The coil has two windings, a primary and a secondary: these names are also used to identify the two circuits which make-up the complete system. Primary is the l.t. circuit supplied by the battery and secondary is the h.t. circuit that incorporates the distributor and sparking plugs. Each circuit must be complete to give current flow, so the end of the secondary winding in the coil is earthed. This is achieved by connecting the winding either to an l.t. coil terminal (normally the negative) or to an additional coil terminal that is linked by an external cable to earth. The latter arrangement is needed on a vehicle that uses an insulated return (I.R.) system, so this type of coil is called an I.R. coil to distinguish it from the common earth return (E.R.) type.

Interruption of the primary d.c. current for the induction of the h.t. voltage into the secondary

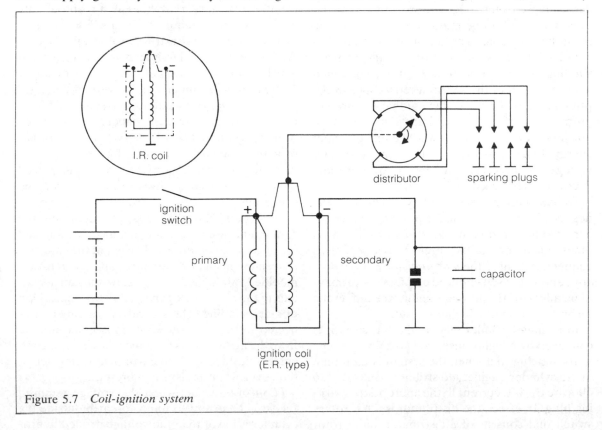

Figure 5.7 *Coil-ignition system*

winding, is made by the contact breaker at the instant the spark is required. Precise timing of the spark is achieved only if the break in the primary circuit is sudden, so to avoid arcing at this critical stage a capacitor is fitted 'across' the contact breaker.

The principles of mutual induction and the transformer are described on pages 25 and 118. These principles are used in the operation of a coil-ignition system and at this stage only the basic operation is considered.

When the ignition switch and contact breaker are both closed, a current of about 3 A flows through the primary circuit. Passage of this current through the primary winding of the coil creates a strong magnetic flux around the winding. At the appropriate time, the contact breaker is opened by a cam driven at the same speed as the engine camshaft, i.e. half crankshaft speed. The breakage of the primary circuit gives a sudden collapse of the magnetic flux in the coil and causes an e.m.f. to be mutually induced into the secondary winding. The secondary winding has about 60 times as many turns as the primary winding, so the transformer action, combined with the effect of the self-induced voltage in the primary, steps-up the voltage to that required to produce a spark at the plug. If the coil were 100% efficient the output energy would equal the input energy. In this case the increase in the secondary voltage is accompanied by a proportional decrease in the current.

By connecting the secondary winding to the negative l.t. coil terminal, the primary and secondary windings will be arranged in series with each other when the contact breaker is open. This connection, called the *auto-transformer connection*, allows the self-induced e.m.f. in the primary to be added to the mutually-induced e.m.f. in the secondary to give a higher output.

In a single-cylinder engine, the h.t. current is conveyed by a highly-insulated lead direct to the sparking plug, but when the system is used on a multi-cylinder engine, a distributor is needed to allocate the h.t. current to the appropriate sparking plug. In effect, the distributor is a h.t. rotary switch that consists of a distributor and a rotor arm, which revolves at camshaft speed. The plug leads are connected, in the firing order of the cylinders, to brass electrodes in the cap and a lead from the coil tower makes contact with a carbon brush that rubs on a brass blade forming part of the rotor arm.

An automatic advance mechanism alters the timing of the spark to suit the engine speed and load. This mechanism is situated adjacent to the contact breaker and it alters the spark timing by moving both the cam and the baseplate on which the contact breaker is mounted.

The unit that incorporates the distributor, contact breaker and automatic advance mechanism is commonly called the *ignition distributor*.

Coil-ignition components

Ignition coil

Sometimes the ignition coil is called a pulse generator because it provides an h.t. output only when a spark is required. Figure 5.8 shows in exploded form the constructional details of a coil. At the centre is a laminated iron core around which is wound a secondary winding of about 20 000 turns of thin enamelled wire of diameter 0.06 mm. Over this winding, and separated from it by layers of varnished paper, is placed the primary winding. For a 12 V system this consists of about 350 turns of enamel-covered wire of diameter 0.5 mm. Varnished paper is placed between each layer of wire to improve the insulation.

A slotted iron sheath is placed inside the aluminium case to localize the magnetic flux and the winding assembly is held clear of the case by a porcelain insulator support and a plastics-moulded, air-sealed cover. Low-tension terminals in the cover are connected to the ends of the primary winding. The secondary winding is connected to the coil tower which is made remote from the l.t. terminals to reduce the risk of the h.t. current flashing-over to earth or tracking across the cover when moisture is present.

Flash-over occurs when the voltage required for the h.t. current to jump-to-earth outside the cylinder is lower than the voltage needed to pro-

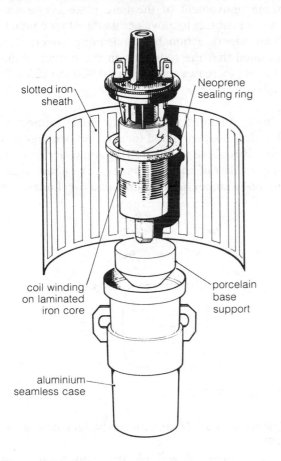

slotted iron sheath

Neoprene sealing ring

coil winding on laminated iron core

porcelain base support

aluminium seamless case

Figure 5.8 *Coil construction*

duce a spark in the cylinder. *Tracking* is the term used when the h.t. current takes an alternative path to earth over the surface of an insulator instead of sparking at the plug: the spark path taken by the current burns the surface and leaves a deposit which then acts as a conductor. To avoid tracking, the insulator surfaces should be non-porous.

The windings of most coils are immersed in oil. This improves insulation, overcomes the corona effect (faint glow of light around the coil) and reduces moisture problems. In addition, the oil improves cooling of the primary winding especially when the ignition is accidently left switched-on for a long period with the engine stationary.

Contact breaker

This cam-actuated switch acts as a *trigger* to signal when an h.t. impulse must be applied to the sparking plug. The cam revolves at half crankshaft speed, so in one revolution of the cam all cylinders are fired.

For 4-stroke engine, the numbers of lobes on the cam are:

4 cylinders	4 lobes
6 cylinders	6 lobes
8 cylinders	8 lobes

Figure 5.9 shows the layout of a contact breaker assembly for a 4-cylinder, 4-stroke engine. The two contacts, or *points* as they are often called, are made of tungsten–steel alloy: this metal resists the electrical burning action that occurs when the contacts are parted. One of the contacts if fixed to the baseplate and the other is attached to a plastics block which rubs on the cam face. A stainless-steel strip spring pushes the heel of the block firmly on to the cam, holds the contacts closed when the heel is free from the cam lobe and also acts as a conductor for the current.

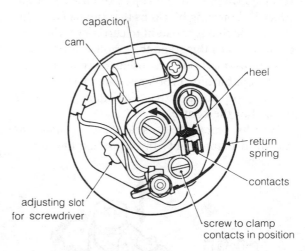

capacitor

cam

heel

return spring

contacts

adjusting slot for screwdriver

screw to clamp contacts in position

Figure 5.9 *Contact breaker assembly*

The cam in Figure 5.9 is positioned at a point where the contacts are just opening; this represents the instant that the spark occurs. Further rotation of the cam opens the contacts wider up to the point where the gap is the greatest. In this position the gap can be checked with a feeler gauge; a typical gap is 0.38 mm (0.015 in).

When the distributor assembly is positioned to give the correct sparking timing, an alteration to the contact gap will change the timing; e.g. a smaller gap causes the cam to strike the contact heel later so the spark is retarded.

Examination of a contact breaker that has been in long service shows that the metal from one contact has vaporized and has been transferred to the other contact (Figure 5.10). The crater nor-

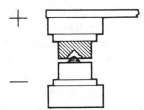

Figure 5.10   *Pitting and piling of contacts*

mally occurs on the positive side, but this is reversed when a smaller capacitor has been used. Electrical burning blackens the contact face; this is an oxide that is resistant to current so when the contacts reach this stage they should be changed.

Various methods are used to overcome the crater and burning problems. One method uses a contact, on the positive side, that has a hole formed in the centre.

Another method uses a *sliding contact* as shown in Figure 5.11. In this design the opera-

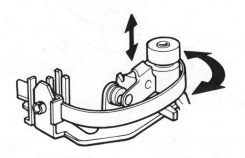

Figure 5.11
Sliding contact type of contact breaker

tional movement of the base plate causes the smaller contact to move across the other contact. This wiping action has a cleaning effect; it is claimed that the reduction in the pitting of the contact increases the life to 40 000 km (25 000 miles).

Dwell   Figure 5.12 shows the angle formed by the closed–open period for a 4-cylinder engine is 90°. This phase angle, or firing angle, depends on the number of engine cylinders. The phase angle is 360/(number of cylinders), so a 6-cylinder cam

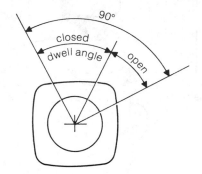

Figure 5.12   *Dwell angle*

has an angle of 60°, and for 8 cylinders the angle is 45°.

The angle moved by the cam during the *contact-closed period* is the *dwell angle* (or cam angle). An instrument called a dwell meter is used to measure the angle and since this method takes the reading whilst the engine is running, a more accurate result is obtained.

A typical dwell angle for a 4-cylinder engine is 54 ± 5°, but the angle depends on the type of distributor and the number of engine cylinders. When the contact gap is set to give the correct dwell angle, the gap measurement (in mm) should be within the specified limits, but this will not be so if the unit is worn.

When the contact gap is increased, the dwell angle is decreased. Furthermore, a decrease in the dwell angle advances the ignition by a similar amount, e.g. reducing the dwell angle from 54° to 51° will advance the ignition by 3°. The affect of dwell on the ignition timing emphasizes the need for the dwell on each cam lobe to be equal; if this is not so, the timing variation between cylinders will make the engine run erratically.

Sometimes the dwell is stated as *percentage dwell*. In this case the dwell angle is related to the phase angle. To find the percentage dwell the following is used:

$$\text{percentage dwell} = \frac{\text{dwell angle}}{\text{phase angle}} \times 100$$

A dwell angle of 54° for a 4-cylinder engine has a percentage dwell of 60%. This is obtained from:

$$\text{percentage dwell} = \frac{54}{90} \times 100 = 60\%$$

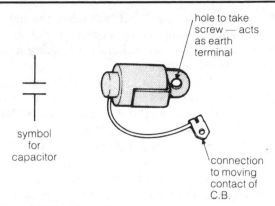

Figure 5.13   *Capacitor*

Capacitor

The importance of a capacitor, or condenser, can be seen when it is disconnected and the action of the contact breaker is observed: as the cam is rotated, severe arcing takes place at the contacts. When the contacts are opened, an induced e.m.f. of over 400 V generated in the primary circuit causes a spark to jump across the contacts as they initially part. The passage of this induced current, in the form of a spark across the contacts, results in a gradual fall in the primary current instead of a sudden fall that is needed. Besides affecting the speed of collapse of the magnetic flux, arcing quickly destroys the surface of the contacts. Therefore the duty of a capacitor is to minimize arcing and as a result speed-up the collapse of the magnetic flux. Details of a capacitor are given on page 33.

A capacitor fitted in an ignition circuit acts as a 'buffer' device. When the contacts have just parted the capacitor gives the surge current an alternative path to take, so instead of jumping the small contact gap, the current flows into the capacitor and charges it up. After a fraction of a second the capacitor discharges, but by this time the contact gap is too wide for the spark to jump across.

Figure 5.13 shows a cylindrical-type capacitor commonly used with a coil-ignition system. It consists of two rolled-up sheets of metallized paper separated from each other by a dielectric insulator. The earthed aluminium-alloy container is joined to one of the sheets and an insulated terminal, attached to a 'pig tail', is connected to the other sheet. A typical capacity is about 0.2 μF.

The capacitor is connected in parallel with the contact breaker and is positioned close to the breaker to minimize inductance and resistance of the lead.

Automatic advance mechanism

Precise timing of the spark is essential if maximum power and economy are to be obtained. If the spark occurs at the incorrect time in relation to the piston position, then problems will occur such as: overheating, pinking, piston damage and exhaust pollution.

These problems are overcome when the spark timing allows the maximum cylinder pressure to always occur about 12° after t.d.c.

A certain time elapses between the production of the spark and maximum cylinder pressure. For a given engine this 'burn-time' is affected by the air–fuel ratio and by the compression pressure; the latter is governed by the throttle opening.

Timing the spark to suit the speed   Taking into account the pressure and mixture factors, it is possible to 'time the spark' to give the required pressure at the correct instant, but this timing will only be suitable for one particular speed. At a faster speed, the crankshaft will move through a larger angle during the burn-time, so the spark must be made to occur earlier, i.e. the ignition has to be advanced.

A simple example of the spark advance requirement is shown in Figure 5.14(a). For this engine the burn-time is 0.004 second, so at 1000 rev/min the spark timing is 10° before t.d.c. A+ maximum pressure is at 12° after t.d.c. After this speed the total burning period is 22°.

Assuming the burn-time is constant at 0.004 second, then the angle moved by the crankshaft and the spark advance is as follows:

Speed (rev/min)	Angle during burn	Advance from t.d.c.
1000	22°	10°
2000	44°	32°
3000	66°	54°

Figure 5.14(b) shows the spark timing for a speed of 2000 rev/min.

In practice the burn-time does not remain constant, so taking the variation into account gives a spark advance requirement as shown in Figure 5.14(c).

Figure 5.15 shows one type of speed-sensitive centrifugal advance mechanism. Although a number of different constructions are made, the basic principle of operation is the same. The rolling contact type shown consists of two flyweights that are pivoted on a baseplate, which is driven by the distributor spindle. A contoured face on the driving side of each flyweight acts against the cam plate on to which is secured the contact breaker cam. This cam bears on the drive spindle but does not receive a direct drive from it: instead the cam is driven via the flyweights. Two tension springs, situated between the base plate and the cam plate, hold the cam plate firmly against the flyweights. The strength of the springs determines the movement of the flyweights in relation to the centrifugal force generated at a given speed.

When the engine speed is raised, the increase in centrifugal force on the flyweights overcomes the resisting force of the springs and causes the

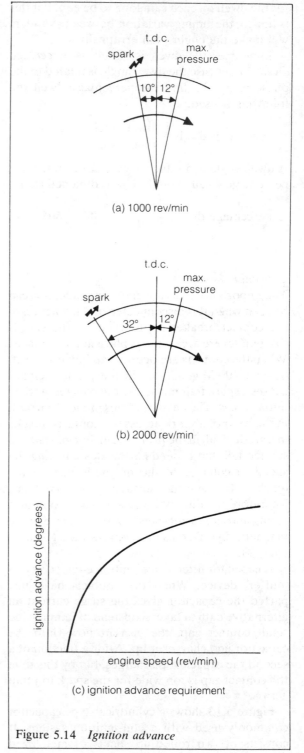

(a) 1000 rev/min

(b) 2000 rev/min

(c) ignition advance requirement

Figure 5.14   *Ignition advance*

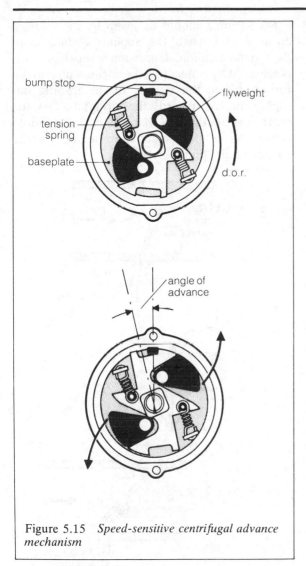

Figure 5.15 *Speed-sensitive centrifugal advance mechanism*

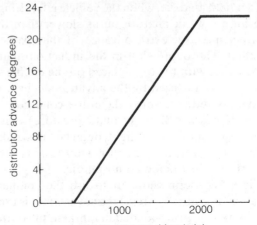

Figure 5.16 *Typical advance given by centrifugal advance mechanism*

degrees. Figure 5.16 shows the relationship between advance and speed.

Some advance mechanisms use unequal strength springs and Figure 5.17 shows this type. The strong spring is slack on its post and the weaker spring is under tension. This arrangement relies on the weaker spring only to resist outward movement of the flyweights until an engine speed of about 1000 rev/min is reached; above this speed both springs come into operation. A unit constructed in this way provides a large rate of advance up to 1000 rev/min and a smaller rate of advance beyond this speed.

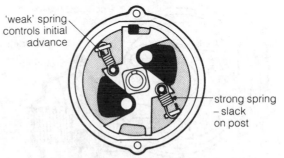

Figure 5.17 *Centrifugal advance unit fitted with unequal strength springs*

Timing the spark to suit the load For economy reasons some carburettors supply a slightly weakened mixture when the engine operates

flyweights to move outwards. As this occurs, the part rotation of the flyweights moves the cam plate forward in relation to the base plate and causes the cam to open the points earlier. This action gives a progressive advance to match the increase in speed up to the point where the full travel of the flyweight is reached.

Altering either the spring strength or the contour of the flyweight changes the angle of advance for a given speed.

The maximum advance for a typical mechanical advance mechanism is about 46 crankshaft under light load, i.e. when the vehicle is 'cruising'. Because a weak mixture burns slower than the correct mixture, extra advance of the spark is needed. The depression in the induction manifold varies with the load placed on the engine so the manifold is used by the advance mechanism and carburettor to sense the cruise condition. A high depression (low absolute pressure) exists when the engine load is light, but when it is under heavy load, the depression is very low; i.e. the pressure is just below atmospheric.

Manifold depression, or to use the common (but technically incorrect) term 'vacuum', is used to operate a spring-loaded diaphragm to control the timing of the spark. This load-sensitive *vacuum-control* operates independently from the centrifugal control mechanism.

Figure 5.18 shows a typical vacuum advance unit which gives an advance of about 13 crank-shaft degrees. The diaphragm chamber is mounted on the side of the distributor unit and a rubber pipe connects the chamber with the induction manifold. A small vent on the non-vacuum side of the diaphragm allows atmospheric pressure to act on that side. A linkage connects the diaphragm to the contact-breaker baseplate, which is allowed to part-rotate in relation to the distributor body. To advance the spark the baseplate is moved in a direction *opposite* to cam rotation, i.e. the contact breaker heel is moved towards the cam lobe.

No advance should be given by the vacuum advance unit when the engine is idling even though the manifold depression is very high. This is achieved by connecting the vacuum pipe to the carburettor in the vicinity of the throttle but slightly biased towards the air intake. As the throttle is nearly closed at idling, the manifold depression is prevented from acting on the advance unit (Figure 5.19).

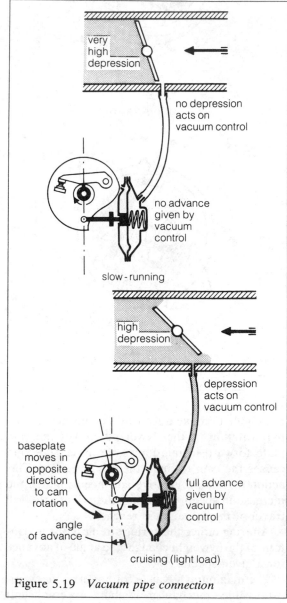

Figure 5.19   *Vacuum pipe connection*

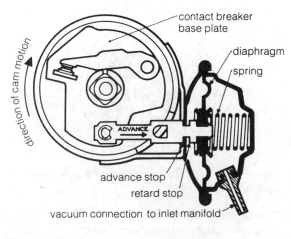

Figure 5.18   *Operation of vacuum control*

At small throttle openings, maximum depression acts on the diaphragm, so the pressure difference on the diaphragm moves it against the resistance of the spring and advances the ignition. At other throttle openings, the reduced depression will give an appropriate advance to suit the cylinder conditions.

When the vehicle is cruising, a sudden opening of the throttle will immediately destroy the manifold depression. This is fortunate because the vacuum advance will retard the ignition and counteract the tendency for the engine to pink under these heavy load conditions. Figure 5.20 shows the advance given by the vacuum advance unit.

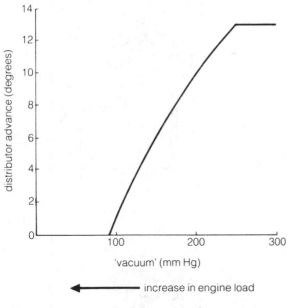

Figure 5.20   *Typical advance given by vacuum advance unit*

Various additions are made to a basic vacuum advance system to satisfy the exhaust emission regulations in force in various countries. Two of these additions are the spark delay/sustain valve and the dual-diaphragm unit.

Spark delay/sustain valve   This dual-purpose valve can be fitted in one of two ways to suit a given engine.

The item shown diagrammatically in Figure 5.21 has a one-way valve and a by-pass bleed orifice. It is fitted in the rubber pipe between the vacuum advance unit and the carburettor.

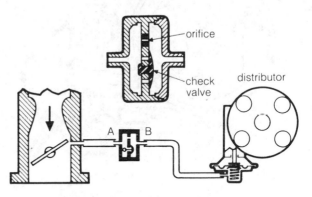

Figure 5.21   *Spark delay valve*

When pipe A is joined to the carburettor the device is a *spark delay* valve. This improves driveability and reduces emissions by delaying the full ignition advance until the air–fuel mixture has stabilized. On some cars the device is called a *spark control* system.

When pipe B is connected to the carburettor the device becomes a *spark sustain* valve. Connected in this way the valve maintains vacuum advance for a short time after the throttle has been operated. Although the valve has little effect on the performance of a warm engine, a considerable improvement in driveability is noticed when the engine is cold.

Since the one valve is used for two different roles, it is essential to fit the valve so that it operates in the manner intended.

Dual-diaphragm vacuum control   This arrangement is used to give an improvement in exhaust emission. It gives extra retardation when the engine decelerates with a closed throttle and when it is idling.

Emission at the idling speed is improved by setting the throttle so that it is opened wider than normal and then the dual-diaphragm is used to retard the ignition so as to offset the increase in speed. The unit shown in Figure 5.22(a) has a

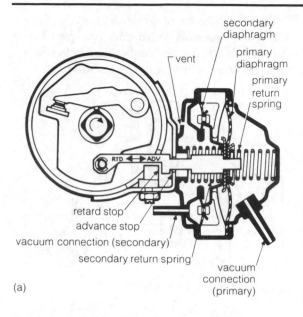

(a)

Figure 5.22 (a)   *Dual diaphragm vacuum*

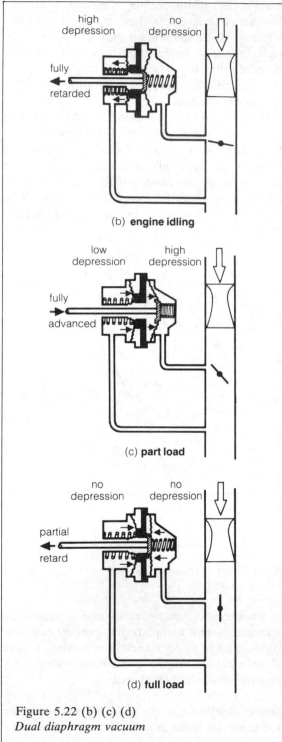

(b) **engine idling**

(c) **part load**

(d) **full load**

Figure 5.22 (b) (c) (d)
Dual diaphragm vacuum

second diaphragm which controls the stop that limits the retardation movement of the first diaphragm. Manifold pressure, at a point far away from the carburettor throttle, is used to sense the idling condition and a rubber pipe connects this point in the manifold with the second diaphragm chamber.

When the engine is idling, the high depression acting on the second diaphragm moves the stop to the left (Figure 5.22(b)). In this condition no depression acts on the main diaphragm so the primary return spring holds the primary diaphragm against the movable stop which is set so that extra retardation of the spark is obtained.

Figure 5.22(c) shows the diaphragm position when the throttle is opened more than one-quarter. In this position, the primary diaphragm gives maximum advance.

In Figure 5.22(d) the throttle is shown full-open and a low depression acts in both chambers. Both diaphragms are in the returned position and the strong secondary spring is pushing the movable stop to the right. When held in this position, the primary diaphragm makes the spark occur earlier compared with the 'idling-timing', but since the engine speed is high, the centrifugal advance system will be providing the main control.

Rotor and distributor cap

The sparking plugs of a multi-cylinder engine must be connected to the secondary winding of the coil when the cylinder is set for firing. The distributor fulfils this task by using a revolving rotor arm to transmit the h.t. impulse to the appropriate fixed electrode in the cap. When the rotor arm tip is adjacent to the cap electrode, the h.t. voltage causes a spark to jump across the small air gap. Each electrode is connected to a highly-insulated cable which conveys the current to the sparking plugs.

Figure 5.23 shows a plan view of the distributor layout. The rotor arm is a press fit on a boss formed on the contact breaker cam. A positive drive, at half crankshaft speed, is achieved by engaging a projection of the rotor with a slot in the driving boss. Electrical contact between the centre king-lead terminal and the brass blade is made by either a spring-loaded carbon brush or a strip spring attached to the rotor arm.

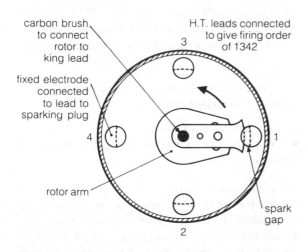

Figure 5.23 *Distributor; plan view*

On some rotor arms, the electrode end of the blade is extended towards the next electrode in the direction of rotation. Use of this feature reduces the risk of the engine running backwards.

When the crankshaft starts to move backwards, the rotor discharges the h.t. current to a plug of a cylinder which has its piston situated in the region of b.d.c. instead of the firing position.

The distributor cap is made of a brittle, anti-tracking, phenolic material that is moulded around the fixed electrodes and cable connections.

The cap is normally secured by quick-action spring clips and provision is made to prevent entry of dust and water. Sparking produces the corrosive gases nitric oxide and ozone, so some form of ventilation or shielding is provided to prevent the gases causing damage to metal surfaces.

Ignition distributor

Figure 5.24(a) shows a distributor which houses the contact breaker, mechanical and vacuum advance mechanisms and the actual h.t. distributor. The shaft is supported in two sintered-

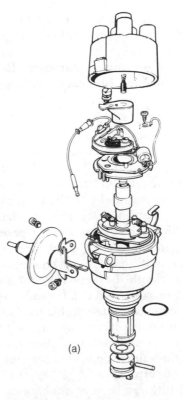

(a)

Figure 5.24 (a) *Distributor*

iron bearings and the drive from the camshaft, at half crankshaft speed, is by a helical skew gear or by an offset male dog (Figure 5.24(b)).

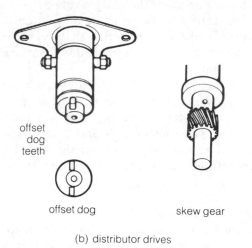

offset dog teeth

offset dog   skew gear

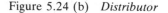

(b) distributor drives

Figure 5.24 (b)   *Distributor*

The distributor is secured to the engine by a plate and clamp, clamp bolt or body flange. Provision is made to enable the distributor body to be partly rotated for timing purposes. Rotation against the direction of rotation advances the ignition.

High-tension cables
In the past, a rubber covered, multi-strand, copper lead was commonly used. But with the advent of plastics, the rubber is often replaced by PVC, since this material gives better protection from oil and water but is less effective than rubber where high temperatures are experienced. Whatever material is used, great care must be taken to prevent the h.t. current shorting to earth.

Each lead should be kept clear of all low-voltage cables and other h.t. leads because mutual induction can cause problems. Any leakage path will reduce the voltage applied to the sparking plug, so cold-starting difficulties can be expected if the voltage applied to the plug is low due to loss of current.

Also moisture can be very troublesome if the leads are porous or when water comes into con-tact with the coil, distributor cap or sparking plugs. In these cases an aerosol-applied silicon spray can be used to repel the moisture. These sprays are extremely effective for dispersing moisture and for sealing against moisture.

Radio–frequency energy is produced by an ignition system and the energy radiated from a metallic h.t. cable causes serious interference to television and radio receivers even when they are situated at a considerable distance from the vehicle. Legislation has been introduced to limit this interference and in order to meet this requirement, the electrical resistance of the h.t. circuit has been increased. This resistance, which reduces the capacity current that discharges each time a plug fires, is obtained by using a special *suppression* cable for all h.t. leads. It consists of a core of graphite-impregnated, stranded and woven rayon or silk, which is insulated by a PVC or Neoprene covering. Special connectors are needed to join the non-metallic cable core to the terminal of the component.

The resistance of a typical cable is about 13 000–26 000 Ω per metre. Engine performance is not affected if the cable resistance is within the limits recommended. Moreover by limiting the discharge current, burning of the distributor and sparking plug electrodes is reduced.

Ballast resistor
A ballast resistor fitted in the primary circuit can fulfil two purposes; these are:

● To improve cold starting
● To reduce the variation in coil output with respect to speed.

Cold-start ballast resistor   In the past the drop in battery p.d. that occurs when an engine is being cranked by the starting motor on a cold morning lowers the coil voltage below that needed to produce a spark at the plug. Under low-temperature conditions the performances of the battery, starting motor and ignition systems are all far from their best, and this deficiency occurs at a time when the ignition requirement is very high.

A low voltage applied to a coil during starting motor operation shows up in a number of ways.

In one case the engine will not start while the starting motor is operating, but it starts easily when it is 'bump-started'. Similarly, some engines will not start until the starter switch is released; at that instant the momentum of the crankshaft gives sufficient movement to allow the engine to fire.

Although these situations are not frequent nowadays, it highlights the need for the driver to limit the duration of time that the starter switch is operated. Beyond a period of about 5 seconds, the voltage output from many batteries falls considerably, whereas if the starter switch is released after a short time and is not re-applied for a few seconds, the battery will then have chance to recover.

Many cold-starting problems have been overcome by the fitting of a ballast resistor, or by using a resistive cable between the battery and the ignition coil. In Figure 5.25 a ballast resistor is shown: this is a 2 Ω resistor connected in series with the ignition switch and a 7.5 V coil.

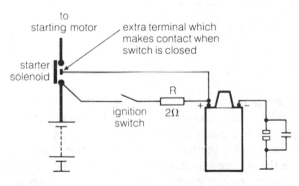

Figure 5.25   *Cold-start ballast resistor*

When the current flow in the circuit is 2.25 A, the voltage drop across the resistor is 4.5 V, so the p.d. applied to the coil is 7.5 V. By designing a coil to suit this voltage (or some other voltage as dictated by the ballast resistor used), the secondary output is kept within the limits required by the engine.

The improvement in the cold starting of the engine is achieved by fitting an extra cable in parallel with the ballast resistor. The ends of this cable are connected to an additional terminal on the starter solenoid switch and the ignition coil. When the starter solenoid is operated, current will be supplied to the coil. Since this current by-passes the ballast resistor, full battery voltage, even though it may be only 10 V at this time, is applied to the coil. Whereas the output from a conventional 12 V coil would be poor during engine-cranking by the starter, the higher-than-normal voltage applied to a 7.5 V coil will actually boost the output to meet the ignition requirement at this time.

Output control ballast resistor   It takes a comparatively long time after the contact breaker has closed for the primary current to build-up to its maximum. Figure 5.26 (overleaf) shows the growth time for a typical ignition coil and in this case a time of about 0.01 second is required before the maximum primary current is obtained.

Current flow starts when the contact breaker closes and continues to build up during the dwell period. Although the dwell angle should remain constant with speed, the dwell time in seconds shortens. When the dwell time is less than about 0.01 second for this coil, the primary current is no longer capable of reaching its maximum, consequently the output from the coil gradually falls when the speed is increased beyond this point.

Figure 5.27 (overleaf) shows the variation in dwell time for 4- and 6-cylinder engines. If the dwell time for a particular speed and engine is read from this graph, it is possible to use Figure 5.26 to ascertain the primary current at that speed.

These graphs show that an engine having 6 or more cylinders suffers a gradual fall-off in coil output when the speed is increased. Fitting an *output control ballast resistor* in series with the primary circuit compensates for this variation in output.

The iron-wire resistor has a high-temperature coefficient such that the 'hot' resistance is about three times as great as the 'cold' resistance. Temperature of the resistor depends on the current passing through it, so when the engine is running at low speed, the long dwell time causes the resistor to run hot; as a result, the average

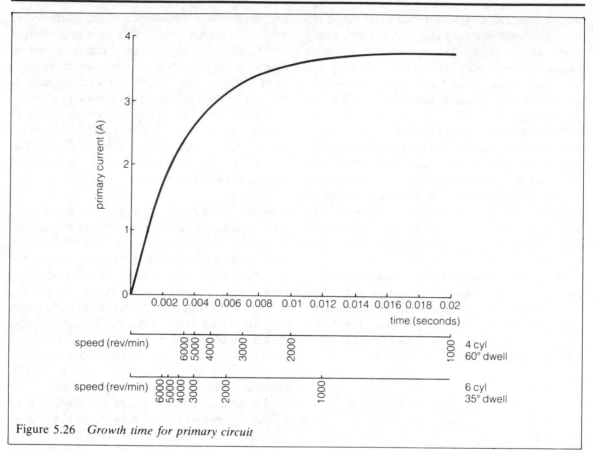

Figure 5.26   *Growth time for primary circuit*

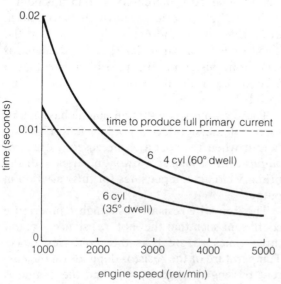

Figure 5.27   *Variation in dwell time*

current in the primary is reduced. This allows the coil to run cooler and also reduces the ill-effects of spark erosion due to high voltage.

Similarly, the natural drop in primary current as the engine speed is increased allows the ballast resistor to run cooler. This causes its resistance value to fall and as a result increases the primary current to offset the fall-off due to speed.

A ballast resistor of this type has a cold value of about 0.25 Ω. The resistor can be fitted either internally in the coil or externally in the circuit.

Low-inductance ignition coil
Current growth in the primary winding of a coil is restricted by the coil's self-inductance. As the primary current gradually increases, self-inductance in the winding produces a back-e.m.f. which opposes any change in the current. This

opposition to current growth is increased as the number of turns on the primary winding is increased.

Engines having a large number of cylinders, especially 8-cylinder units, need a coil which gives a quicker growth than a conventional coil, so a high-output coil, called a *low-inductance coil* is often fitted to these engines. Since the primary winding of a low-inductance coil has fewer turns and therefore the wire is shorter in length, the current flowing when the engine is in its stalled condition is about three times as great. Naturally the erosion wear, due to this high current, on a conventional contact breaker is severe, so the low-inductance coil is often fitted in conjunction with a transistorized breaker system.

Twin contact breaker

The short dwell time associated with high-speed operation of 8-cylinder engines can be improved by using twin contact breakers. This system reduces the time that the primary circuit is broken because it arranges for one set of contacts to 'make' the circuit immediately after the spark has occurred.

In the twin contact arrangement shown in Figure 5.28 the contact set A is connected in parallel with contact set B. Mounted in this way, the circuit is interrupted only when both contacts are opened simultaneously. As soon as contact set A opens to give a spark at the plug, the other set closes and starts the build up of the primary current.

The introduction of transistorized systems has made the twin contact arrangement obsolete.

Magneto ignition

A magneto is a self-contained unit which generates its own electricity and at the right time steps-up the voltage to give a spark at the plug. A magneto has two advantages over a coil-ignition system: it does not require a battery and the voltage output improves as the engine speed is increased.

Its main disadvantage is its low performance at cranking speed. This drawback has made the coil-ignition system universal for cars, but the

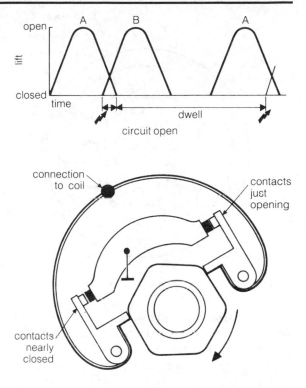

Figure 5.28 *Twin contact breakers*

magneto is still used on small engines such as motor cycles, mowers, etc.

Rotating–magnet magneto   Small engines have used this type of magneto since improved magnetic materials were introduced for permanent magnets.

A flywheel magneto has the magnet cast into a non-ferrous flywheel, hence it is classified as a rotating magnet type. Figure 5.29 (overleaf) shows the basic construction.

In this design a laminated soft-iron armature containing the coil windings, is stationary, and a cam formed on the flywheel hub operates a contact breaker.

Rotation of the magnet with the flywheel causes an alternating magnetic flux to pass through the armature. As the primary coil is wound on this armature, a current is induced into the coil every time a change occurs in the magnetic flux. Movement of the magnet across the

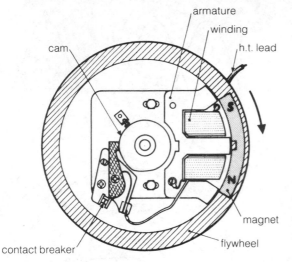

Figure 5.29   *Rotating magnet magneto*

complete armature will give a full reversal of the flux, so an alternating current is generated which peaks every time the flux reversal occurs.

Having generated its own primary current, the magneto transforms the low voltage to a voltage sufficient to produce a spark at the plug. This is achieved by using a circuit as shown in Figure 5.30. It consists of a contact breaker and two windings, a primary and a secondary, interconnected similar to that used in a coil-ignition circuit but without a battery. A capacitor is used to

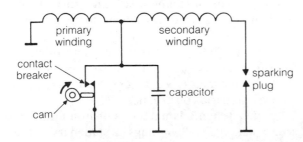

Figure 5.30   *Circuit of magneto*

speed up the collapse of the magnetic flux which it does by reducing arcing at the contacts.

A complete circuit is required for the primary current to build up to its maximum, so during this period the contact breaker is kept closed. Just

before the primary starts to fall in order to build up in the other direction, the contact breaker is opened. This interruption mutually induces a high voltage into the secondary winding which is connected to the sparking plug.

Many magnetos have a safety spark-gap which is used to protect the insulation of the magneto coil when either a plug lead becomes disconnected or the h.t. is open-circuited.

5.3 Sparking plugs

The sparking plug provides the gap across which the high-tension current jumps to give the spark for ignition of the petrol–air mixture.

Since the Frenchman Etienne Lenoir invented the sparking plug in 1860, many detailed changes have been made but the basic construction has remained the same; a highly-insulated electrode connected to the h.t. cable and an earth electrode joined to the plug body.

Sparking plug requirements   The basic requirement is that a spark of sufficient energy should be produced across the electrodes at all times irrespective of the pressure and temperature of the gases in the combustion chamber. Consideration of these two factors show that the plug operating conditions are severe.

Pressure   Besides withstanding a pressure of about 70 bar during combustion, the plug must be able to produce a high-energy spark when the gas pressure is about 10 bar. The voltage required to do this may be as high as 30 kV, so adequate insulation is needed to prevent leakage of the electrical energy to earth.

Temperature   The plug must be capable of withstanding an electrode temperature of between 350°C and 900°C for long periods of time. Its construction should keep the electrode temperature between these limits, because if these limits are exceeded the plug will fail.

Above 900°C the high temperature of the electrodes causes pre-ignition, whereas below 350°C carbon will form on the insulator. This will cause fouling, which is a term used when the plug allows the electrical charge to short to earth instead of jumping the spark gap.

Also in addition to these basic requirements, the plug must be: resistant to corrosion, durable, gas-tight and cheap.

Construction   Figure 5.31 shows the main parts of a typical plug. It consists of an alloy steel centre electrode and an aluminium oxide ceramic insulator, which is supported in a steel shell. Gas leakage past the insulator is prevented by 'sillment compressed powder seals' and leakage between the cylinder head and the shell is by means of a gasket or tapered seat. A single-earth electrode of rectangular cross-section is welded to the

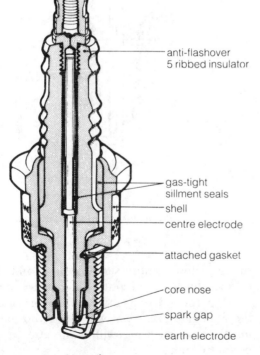

- terminal
- anti-flashover 5 ribbed insulator
- gas-tight sillment seals
- shell
- centre electrode
- attached gasket
- core nose
- spark gap
- earth electrode

Figure 5.31   *Sparking plug*

shell. A hexagon is machined on the shell for the purposes of installation and removal of the plug.

Ribs formed on the outside of the insulator increase the length of the flashover path and also

improve the grip of the lead covers which are fitted to prevent the ingress of moisture.

Plug terms   The length of the thread that screws into the cylinder head is called the *reach* and the diameter of the threaded part indicates the plug size; common sizes used are 10, 12, 14 and 18 mm. Manufacturers use a code to identify their products; the letters and numbers stamped on the insulator give the following information:

- Diameter and reach
- Seat sealing and radio interference features
- Centre electrode features such as the incorporation of a resistor or auxiliary gap
- Heat range
- Firing end configuration

Figure 5.32 shows the position of the plug in a cylinder. The seating height A is less than the reach to the extent of the thickness of the gasket.

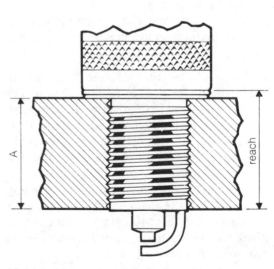

Figure 5.32   *Sparking plug in cylinder head*

Heat range   This term indicates the temperature range in which a plug operates without causing pre-ignition or plug fouling due to carbon or oil deposits on the insulator.

Figure 5.33 (overleaf) shows the heat limits and the effect of road speed on the temperature of a typical plug. In addition to pre-ignition and

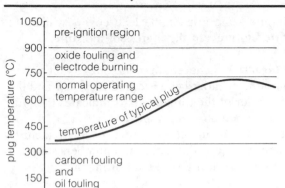
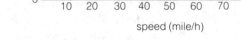

Figure 5.33   *Plug operating temperature*

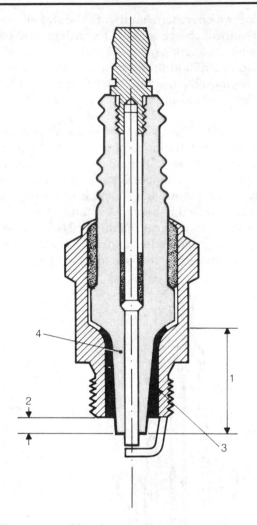

Figure 5.34   *Plug feature affecting temperature*

carbon fouling the graph shows that an operating temperature in excess of about 750°C causes oxide fouling of the insulator and excessive burning of the electrodes.

The operating temperature of a plug depends on the four features shown in Figure 5.34. These are:

1. *Insulator nose length.* This is the distance from the tip of the electrode to the body. The length of this heat-flow path governs the temperature of the insulator nose, so if the path is made short, the plug will run relatively cool.
2. *Projection of insulator.* The amount that the insulator protrudes into the combustion chamber governs the amount of cooling obtained from the incoming air–fuel charge. It is claimed that the 'turbo-action' of the charge gives the plug a better resistance to carbon and oxide fouling.
3. *Bore clearance.* The clearance between the insulator and the shell governs the amount of deposit that can be accepted before the plug electrodes are shorted-out.
4. *Material.* Rate of heat transfer depends on the thermal conductivity of the materials used, especially the material used for the insulator.

Figure 5.35 shows two plugs with different heat ranges. The hot (or soft) plug has a long heat-transfer path and is recommended for cool-running, low-compression engines, and other engines that are used continually at low speed for short journeys. Unless this type of plug is used on these engines, carbon will build up on the insulator so misfiring will occur after a short period of time.

At the other end of the heat range is the cold (or hard) plug. This plug has good thermal conductivity and is used in engines that have a high power output for their size.

To summarize the application of each type, the following general rules are used:

Hot plug for a cold engine
Cold plug for a hot engine

Manufacturers offer a range of plugs, so it is possible to match the heat range of a plug to the engine.

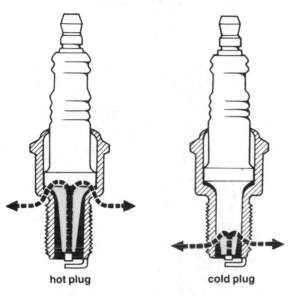

hot plug   **cold plug**

Figure 5.35   *Sparking plug heat range*

Electrode features   A conventional plug uses nickel alloy for the electrodes to give resistance to corrosive attack by combustion products or erosion from high-voltage discharges. In engines where corrosion and erosion are severe, a special material, such as platinum, is sometimes used.

Both electrodes must be robust to withstand vibration from combustion effects and also they must be correctly shaped to allow a spark to be produced with minimum voltage (Figure 5.36(a)). Under normal conditions, erosion eats away the electrodes, so after a period of time the earth electrode becomes pointed in shape (Figure 5.36(b)): in this state it requires a higher voltage to produce a spark.

Between the period of the maintenance checks, the increase in the voltage requirement of the plug is accompanied by a general deterioration in the output voltage of the ignition-generator system. Unless regular attention is given, the system will fail: most probably this will

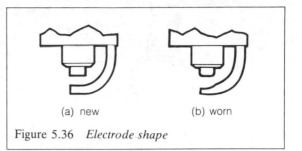

(a) new   (b) worn

Figure 5.36   *Electrode shape*

show up when attempting to start the engine on a cold, damp day.

A typical sparking plug gap is 0.6 mm (0.024 in), but wider gaps are sometimes used for engines which run on a mixture weaker than normal. These mixtures are more difficult to ignite so a higher voltage is required.

An *auxiliary gap* (or booster gap) is used on some types. This series gap is formed between the terminal and the end of the electrode. Its purpose is to reduce the build up of carbon on the insulator nose and so improve the plug performance when the engine is operated at low power for a considerable time.

The introduction of an extra series gap in the circuit raises the h.t. voltage and as a result causes the spark to jump the gap rather than short down the carbon on the insulator.

This is similar to the method used in the past to restore normal sparking from a plug that has been fouled by petrol. The plug lead is held, by suitably-insulated tongs, about 6 mm from the plug terminal and the engine is started. After a short time the plug begins to operate normally.

Copper-cored electrode   Increasing the insulator nose length reduces the risk of carbon fouling when the vehicle is operated on short journeys, but when a high vehicle speed is maintained for a long period, the spark plug seriously overheats.

This temperature problem can be overcome by using expensive electrode materials such as platinum, iridium, silver or gold–palladium, but a cheaper alternative is to use a copper-cored electrode to improve thermal conductivity (Figure 5.37) (overleaf).

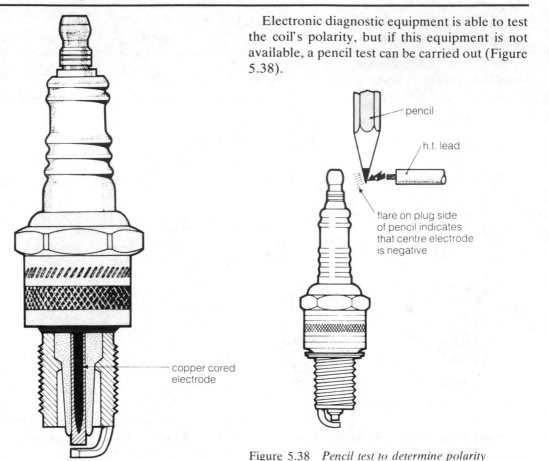

Electronic diagnostic equipment is able to test the coil's polarity, but if this equipment is not available, a pencil test can be carried out (Figure 5.38).

pencil

h.t. lead

flare on plug side of pencil indicates that centre electrode is negative

Figure 5.38   *Pencil test to determine polarity*

copper cored electrode

Figure 5.37   *Copper cored electrode*

Electrode polarity   A lower voltage is needed to produce a spark at the plug when the centre electrode is negative in relation to the h.t. circuit polarity. A hot surface emits electrons and since the centre electrode is the hotter of the two, the natural flow of electrons is from the centre electrode to the earth electrode. If the circuit is connected to give this direction of flow, then the natural flow of electrons will aid, rather than oppose, the electron movement given by the ignition coil.

The direction of electron flow in the secondary depends on the polarity of the primary winding. Nowadays the l.t. terminals on most coils are marked (+) and (−) to indicate the connections required to give a 'negative spark'.

Inspection of a plug which has been in use for a long time shows that more erosion occurs on the earth electrode when the centre electrode is negative.

5.4 Electronic ignition

Since 1960, new requirements for ignition systems have been made that could not be met by the conventional inductive ignition system. The introduction of new exhaust emission criteria in 1965 and the demand for improved fuel economy in 1975, has forced designers to turn to electronics to provide a system to satisfy the statutory requirements for a vehicle. Legislative requirements and driver demand for better engine performance, added to the manufacturer's need to offer a more sophisticated vehicle to counter a competitor's product all show why electronic innovation in this field is taking place.

The full impact of ignition development will take a few years to unfold, but it is expected that the changes in this decade will be far greater than for many years past.

Drawbacks of a conventional system

The basic principle of a conventional inductive ignition system has remained unchanged for over sixty years, but the time has come where it is unable to meet present and future needs as regards energy output and contact breaker performance.

Whereas an ignition output of 10–15 kV was sufficient in earlier days, the modern high-speed engine demands an output of 15–30 kV to ignite the weaker mixtures needed to give good economy and emission. To achieve this improvement a low-inductive coil (see page 118) is often fitted, but because the current through this coil is much higher than a normal coil, the erosive wear of the contact breaker is unacceptable.

This reason alone suggests a need for an electronic replacement for the mechanical breaker, but other drawbacks of the breaker show why it should be superseded. These are:

- Ignition varies from specification as the speed is varied. This is due to (a) wear at the contact heel, cam and spindle and erosion of the contact faces; (b) contact bounce and the inability of the heel to follow the cam at high speed.
- Adverse effect on the dwell time for the coil current due to dwell angle variation.
- Frequent servicing is necessary.

Although many of these breaker drawbacks affect the energy output of the coil, the variation of the spark timing between service intervals has meant that engines fitted with mechanical breakers cannot meet current emission regulations in force in many parts of the world. When statutory requirements insist that exhaust emission levels must be maintained for 80 000 km (50 000 miles), manufacturers have turned away from the mechanical breaker to alternative systems which do not suffer the aforementioned drawbacks. Electronic systems fill this requirement.

The following descriptions cover the basic principles of the main ignition systems used during the period from the start of the change-over to the present day.

Personal safety

Before considering the various systems, it must be appreciated that the voltage output from electronic ignition systems, *even from the l.t. circuit terminals and cables*, is far greater than that developed from a conventional system used in the past.

A shock received from live components can prove fatal, so extra care is needed with this type of ignition system!

Breaker-triggered systems

Transistor assisted contacts (T.A.C.)

This system uses a normal mechanical breaker to 'drive' a transistor which controls the current in the primary circuit. By using a very small breaker current, erosion of the contacts is eliminated and, as a result, the cleaner contact faces maintain good coil output; it also gives accurate spark timing for a much longer period. The use of a low-inductive coil and ballast resistor with this system, allows the benefits of this type of coil to be obtained without suffering the ill-effects of excessive contact arcing produced by the high primary current.

Figure 5.39 (overleaf) shows the basic principle of a breaker-triggered, inductive, semiconductor ignition system. In this system, a transistor performs the duty originally undertaken by the contact breaker, namely to act as a power switch to 'make and break' the primary circuit. The transistor (as described on page 43) in this circuit acts as a relay which is operated by the current supplied by a cam-operated control switch; hence the term *breaker-triggered*.

When the contact breaker is closed, a small control current passes through the base-emitter of the transistor. This switches-on the collector–

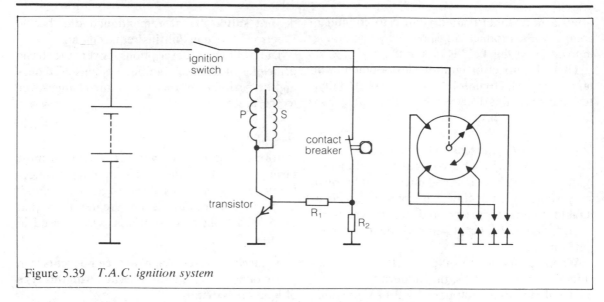

Figure 5.39 *T.A.C. ignition system*

emitter circuit of the transistor and allows full current to flow through the primary circuit to give energization of the coil. At this stage current flow in the control circuit and transistor base is governed by the total and relative values of the resistors R_1 and R_2. Values are chosen to give a control current of about 0.3 A, because this small current will not overload the breaker but is sufficient to give a self-cleaning action of the contact surfaces.

At the instant the spark is required, the cam opens the contact breaker to interrupt the base circuit: this causes the transistor to switch-off. As the primary circuit is suddenly broken, a high voltage is induced into the secondary to give a spark at the plug. This sequence is repeated to give the appropriate number of sparks per cam revolution (Figure 5.40).

Compared with a non-transistorized system, the T.A.C. arrangement gives a quicker break of the circuit and, in consequence, a more rapid collapse of the magnetic flux. Since secondary output depends on the speed of collapse of the flux, a high h.t. voltage is obtained from the T.A.C. system.

With the exception of the extra control module containing the power transistor, the other units of this ignition system are similar to those used with a conventional system.

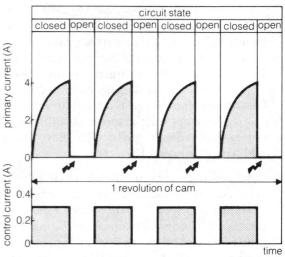

Figure 5.40 *Primary current control (4-cylinder engine)*

Although the basic circuit shown in Figure 5.39 illustrates the principle of a T.A.C. system, extra refinements are needed to protect the semiconductors from overload due to self-induction and to minimize radio interference. In addition it will be seen that the circuit shown is unsuitable for use with a conventional contact breaker that has a fixed-earth contact. To overcome this problem an additional transistor is used (Figure 5.41).

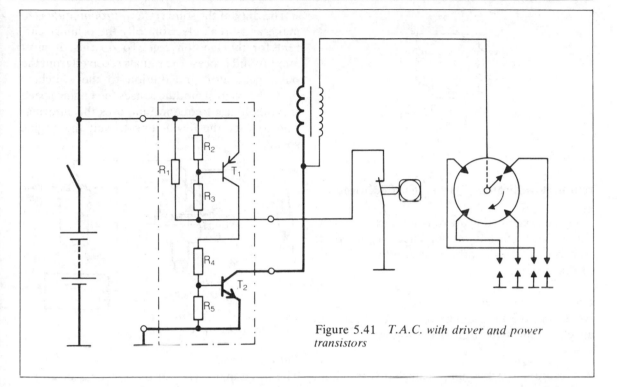

Figure 5.41 *T.A.C. with driver and power transistors*

In this layout the transistor T_1 is fitted in series with the contact breaker in the control circuit and its duty is to act as a driver for the power transistor T_2. As in the previous systems, resistors are fitted to limit the base current in T_1 and T_2 as well as the contact-breaker current.

When the contact breaker is closed a small current flows in the control circuit. Most of this current passes through R_1, but a very small proportion is passed through the base of T_1 which causes the transistor to switch-on. This sensitive transistor then supplies a current to the base of the power transistor T_2 and causes this transistor to switch-on. When this happens, the collector–emitter of T_2 conducts and completes the primary circuit to allow the build up of magnetic flux in the coil.

At the time of the spark, the contact breaker is opened; this interrupts the current in the control circuit and base circuit of T_1. With T_1 switched-off the current is cut-off from the base of T_2, so this causes T_2 to switch-off and break the primary circuit.

The power transistor T_2 must be robust to handle the large current and high voltage due to self-induction. These conditions are particularly severe when a low-inductance coil is used, especially when the engine is being started. In this case a current of about 9 A built up in the primary circuit during normal operation can increase to about 16 A when cold-starting. As a conventional breaker can handle a maximum of only 5 A, the high currents associated with this type of coil can be switched effectively only by electronic means. Even so, this high load would reduce the reliability of a normal power transistor so a special dual-transistor called a *Darlington amplifier* is used.

Figure 5.42 (overleaf) shows the circuit of a Darlington amplifier: this consists of two transistors formed in an *integrated circuit* (I.C.) to make one unit with three terminals, E, B, and C.

When a small current is supplied to the base of T_1, it switches-on and causes a proportionally larger current to flow to the base of T_2. This, in turn, switches-on T_2 which allows the main cur-

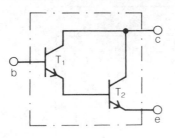

Figure 5.42   *Darlington amplifier*

rent to flow through T_2 from the collector to the emitter.

In the system shown as Figure 5.41, the substitution of a Darlington amplifier for the power transistor T_2 considerably improves the reliability of the system.

Breakerless systems
Replacing the mechanical contact breaker with an electronic switch gives the following advantages:

- Accurate spark timing achieved throughout the speed range.
- No contacts to erode and wear. This eliminates maintenance in respect of constant replacement, dwell adjustment and setting of the spark timing. Furthermore, the timing remains correct for a very long period.
- Build-up time for the ignition coil can be varied by altering the dwell period to suit the conditions. This gives a higher energy output from the coil at high speed without the risk of h.t. erosion at low speed.
- No bouncing of contacts at high speed to rob the coil of its primary current.

Figure 5.43 shows the main layout of a breakerless, electronic ignition system. The distributor unit is similar to a conventional unit with the exception that the contact breaker is replaced by an electronic switch called a *pulse generator*. As the name suggests, this device generates an electrical pulse to signal when the spark is required. This action is similar to the trigger of a gun – when the trigger is operated the coil fires its h.t. charge.

The duty of the solid state *control module* is to make-and-break electronically the primary current for the ignition coil. To do this, it must amplify and process the signals received from the pulse generator. In addition to the switching duty, the control module senses the engine speed from the pulse frequency and uses this information to vary the dwell time to suit the engine speed.

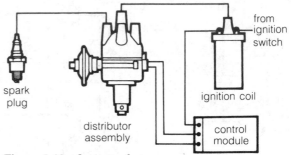

Figure 5.43   *Layout of breakerless electronic system*

Pulse generator
The three main types of pulse generator are:

1. Inductive
2. Hall generator
3. Optical

Inductive type   Figure 5.44 shows one type of inductive pulse generator. The permanent magnet and inductive winding are fixed to the baseplate and an iron trigger wheel is driven by the distributor shaft. The number of teeth formed on the trigger wheel or reluctor matches the number of engine cylinders. When a tooth is positioned

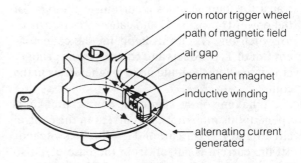

Figure 5.44   *Inductive pulse generator*

close to the soft iron stator core, the magnetic path is completed and this gives a flux flow as shown in the diagram. When the trigger wheel is moved away from the position shown, the air gap between the stator core and the trigger tooth is increased. This larger gap increases the magnetic resistance or reluctance, so the flux in the magnetic circuit is decreased.

Generation of an e.m.f. in the inductive winding fitted around the iron stator core occurs as a result of the *change* in the magnetic flux, so maximum voltage is induced when the rate of change in flux is greatest. This occurs just before, and just after, the point where the trigger tooth is closest to the stator core.

Figure 5.45 shows the variation in voltage as the trigger wheel is moved through one revolution. This shows that the build up of flux gives a positive peak and the decay of flux gives a negative peak. In the trigger position of greatest flux, no e.m.f. is induced into the winding. It is this mid-point of change between the positive and negative pulses that is used to signal that the spark is required.

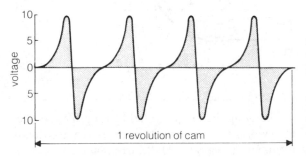

Figure 5.45 *Voltage output from pulse generator*

Rotational speed of the trigger wheel governs the rate of change of the flux, so the output of the pulse generator varies from about 0.5 V to 100 V. This voltage variation, combined with the frequency change, acts as sensing signals that can be used by the control module for purposes other than spark triggering.

The size of the air gap varies the reluctance of the magnetic circuit, so the output voltage depends on the size of the air gap. Due to the magnetic effect, the gap is checked with a non-magnetic feeler gauge, e.g. a plastics gauge.

A Bosch pulse generator operates on a similar principle but uses a different construction (Figure 5.46) (overleaf). This type uses a circular disc magnet with the two flat faces acting as the N and S poles. On the top face of the magnet is placed a soft iron circular pole piece; this has fingers bent upwards to form four stator poles in the case of a 4-cylinder engine. A similar number of teeth formed on the trigger wheel make a path for the flux to pass to the carrying plate that supports the magnet. The inductive coil is wound concentrically with the spindle and the complete assembly forms a symmetrical unit that is resistant to vibration and spindle wear.

Some manufacturers do not use a conventional distributor. Citroën use a single metal slug called a target that is bolted on the periphery of the flywheel and a target sensor mounted on the clutch housing (Figure 5.47) (overleaf). The target sensor comprises an inductive winding arranged around a magnetic core and set so that the core is 1 mm ± 0.5 mm from the slug when No. 1 piston is just before t.d.c.

The voltage output is similar to other pulse generators with the exception that in this case the control module (computer) receives only one signal pulse per revolution.

For control purposes, Citroën fit a second target sensor, of identical construction to the other sensor, adjacent to the starter ring teeth on the flywheel. This sensor signals the passage of each flywheel tooth so that the computer can count the teeth and determine the engine speed; this sets the ignition advance to suit the conditions.

Hall generator The operation of this type of pulse generator is based on the Hall effect as described on page 26.

When a chip of semiconductor material, carrying a signal current across it, is exposed to a magnetic field, a small voltage called the Hall voltage is generated between the chip edges at 90° to the path taken by the signal current.

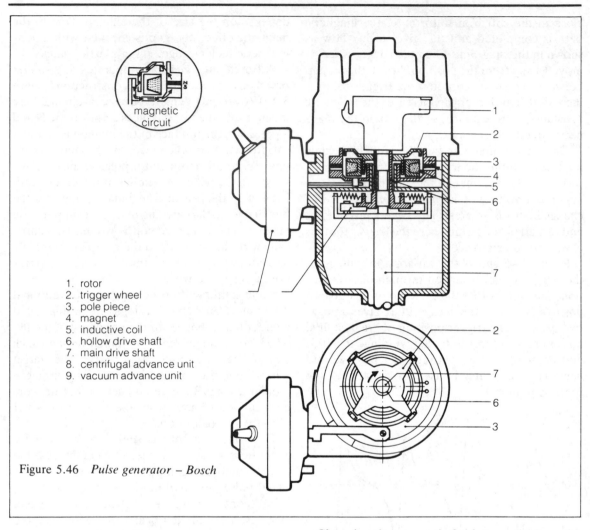

1. rotor
2. trigger wheel
3. pole piece
4. magnet
5. inductive coil
6. hollow drive shaft
7. main drive shaft
8. centrifugal advance unit
9. vacuum advance unit

Figure 5.46   *Pulse generator – Bosch*

Figure 5.47   *Pulse generator – Citroën*

Changing the magnetic field strength alters the Hall voltage, so this effect can be used as a switching device to vary the Hall current and trigger the ignition point.

The principle is shown in Figure 5.48. A semi-conductor chip retained in a ceramic support, has four electrical connections; an input signal current is supplied to AB and an output Hall current is delivered from CD. Opposite to the chip, and separated by an air gap, a permanent magnet is situated. Switching is performed by vanes on a trigger wheel which is driven by the distributor spindle.

When the metal vane is clear of the air gap, the chip is exposed to the magnetic flux and the Hall

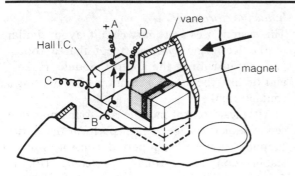

Figure 5.48   *Hall effect*

voltage is applied to CD: at this stage the switch is turned-on and current flows in the CD circuit.

Moving the vane into the air gap between the magnet and the chip blocks and diverts the magnetic flux away from the chip: this causes the Hall voltage to drop to zero. When the vane is in this flux-blocking position, the switch is off; i.e. no Hall current flows in the CD circuit.

The control module used with this system switches-on the primary current for the ignition coil when the pulse generator's trigger vane is passing through the air gap. Therefore the dwell period is governed by the angular spacing of the vanes; the smaller the space between the vanes the greater is the time that the primary circuit is closed. This closed period is terminated and the spark is produced at the instant when the Hall switch is closed, i.e. *the spark occurs when the vane leaves the air gap*.

This may be summarized as follows:

Vane position	Hall switch position	Ignition coil primary
In air gap	off	on
Outside air gap	on	off

Figure 5.49 shows the layout of a Hall generator used in a Bosch distributor. The semiconductor chip in this model is incorporated in an integrated circuit which also performs the duties of pulse shaping, pulse amplification and voltage stabilization. The number of vanes on the trigger

wheel equals the number of engine cylinders. In this design the trigger wheel and the rotor arms form one integral part.

A three-core cable connects the Hall generator with the control module; the leads are signal input, Hall output and earth.

It is impossible with the inductive pulse generator to obtain a spark when the engine is stationary, but this is not so with the Hall generator.

Care should be exercised when handling this system due to the risk of receiving an electric shock!

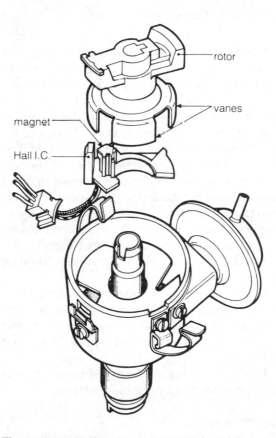

Figure 5.49   *Hall generator – Bosch*

Optical pulse generator   This system senses the spark point by using a shutter to interrupt a light beam projected by a light-emitting diode (LED) on to a phototransistor (see pages 43 and 46).

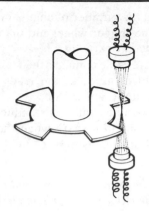

Figure 5.50 *Optical pulse generator*

Figure 5.50 shows the principle of this type of trigger. An invisible light, at a frequency close to infra-red, is emitted by a gallium arsenide semi-conductor diode and its beam is focused by a hemispherical lens to a width at the chopping point of about 1.25 mm (0.05 in). A steel chopper, having blades to suit the number of cylinders and dwell period, is attached to the distributor spindle; this controls the time periods that the light falls on the silicon phototransistor detector. This transistor forms the first part of a Darlington amplifier which builds up the signal and includes a means of preventing timing change due to variation in line voltage or due to dirt accumulation on the lens.

The signal sent by the generator to the control module switches-on the current for the coil primary so when the chopper cuts the beam the primary circuit is broken and a spark is produced at the plug.

This opto-electrical method of triggering was developed for the Lumenition system.

Control modules
The control module, or trigger box, is responsible for switching the current of the primary winding of the ignition coil in accordance with the signal received from the pulse generator.

Two control systems in use are:

1. Inductive storage
2. Capacity discharge

Inductive storage
This system uses a primary circuit layout similar to the Kettering system except that a robust power transistor in the control module 'makes-and-breaks' the primary circuit instead of using a contact breaker.

A typical control module has four important semiconductor stages which perform the duties of pulse shape, dwell period control, voltage stabilization and primary switching (Figure 5.51).

Pulse shaping The full line in Figure 5.52 represents the output voltage from an inductive-type pulse generator when it is connected to a control module circuit. It should be noted that the full negative wave is achieved only when the generator is tested on open circuit.

After feeding the a.c. signal to the trigger circuit stage, the pulse is shaped into a d.c. rectangular form as shown in Figure 5.52. The width of the rectangular pulse depends on the duration of the pulse output from the generator, but the height of the rectangle, or the current output from the trigger circuits, is independent of engine speed.

Dwell period control and voltage stabilization This stage normally varies the dwell period by altering the start of the dwell period. Secondary output is reduced when the dwell period is decreased, so by means of this control feature, the period of time that current flows through the primary winding of the coil is altered to suit the engine speed.

The voltage supplied to this resistor–capacitor (R.C.) network must not vary, even though the supply voltage to the control module alters due to changes in charging output and consumer loads. This control duty is performed by a voltage stabilization section of the module.

Primary switching Switch control of the primary circuit current is normally performed by a Darlington amplifier. Pulse signals received from the dwell period control stage are passed to a driver transistor which acts as a control current amplifier. At the appropriate times, current from

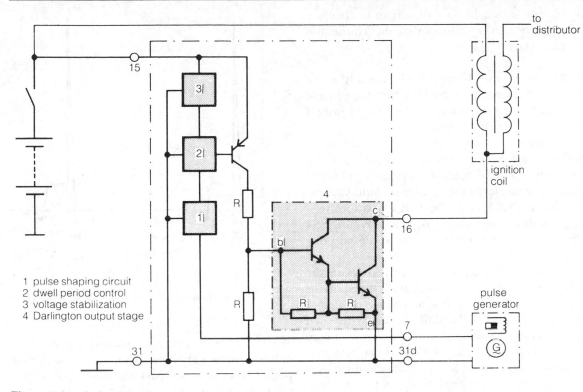

Figure 5.51   *Inductive storage control module*

1 pulse shaping circuit
2 dwell period control
3 voltage stabilization
4 Darlington output stage

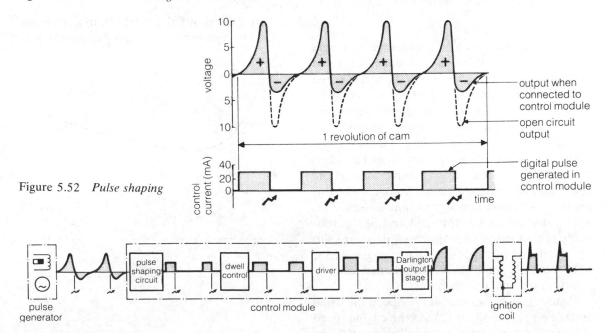

Figure 5.52   *Pulse shaping*

Figure 5.53   *Pulse processing*

the driver is switched-on or -off to control the heavy-duty power transistor of the Darlington output stage.

Pulse processing   Figure 5.53 shows the sequence of events from the time that the original pulse generator's signal is received to the instant of the spark in the cylinder.

The secondary output patterns show the image given when a cathode ray oscilloscope (C.R.O.) is connected to the output of an ignition coil forming part of an electronic ignition system. Vertical and horizontal axes of the C.R.O. pattern represent voltage and time respectively. Figure 5.54 shows the main features of one secondary discharge.

At the instant the primary circuit is broken, the secondary voltage increases until the spark is initiated; when this occurs the voltage needed to sustain the spark falls until a value is reached which is then maintained until the output energy is no longer sufficient to support the sparking process. At this point the secondary voltage rises slightly; it then falls and oscillates two or three times as the remaining energy is dissipated in the coil.

Secondary output control   Disregarding changes due to mechanical defects, a breaker-triggered system has a constant dwell over the whole speed range. This means that at high speed the dwell period is too short; as a result the secondary output is poor owing to the comparatively low primary current. Use of a low-inductance coil improves the output in the upper speed range, but this type of coil causes erosive wear at the lower end of the speed range. To overcome this problem a *constant energy* system is used. This system uses a high-output coil and is electronically controlled to vary the dwell period to suit all speeds. At low speed the percentage dwell is kept relatively small but it is lengthened progressively as the speed is increased.

Figure 5.54 shows that at low speed the dwell starts at (1) and ends at (2). As the engine speed is increased, the start of the dwell period (i.e. the point at which current starts to flow in the prim-

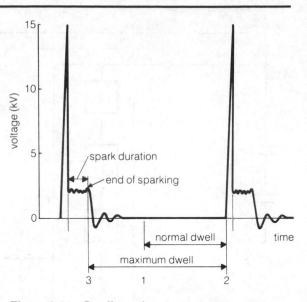

Figure 5.54   *Dwell in relation to secondary voltage*

ary winding) is gradually moved towards the extreme limit (3). Any increase in dwell past point (3) reduces the spark duration because this point represents the end of the sparking discharge period.

Figure 5.55 shows the variation in percentage dwell with engine speed. At idling speed the per-

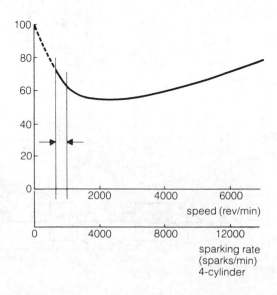

Figure 5.55   *Alteration in dwell to suit engine speed*

centage dwell is set large to provide a high-energy spark to limit exhaust gas emissions, but between idling and 4000 rev/min the increase in percentage dwell prevents a reduction in the stored energy. As a result, this gives a near-constant secondary available voltage up to the system's maximum; this is generally about 15 000 sparks/min.

When the system is used on 6- and 8-cylinder engines it is necessary to reduce the percentage dwell at speeds beyond 5000 rev/min. This is because the start of the dwell would otherwise occur before the end of the spark discharge period. To overcome this problem, the control system uses a transistor to switch-on the primary current a given time after the spark has been initiated; a duration of 0.4 millisecond is normally sufficient to meet most combustion requirements.

Constant-energy systems using dwell-angle control gives an output as shown in Figure 5.56. (The use of the terms dwell angle, percentage dwell and dwell period should be noted. In many electronic systems the term 'dwell angle' is inappropriate.)

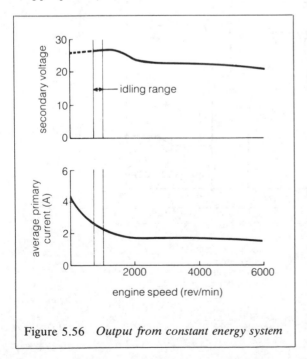

Figure 5.56 *Output from constant energy system*

Control-module circuit Having identified the functions performed by a control module, it is now possible to examine a typical circuit.

Figure 5.57 (overleaf) shows a simplified circuit laid out in a manner to identify the four main sections, A, B, C and D.

A. *Voltage regulation* Voltage stabilization is performed by the Zener diode (ZD). This ensures that the voltage applied to control sections B and C is kept constant and is not affected by voltage variations that take place in other circuits of the vehicle.

Voltage drop across a diode is constant so this feature is utilized to provide the regulated voltage to drive the control circuit (see page 43).

B. *Pulse shaping* In the form shown, the two transistors, T_1 and T_2, produce an arrangement called a *Schmitt Trigger*. This is a common method for converting an analogue signal to a digital signal, i.e. an A/D converter for forming a rectangular pulse.

Transistor T_1 is switched-on when the pulse generated by the external trigger is of a potential that opposes current flow from the battery to the trigger via the diode D. This causes current to flow through the base-emitter of T_1 which switches-on the transistor and diverts current away from the base of T_2. The action of this Schmitt Trigger causes T_2 to be 'off' when T_1 is 'on' and vice versa; the voltage at the time of switching is governed by the threshold voltage required to switch-on T_1.

In this application the switching of T_1 occurs at a very low threshold voltage so, for practical purposes, the switching is considered to take place at a point when the trigger potential changes from positive to negative.

C. *Dwell control* Coil primary current flows when the p–n–p transistor T_4 is switched-on. This is controlled by T_3 in a way that when T_3 is 'on', T_4 is also 'on'. The switching of T_3 is controlled by the current supplied via R_5 and the state-of-charge of the capacitor C. All the time the capacitor is being charged with current from R_5

no current passes to the base of T_3; during this stage, T_3 is switched-off. Only when the capacitor is fully charged will current pass to the base of T_3 and switch it on to start the dwell period, i.e. to initiate current flow in the primary winding of the coil.

The time taken to charge the capacitor dictates the dwell period, so in this case the *R-C Time Constant* is determined by the amount that the capacitor is discharged prior to receiving its charge from R_5.

At low engine speeds, the transistor T_2 will be switched-on for a comparatively long time, so this allows the capacitor plate adjacent to T_2 to pass to earth the charge it received from R_4 when T_2 was switched-off. At this slow speed, there is sufficient time for the capacitor to fully discharge to a point where the plate potential becomes

similar to earth; this causes the capacitor to attract a large charge from R_5 when the transistor T_2 switches 'off'. Since the time taken to provide this charge is long, the switch-on point of T_3 will be delayed and a short dwell period will result.

At high speed, the duration that T_2 is switched-on is short so this only allows partial discharge of the capacitor. As a result, the time taken to charge the capacitor is shorter and the dwell commences at an earlier point and gives a longer period.

Interruption of the coil primary takes place when T_2 is switched-on. This is dictated by the trigger signal so the end of the dwell period always takes place at the same time. At the instant T_2 switches-on, the capacitor starts to discharge; this causes T_3 to switch-off to trigger the spark.

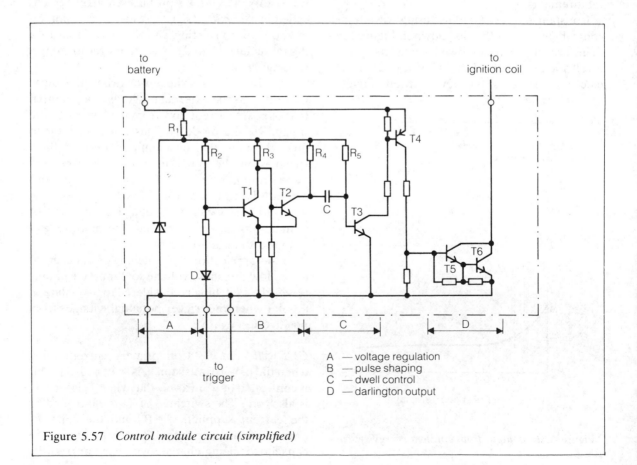

Figure 5.57 *Control module circuit (simplified)*

D. Darlington output   A 'Darlington pair' is a common power transistor array used for switching large currents. The pair consists of two robust transistors, T_5 and T_6, which are integrally constructed in a metal case having three terminals – base, emitter and collector.

When a forward-biased voltage is applied to the base–emitter circuit of T_5, the transistor is switched-on. This increases the voltage applied to the base of T_6 and when it exceeds the threshold value, this transistor also switches-on. When T_5 and T_6 are switched-on, the coil primary is energized, but when T_5 is switched-off by the switching-off of T_4, the primary circuit is broken and a spark is generated.

The switching of transistors is summarized as follows:

Trigger pulse	T_1	T_2	T_3	T_4	Primary
+	on	off	on	on	on
					← spark
−	off	on	off	off	off

To make the system suitable for a vehicle, additional capacitors and diodes are fitted to the circuit shown in Figure 5.57. These extra items prevent damage to the semiconductors from high transient voltage and they also reduce radio interference.

Alternative method of dwell control   Another method of achieving dwell-angle control is to superimpose a reference voltage on to the output signal supplied by the pulse generator (Figure 5.58(a)).

In this system the triggering of the spark at the end of the dwell period occurs at the change-over point between the positive and negative waves, whereas the start of the dwell period is signalled when the pulse voltage exceeds the reference voltage.

At low speed a reference voltage of 1.5 V acts on the dwell-control stage and this rises to 5 V at high speed. Figure 5.58(b) shows that the stronger pulse signal, combined with the higher reference voltage, gives a longer dwell period.

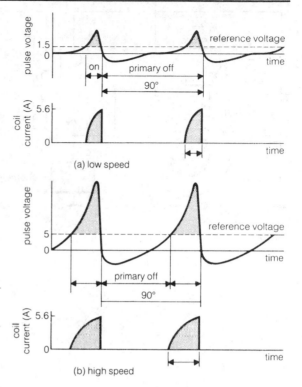

Figure 5.58   *Use of reference voltage to control dwell*

No pulse signal is generated when the engine is stationary, so the dwell control cannot operate. This feature ensures that no current can flow through the coil when the engine is stalled.

Ford Escort electronic ignition   Since 1981, Ford 1300 and 1600 engines have been fitted with electronic ignition systems. In this layout the control module is mounted on the side of the distributor assembly and connection is made by a four-pin multi-plug built into the distributor body. External l.t. cables from the distributor are limited to two leads; these connect with the coil and ignition switch (Figure 5.59) (overleaf). A tachometer, connected to the '−' side of the coil, uses the l.t. charge pulses of the coil to sense the engine speed.

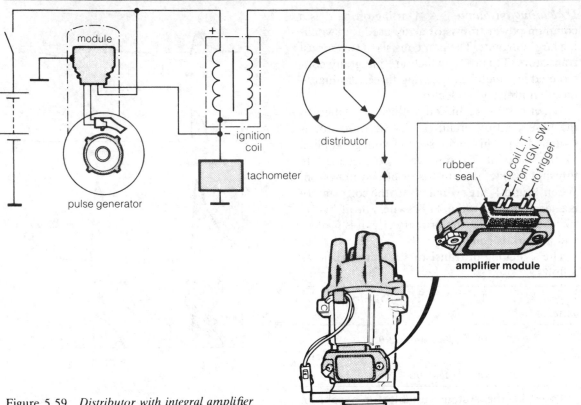

Figure 5.59   *Distributor with integral amplifier*

When fitted initially, the distributor is accurately timed to the engine and, because it has a breakerless construction, no further check of the timing is necessary when the vehicle is being serviced. Because the 'dwell angle' is governed by the control module, no check or adjustment is necessary.

Honda electronic ignition   This system, which is used on the Accord, has an inductive-type pulse generator and a control module called an *igniter* (Figure 5.60).

In this system, the switching of the coil's primary current is performed by two transistors, a driver transistor T_1 and a power transistor T_2. The pulse generator has a reluctor, shaped in the form of a saw tooth, to produce the a.c. wave form.

When the ignition switch is closed with the engine stationary, a voltage is applied by R_2 to the base of T_1. This voltage is above the T_1 trigger voltage and, since the resistance of the pulse generator's winding is about 700 Ω, the transistor

T_1 will be switched-on. At this stage, T_1 conducts current 'a' to earth instead of passing the current to the base of T_2, so T_2 is swiched-off and the primary circuit is open.

Cranking the engine causes an e.m.f. to be generated by the movement of the reluctor. When the polarity of the generator's e.m.f. at the T_1 end of the winding is negative, the resistor R_2 supplies current which flows through the winding and diode D_1 to earth. At this stage, voltage applied to the base of T_1 is less than the trigger voltage, so T_1 is switched-off. Current 'a' from R_3 is now diverted from T_1 to the base of T_2, so T_2 is switched-on and current passes through the primary winding of the ignition coil.

When the e.m.f. from the pulse generator is reversed, the combined effect of the voltage from R_2 and the e.m.f. from the pulse generator triggers T_1. This causes T_1 to switch-on, T_2 to switch-off to interrupt the primary current and give a spark at the plug.

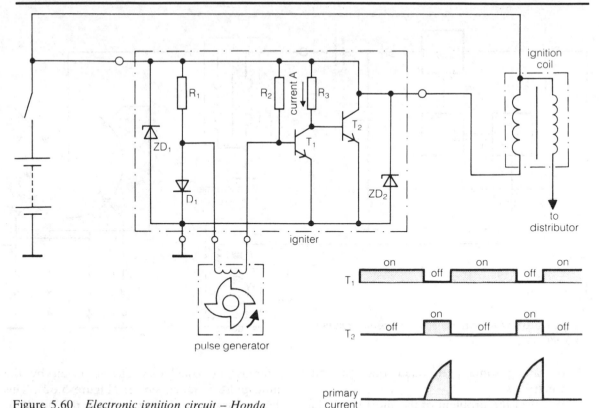

Figure 5.60   *Electronic ignition circuit – Honda*

Zener diodes ZD_1 and ZD_2, fitted at each end of the primary winding, conduct to earth the high-voltage oscillatory currents caused by self-induction. These diodes protect both transistors from high-voltage charges.

Capacity discharge (C.D.)

The module of this system stores electrical energy of high voltage in a capacitor until the trigger releases the charge to the primary winding of a coil. In this system the coil is a *pulse transformer* instead of being an energy-storage device as is normal (Figure 5.61) (overleaf).

To obtain a voltage of about 400V for the capacitor, the battery current is first delivered to an *inverter* (to change d.c. to a.c.) and then it is passed to a transformer to raise the voltage. When the spark is required, the trigger releases the energy to the coil primary winding by 'firing' a thyristor (a type of transistor switch which, once triggered, continues to pass current through the

switch even after the trigger current has ceased). Sudden discharge of the high-voltage energy to the primary winding causes a rapid build up in the magnetic flux of the coil and induces a voltage in excess of 40 kV in the secondary circuit to give a high-intensity, short-duration spark.

Although the C.D. system is particularly suited to high-performance engines, a spark duration of about 0.1 ms given by this system is normally too short to ignite reliably the weaker petrol–air mixtures used with many modern engines.

Advantages of a C.D. system are:

- High secondary voltage reserve.
- Input current and output available voltage are constant over a wide speed range.
- Fast build up of output voltage. Since the speed of build up is about ten times faster than the inductive type of electronic ignition, the C.D. system reduces the risk of the h.t. current shorting to earth via a fouled plug insulator

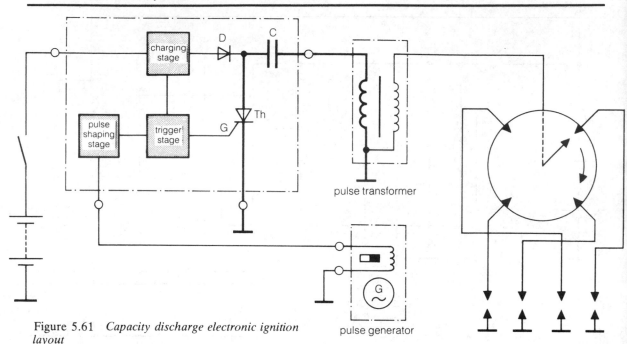

Figure 5.61   *Capacity discharge electronic ignition layout*

pulse transformer

pulse generator

or taking some path other than the plug electrodes.

To offset the problem of the short spark duration, advantage is sometimes taken of the high secondary output by increasing the sparking plug gap to give a larger spark.

Although the system can be triggered by a mechanical breaker, the advantages of a pulse generator, using either inductive or Hall effect, makes this type more attractive. The a.c. signal from the generator is applied to a pulse-shaping control circuit which converts the signal into a rectified rectangular pulse and then changes it to a triangular trigger pulse to 'fire' the thyristor when the spark is required.

Charging of the 1 μF capacitor by the charging stage to the voltage of about 400 V is performed by a voltage transformer which gives either a single- or multi-pulse output. In both cases, a diode is fitted between the charging stage and the capacitor to prevent the current flowing back from the capacitor.

Single-pulse charging of the capacitor is preferred because it enables the build up to maximum voltage to be achieved in about 0.3 ms,

whereas the oscillatory charge given by the multi-pulse is much slower (Figure 5.62). This short charge-up time overcomes the need for 'dwell-angle' control because the charge time of a C.D. system is independent of engine speed. Since the primary winding of the ignition transformer (coil) always receives a similar energy discharge from the capacitor, the secondary available voltage is constant throughout the speed range (Figure 5.63).

Although the external appearance of an ignition transformer of a C.D. system is similar to a normal ignition coil, the internal construction is quite different. Besides being more robust to withstand higher electrical and thermal stresses, the inductance of the primary winding is only about 10% of that of a normal coil. Since its impedance is only about 50 kΩ, the C.D. coil will readily accept the energy discharged from the capacitor. In view of this, the rise in secondary voltage is ten times faster. It is this feature that reduces the risk of misfiring due to the presence of h.t. shunts, i.e. leakage paths such as a fouled sparking plug, which has a resistance of 0.2–1.0 MΩ.

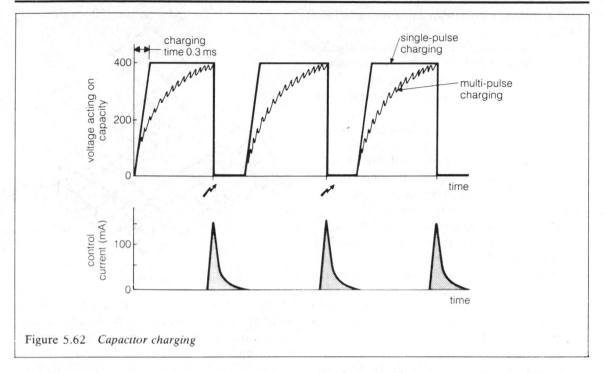

Figure 5.62 *Capacitor charging*

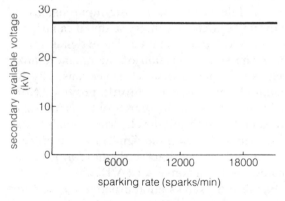

Figure 5.63 *Secondary output from C.D. system*

As the ignition transformer and capacitor forms an electrically-tuned circuit, it is necessary to fit the recommended type of transformer when replacement is required. Although a standard coil used in place of an ignition transformer will operate without damaging the system, many of the advantages of a C.D. system are lost. Conversely, if an ignition transformer is used with a non-C.D. system, damage to control module and transformer will occur immediately the 'system' is used.

The C.D. principle is also used in some small engines as fitted to motor cycles, lawn mowers, etc. A battery is not used in these cases, so the energy needed by the C.D. system is generated by a magneto.

Digital electronic ignition system
This system provides the next step in the evolution from the breakerless arrangements first used on vehicles in the late 1970s.

After a time, moving parts in a conventional distributor drive wear and when any slackness in the drive affects spark timing, some of the advantages of a breakerless electronic system are lost. The timing variation due to component wear, together with the introduction of more stringent emission and fuel-economy regulations, has forced many manufacturers to use a distributor assembly and ignition control unit which satisfies the following requirements:

- To provide an optimum spark timing to suit all load and speed conditions. In particular, it must provide a high degree of advance at light load.

- To give a constant energy output over the full speed range as needed by lean-burn engines.
- To be able to give a spark timing which allows the engine to operate just clear of the detonation region.

These requirements are achieved by using a solid-state, digital-control unit to perform the duties undertaken originally by the mechanical advance mechanism; i.e. an electronic system replaces the centrifugal and vacuum control units used in a distributor for the last 50 years.

Microelectronic memories in which the stored data cannot be changed are called *Read-Only Memories* (R.O.Ms). Sometimes the manufacturer wishes to buy a standard memory chip and then program it to suit a given application: this is called an EPROM (An Erasable Programable R.O.M.) and is used when the number of units required is limited.

Electronic spark advance (E.S.A.) Besides controlling the dwell period to suit the engine speed, this unit also varies, by electronic means, the angle of advance. A memory chip is programmed with data obtained from prototype engine tests to give the optimum advance needed for the best performance in respect of power, economy, acceleration and emission.

Two main factors dictate the angle of advance; they are speed and engine load. Figure 5.64 shows a typical map of spark advance that is programmed into a memory chip. It is normally shown as a three-dimensional 'graph' having three axes – x, y and z, representing speed, advance and engine load respectively.

Angle of advance is the 'contour height' from the axis or base of the map to the intersection point obtained from set conditions of speed and engine load. For each step of speed and load, the map shows the spark advance that should be given.

A typical control grid is sub-divided into 16 throttle positions (engine load) and 16 engine speed positions, thus providing 16 × 16 or 256 memory calibration points. On more expensive

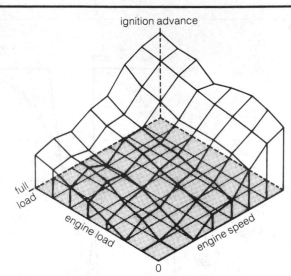

Figure 5.64 *Spark advance map*

systems an additional computer program allows each one of these positions to be divided further into 8 more points to give even more control of the timing.

The single-chip minicomputer fitted in the control module needs three basic input signals to allow it to search its memory so that it can trigger the spark at the correct time. These signals relate to engine speed, position of the engine crankshaft, and engine load, which can be sensed by the manifold depression or throttle position. Normally analogue sensors search-out this information and then transmit the electrical signals to the control module. After the signal has been processed, it is converted to a digital form by an analogue–digital converter (A/D).

Figure 5.65 shows an alternative method for displaying data stored in a memory unit; the two axes represent the digital signals needed to indicate the correct angle of advance for the engine. For example, when the digital signals received from the speed and load sensors are 3 up and 2 across, the memory output will read 33. After this memory output has been processed, the spark will be triggered 33° in advance of the time when the crank angle sensor sends its reference signal.

This advance setting is called-up after each revolution of the engine, so when the sensors detect a slight change in either the speed or

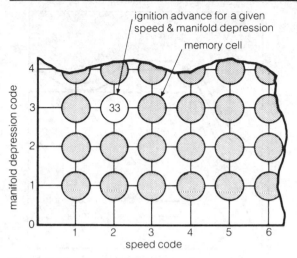

Figure 5.65 *Storage of ignition advance data*

engine load, a suitable alteration in the spark timing is made.

Figure 5.66 shows a first generation digital electronic system. An E.S.A. unit can be made as a discrete (separate) or hybrid module. Hybrid systems lend themselves particularly well to mass production and since they fulfil other duties besides ignition control, the cost, weight and size are all reduced. In the hybrid system the conductors, contact surfaces and resistors are printed, but most semiconductors and capacitors are soldered in place.

The E.S.A. system uses two speed-based inductive sensors which are mounted either in the distributor, or, if greater precision is required, fitted adjacent to the flywheel teeth. One of these

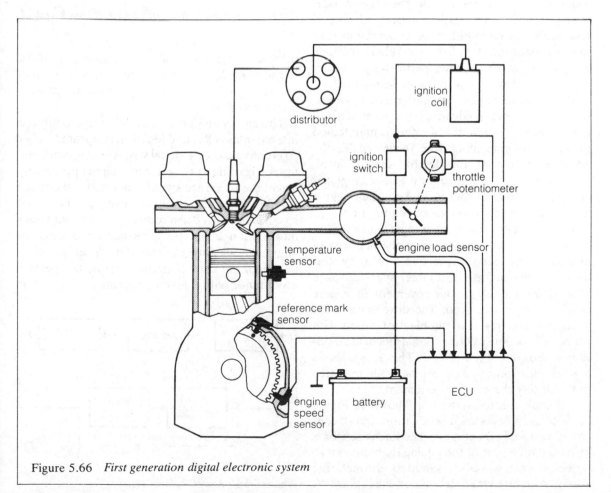

Figure 5.66 *First generation digital electronic system*

sensors monitors engine speed and the other provides a given reference mark, such as 10° before t.d.c., for the firing impulse.

Engine load can be sensed by a throttle potentiometer or by an inlet manifold pressure transducer such as a silicon strain-gauge type.

Extra sensors may be added to this basic system to refine further the operation of the Electronic Control Unit (E.C.U.). These additions often include special transducers to signal the engine temperature and detect the onset of combustion knock (detonation).

Electronic spark advance with knock control　It has long been known that a large spark-advance is needed to obtain maximum power and economy from an engine, but when the spark is over-advanced, combustion knock will occur. Knock should be avoided for two reasons: the sound is undesirable; also it is likely that knock will cause engine damage particularly when it occurs when the engine is under heavy load.

Modern lean-burn engines are prone to knock, so when the ignition advance is programmed into a memory unit, a margin of safety is maintained to keep the engine knock-free. In view of this, the advance angle is set considerably less than ideal. Even so, the engine still enters the knock region if an inferior grade of fuel is used or when engine wear alters the spark-advance requirement.

This can be overcome by using a *knock limiter* to slightly retard the ignition at the first sign of detonation. Engine protection given by this device allows the use of a larger spark-advance than is normal, so an improvement in engine performance is achieved. The difference in the spark-advance given by an electronic unit with spark control, and a conventional mechanical system is shown in Figure 5.67. This graph shows how an electronic system can be made to match more closely the ideal requirement.

A knock-control system, as shown in Figure 5.68, consists of a sensor, evaluation circuit, control circuit and an actuator. The knock sensor is mounted on a part of the engine that allows it to detect sound waves transmitted through the engine structure when detonation takes place. A

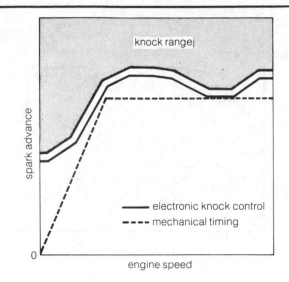

Figure 5.67　*Graph shows advantage of knock control*

piezo-ceramic disc is the active component of the sound-transducer sensor. This delivers a small voltage signal when it is triggered by mechanical oscillations.

The analogue signal from the sensor is filtered in a bandpass filter and fed to an integrator. After A/D conversion the signal flow is split up with one branch leading to the reference signal processing stage; the reference signal generated is the mean value of the previous power strokes. The actual signal is compared in a comparator; this furnishes information about the presence or absence of engine knock for each cylinder. When knock is detected, the control circuit retards the spark in accordance with a given program.

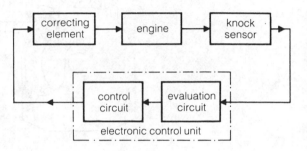

Figure 5.68　*Knock control system*

A safety circuit is built into the system to protect the engine in the event of a cable break, sensor failure or evaluation circuit fault. To cover these situations, the spark is retarded and an instrument light comes on to warn the driver that a fault is present.

When a knock-control system is used on a turbo-charged engine, it can also be arranged to control the turbo-charger boost. When the sensor detects detonation, the spark-advance is quickly reduced to take the engine out of the knock region. This is followed by the charging pressure being reduced by the opening of the waste gate valve. When this slower-acting control has taken effect, the spark-advance is restored to its optimum setting.

Figure 5.69 shows a digital system manufactured by Austin-Rover. This system is called Programmed Ignition.

Distributorless electronic ignition

As progress is made into electronic systems, attention is drawn to the only moving part left – the h.t. distributor. When this is eliminated, a fully electronic system is obtained; this is compact, does not require a mechanical drive and does not suffer from wear problems.

Distributorless systems in use incorporate one ignition coil for every two cylinders, so a 4-cylinder engine requires two coils. Whereas an ordinary coil has one end of the secondary winding connected to the primary, a distributorless system has its secondary winding connected to two sparking plugs, one at each end of the winding (Figure 5.70) (overleaf).

When the primary is interrupted in the normal way, the h.t. voltage produces a spark at both sparking plugs. Since only one cylinder is primed for ignition, the spark in the other cylinder, which

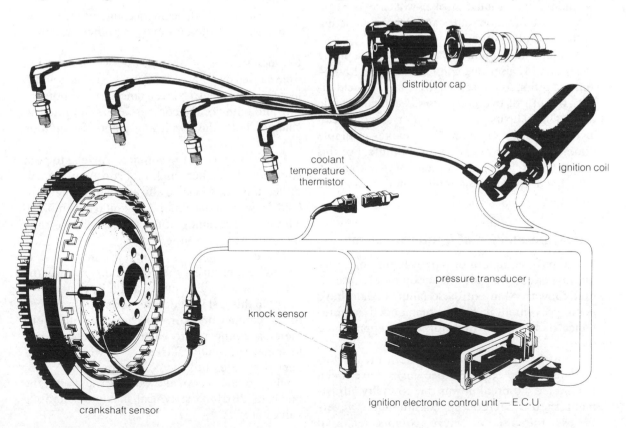

distributor cap

ignition coil

coolant temperature thermistor

pressure transducer

knock sensor

ignition electronic control unit — E.C.U.

crankshaft sensor

Figure 5.69   *Programmed ignition system   (Austin-Rover)*

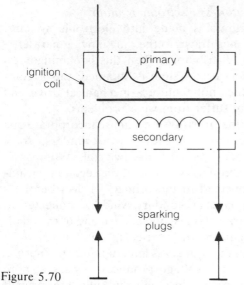

Figure 5.70
Distributorless ignition system

is ending its exhaust stroke, will not perform any useful function other than the completion of the secondary circuit.

A 4-cylinder engine operating with a firing order of 1342 uses one coil to serve Nos. 1 and 4 and the other coil to serve Nos. 2 and 3. Incidentally the use of two coils allows each coil ample time to build up its primary current, so a dual-coil layout provides a system which makes available double the number of sparks per minute. For this reason, this type of system was originally developed for use on multi-cylinder 2-stroke engines.

5.5 Maintenance of ignition systems

In the past, complete or partial failure of an ignition system was the most common cause of breakdowns. Many of these failures could have been prevented if the recommended maintenance had been carried out at the appropriate time.

With the introduction of modern breakerless systems, many maintenance tasks have been eliminated, so breakdowns due to faulty adjustment should be reduced considerably. Nevertheless the use of more complicated and sophisticated systems, which incorporate extra devices to improve performance, will increase the risk of breakdown unless the manufacturers achieve a high level of quality control of their products.

Since maintenance tests of breakerless systems require special techniques, the conventional Kettering system is considered first.

Maintenance of a conventional ignition system
Many manufacturers recommend that the ignition system is checked every 10 000 km (6000 miles). At this time the following tasks are performed:

- Contact breaker replaced and adjusted.
- Sparking plugs cleaned and tested.
- Wiring is checked for condition and security.
- Dirt and moisture is removed from the coil and any other surface which is exposed to h.t. charges.
- Lubrication of the cam face and also the moving parts of the automatic advance system.

Contact breaker Constant use over a period of time causes burning of the contact surfaces which increases the electrical resistance. In addition pitting and piling of the contacts alters the gap: this changes both the dwell angle and the ignition timing.

Failure to service the contacts will lead to poor engine performance, high fuel consumption and difficult starting, especially when the engine is cold. Low available secondary voltage, combined with incorrect timing of the spark, will also raise the exhaust emission levels.

The condition of the contacts can be ascertained by measuring the voltage drop across the contacts when the contacts are fully *closed*.

Often this test includes the wiring from the coil to the contact breaker. To check this, the voltmeter is connected to earth and to the l.t. coil terminal (normally marked '−') which is connected by a lead to the contact breaker.

The voltage drop allowable varies with the make of distributor; typical maximum limiting values are:

Ducellier and Delco   0.25 V

Lucas, Bosch and most other types	0.35 V
Ford Motorcraft	
without tachometer	0.4 V
with tachometer connected	0.6 V

Nowadays it is rare that the contact points are cleared or reground; instead a new contact set is fitted at the appropriate time. This task is simplified by the use of 'quick-fit' assemblies that are held in place and adjusted by only one screw. With all types, attention is needed to ensure that the insulated lead from the coil makes contact with the strip spring attached to the movable contact, but it must not make contact with the casing.

The correct contact gap is set by using a feeler gauge or dwell meter.

Feeler gauge method This is shown being used in Figure 5.71. The cam is rotated until the contacts are *full open* and the gap is checked with a *clean* feeler gauge. A typical gap is 0.35–0.412 mm (0.014–0.016 in): this is obtained by slightly slackening the contact set clamp screw and moving the contact plate until the correct setting is obtained. Generally a niche is cut in the plate to enable a screwdriver to lever the plate in the required direction.

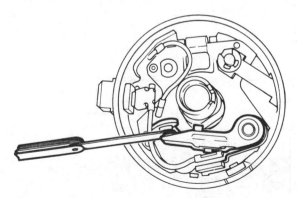

Figure 5.71 *Adjustment of contact breaker gap*

After tightening the contact adjusting screw and checking the gap, a smear of high melting-point grease is applied to the cam face before refitting the rotor arm and distributor cap.

The correct gap is essential because an incorrect gap causes the following faults:

Gap too wide	Timing is advanced and the time for the primary current to build-up is reduced
Gap too narrow	Timing is retarded

Dwell meter method This takes into account the wear on the distributor bushes and cam eccentricity. This wear causes the cam to take up a different position when it is in motion compared with the static condition used for the feeler gauge adjustment.

Dwell is expressed either as an angle or as a percentage; it indicates the time that the contacts are *closed* (Figure 5.72). For a 4-cylinder engine, a typical dwell angle is $54° \pm 5°$, i.e. 49°–59°. A dwell angle of 54° gives a percentage dwell of 54/90 = 60%. This means that the contacts are closed for 60% and open for 40% during the phase in which the spark for one cylinder is being produced.

A dwell meter is similar to a voltmeter, so when it is connected between the contact-breaker lead and earth, it measures the average voltage. Since the voltage fluctuates as the contacts quickly

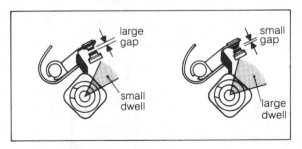

Figure 5.72 *Dwell*

open and close, the instrument needle is made to give a steady reading (Figure 5.73) (overleaf). The reading obtained will depend on the ratio between the open and closed periods.

Dwell measurement is taken when the engine is slow-running. If the reading is incorrect, the dwell meter can be used while the engine is being cranked with the starting motor and the contacts are being adjusted.

Contact gap and dwell are interrelated: a larger gap causes the contacts to open earlier, so this gives a shorter dwell.

Similarly, dwell-angle alteration changes the spark timing; a dwell angle change from 54° to 59° causes the spark to occur 5° later.

Spindle wear and other drive defects can be diagnosed by observing the change in dwell angle as the speed is increased from 1000 rev/min to 3000 rev/min. Normally the *dwell variation* should be less than 3°.

Sparking plugs After being used for a long period of time, erosion causes the plug gap to increase, and carbon, oil or petrol additive deposits on the internal surfaces decrease the plug's resistance to misfire.

Any shunt path to earth decreases the energy available for the spark, so this results in poor engine performance, especially power and economy, at times when maximum energy is required.

Alteration of the spark gap varies the voltage supplied by the coil for a given compression pressure. The effects are:

Gap too small	Low-voltage spark which may be insufficient to fully ignite the air–fuel mixture. This shows up when cold-starting, slow-running and when the engine is cruising with a weakened mixture.
Gap too wide	High-voltage spark which may demand more voltage than that produced by the coil. Consequently high-speed and cold-starting performances are poor, and erosion of the distributor and sparking-plug electrodes will be severe.

Before removing the plugs the h.t. leads should be removed by withdrawing the connectors rather than pulling on the leads. Plug removal is achieved by using a suitable plug socket and care must be exercised to avoid cracking the ceramic insulator.

Cleaning is performed normally by sand-blasting, followed by air-blasting to remove all particles of sand from the interior of the plug. Deposits on the electrodes at the spark gap are removed by a contact file and afterwards the gap

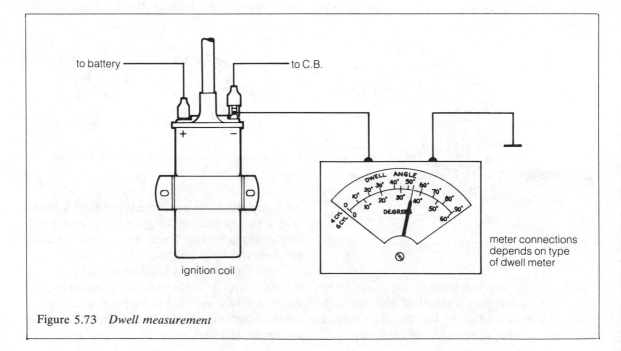

Figure 5.73 *Dwell measurement*

should be reset to the recommended value by bending the earth electrode. A wire-type feeler gauge allows more accurate setting of the gap because a flat feeler blade does not allow for the contour of the earth electrode when erosion has occurred.

Attention should be given to cleaning all sand from the threads and deposits from the external surface of the insulator.

Testing is carried out normally in a chamber that allows the air pressure to be varied while the plug is supplied with an h.t. current at a pre-determined maximum voltage. This voltage is set to give good regular sparking by a new plug up to a given pressure.

The plug is screwed into the chamber and left finger-tight to allow slight escape of air so as to improve ionization. After clipping on the h.t. test lead, the switch is operated and the air pressure is gradually increased. Sparking is observed through a window in the chamber and the pressure registered on the gauge is noted. A good plug should give regular sparking at the electrodes up to the set pressure.

Manual cleaning by a wire brush is a poor substitute for a sand-blast. Care must be taken when using a brush to avoid damaging the insulator, but sufficient pressure must be applied to remove all deposits.

Inspection of the plug prior to cleaning often acts as a guide to the state of tune of the engine. Figure 5.74 (overleaf) shows the appearance for the following conditions:

- *Normal* The core nose is lightly coated with a grey-brown deposit and the electrode is not unduly eroded. Typical erosion wear causes the gap to increase by about 0.016 mm per 1000 km (0.001 in/1000 miles).
- *Carbon fouling* A matt black sooty appearance which can cause the h.t. charge to short to earth without jumping the spark gap. This deposit originates from an over-rich mixture; also it occurs when the vehicle is engaged on short journeys that require the repeated use of the choke as well as other conditions that cause the engine to operate below its normal running temperature.

- *Split core nose* Initially a hair-line crack appears which soon causes a part of the insulator to break away. This failure is often promoted by detonation.
- *Overheating* This condition can be caused by any factor which gives overheating of the combustion chamber. Plug temperature increases with ignition advance so if over-advance is given, plug failure will result. A plug in this condition should be discarded and the cause rectified.

When refitting a plug to the engine the correct torque should be applied especially with plugs having a taper seat. If a torque wrench is unavailable the following technique can be used:

Gasket seat	Tighten 'finger-tight' and then rotate the plug 1/4 turn with a plug spanner
Taper seat	Ensure the seat is clean, tighten 'finger-tight' and then use a plug spanner to rotate the plug 1/16 turn

Ignition timing The need for correct timing cannot be over-emphasized, since incorrect spark timing can cause many problems as stated previously in this book (page 100).

When a distributor unit is refitted to an engine the following method may be used assuming detailed manufacturer's instructions are unavailable.

1. Set No. 1 piston to t.d.c. 'compression'.
2. Connect the drive so that the contacts are just *opening* when the rotor arm is pointing to the distributor segment 'feeding' No. 1 cylinder. (The contact position may be found by using a lamp as shown in Figure 5.75.) (overleaf).
3. Fit h.t. leads to the distributor in the order that the cylinders fire.
4. Start engine and make final adjustments with a timing light.

This method is sometimes called *static timing* and manufacturers often quote the crankshaft position at which the spark should occur when the

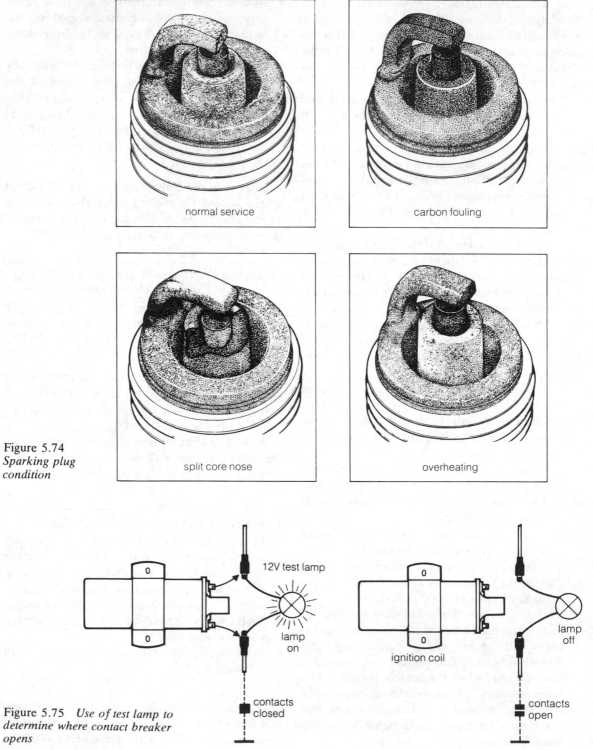

Figure 5.74
*Sparking plug
condition*

normal service

carbon fouling

split core nose

overheating

12V test lamp

lamp
on

contacts
closed

ignition coil

lamp
off

contacts
open

Figure 5.75 *Use of test lamp to
determine where contact breaker
opens*

timing is set in this way, e.g. a static timing of 4° indicates that when the crankshaft is set to 4° before t.d.c., the test lamp should show that the contact points have just opened.

Stroboscopic timing   A strobe-type timing light gives a sudden flash of light at the instant the spark occurs. The high-intensity, short-duration flash is triggered by the h.t. impulse from No. 1 plug lead. When the lamp is held close to the member that carries the timing marks, the instantaneous flash illuminates the marks and 'freezes' the motion to enable the spark timing to be ascertained (Figure 5.76).

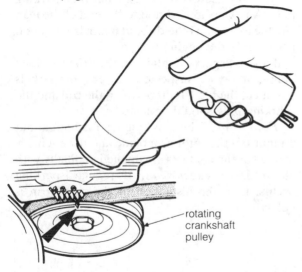

rotating crankshaft pulley

Figure 5.76   *Stroboscopic timing*

Care must be exercised during this test because the timing-light illumination gives a false impression that the fan and other moving parts are stationary. Also, *loose clothing must be kept clear of rotating parts!*

To set the timing, the recommended advance at a certain speed should be known, e.g. advance is 10° at 900 rev/min with the vacuum pipe disconnected. This example requires the engine speed to be set to 900 rev/min, the vacuum advance pipe disconnected and the mark on the crankshaft pulley (or flywheel) should then show an advance of 10° when the timing is correct. If the timing is incorrect, the distributor clamp

should be slackened and the distributor body rotated until the correct setting is obtained. Slight rotation of the distributor body in the same direction as the movement of the rotor arm retards the ignition. When the timing is correct, the distributor must be clamped securely, the timing rechecked and the vacuum pipe reconnected.

A development of this method allows the automatic advance mechanism to be tested. Special timing lights are available which incorporate a control to make the flash occur a set time after the spark is produced. The extent of this delay is indicated to the operator by an analogue-type meter or digital read-out.

After setting the normal timing, the engine speed is raised through set increments. At each speed the timing-light control is altered to make the original timing marks coincide: the angle of advance is shown on the meter.

Ignition circuit testing
If an engine fitted with a conventional ignition system fails to start, the following checks can be made to test the operation of the system.

1. *Visual check*   Check all cables and connectors for security and ensure that the battery is in good condition.
2. *Coil output*   Remove king lead from distributor cap, fit extension (e.g. centre electrode from old sparking plug) and hold lead with insulated pliers so that the end of the extension is about 6 mm from a good earth such as the engine block (Figure 5.77) (overleaf). Switch-on ignition and crank the engine. If a good spark is obtained, then the fault, if present, is in the secondary circuit beyond the coil h.t. lead, so proceed to Test 7.
3. *Contact breaker*   If no spark, or a poor spark from the coil h.t. lead is evident, then examine the contact breaker for the condition of the surfaces and the gap setting.

The surface condition is best checked with a voltmeter, but a 12 V, 3 W test lamp can also be used in parallel with the contacts: the lamp should go out when the contact close.
4. *Primary circuit*   A high resistance in the

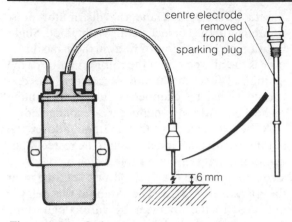

Figure 5.77 *Coil output test*

primary circuit can be pin-pointed by measuring the p.d. at various parts of the circuit (Figure 5.78). This test requires the contact breaker to be *closed*, so that current flows in the circuit. Unless the contacts are set in this position, the voltage at the points 1–4 in Figure 5.78 will correspond to the battery e.m.f. unless the circuit is completely broken at some point between 1 and 4.

5. *Capacitor* Unless special test equipment is available, the capacitor condition is assessed by substituting a test capacitor. Failing this, the contact breaker can be observed when the engine is cranked with the distributor cap removed so as to assess the degree of arcing. If arcing is severe, the capacitor should be changed.

6. *Ignition coil* The resistance of each winding can be checked by an ohmmeter. Typical values for a conventional coil are:

Primary winding   1.2–1.4 Ω
Secondary winding   5–9 kΩ

A visual inspection of the insulated surfaces should be made for signs of tracking. Breakdown of insulation, internal and external, often shows up when the coil is 'loaded'.

7. *H.T. leads* The coil h.t. lead is reconnected and the h.t. lead is removed from the sparking plug. An extension is fitted to the end of the lead and the extension is held about 6 mm from a good earth while the engine is cranked.

If a good spark is obtained, then the sparking plugs should be suspected. When no spark is obtained, the h.t. circuit between the coil and the sparking plug should be tested.

8. *Rotor arm* The coil king lead is held about 3 mm from the rotor arm blade and the engine is rotated or the contacts are opened manually with an insulated screwdriver. If regular sparking occurs, then the insulation of the rotor arm is defective.

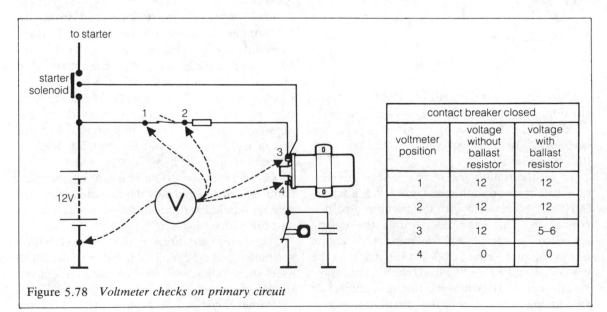

	contact breaker closed		
voltmeter position	voltage without ballast resistor	voltage with ballast resistor
1	12	12
2	12	12
3	12	5–6
4	0	0

Figure 5.78 *Voltmeter checks on primary circuit*

9. *Distributor cap* The cap is examined visually for cracks and signs of tracking: this is a thin carbon line which allows the h.t. current to take an easier path than that required. Tracking can cause misfiring, but normally the engine will start with the cap in this condition.

If the weather is damp, then h.t. leads should be checked at a very early stage in the test program. Moisture causes the electrical charge to 'leak' to another lead or directly to earth, so to prevent this, the h.t. leads should be sprayed with a moisture repellent or they should be separated from one another.

Complete breakdown of the king lead will prevent the engine from starting whereas a similar failure of a plug lead will cause regular misfiring under load or poor engine performance in general. An ohmmeter should be used to check the resistance of the h.t. leads, which, in the case of a supression lead, will have a value of 13 000–26 000 Ω/metre.

Diagnostic equipment – engine analysers
The need for quick and accurate diagnosis of engine faults makes it necessary for the automotive engineer to possess and use modern electronic equipment. This equipment has been designed to pin-point common and obscure faults in the engine and associated systems, and, where necessary, provide the vehicle operator with detailed information on the condition of the engine.

A number of analysers are available and these can be divided broadly into two main categories:

- Oscilloscopes for the display of test patterns.
- Computers which test the engine systems and display the results on a V.D.U. (visual display unit).

Oscilloscope Figure 5.79 shows a Crypton Diagnostic Centre which comprises a cathode-ray oscilloscope (C.R.O.) together with analogue and digital meters for testing and adjusting where necessary:

- Ignition circuits and components.
- Spark timing.

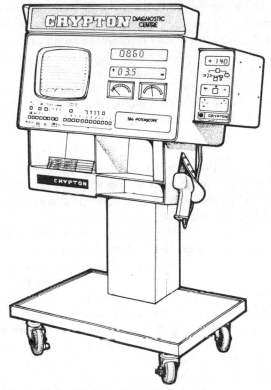

Figure 5.79 *Ignition diagnostic equipment*

- Relative output from cylinders (cylinder balance).
- CO emission from exhaust.
- Electrical equipment involving the use of basic test meters.

C.R.O. display The pattern formed on the screen is obtained by a spot of light which sweeps horizontally across the screen in step with the engine speed. Each time a given sparking plug fires, the spot is made to return to the left-hand side of the screen. As the phosphorus coating on the inside of the cathode-ray tube retains the image for a short time, the human eye sees a continuous line formed on the screen instead of a moving spot of light.

Vertical deflection of the line corresponds with the voltage sensed at a given point in the ignition system. Switches are provided to select various meter scales and voltage pick-up points in the circuit; these enable the operator to analyse the complete process of ignition.

In this form the C.R.O. is a high-speed voltmeter. Whereas the sluggish movement of an analogue meter needle only gives a mean value of the voltage, a C.R.O. shows the actual voltage at all stages of the operating cycle.

Before using a C.R.O. to examine ignition patterns, the primary circuit is normally tested with a voltmeter to eliminate the effects of basic circuit faults from the patterns.

Primary pattern   The trace shown in Figure 5.80 represents an ignition system in good order. It shows the voltage variation in the primary circuit for the production of one spark.

At point (1) the contacts open and self-induction causes the voltage to increase. The oscillations at (2) are due to the charge–discharge action of the capacitor. A good capacitor should give four or five oscillations at this stage.

After the rise in the voltage at point (3) at the end of the sparking period, the energy remaining in the primary winding is dissipated. This phase (4) should produce four or five oscillations as the energy fades away.

At point (5) the contacts close and the trace should show a clean line joining the two horizontal lines. Although preliminary tests should pinpoint a dirty contact condition, the C.R.O. will show up this condition as a 'hash' at the point where the contacts close (Figure 5.81(a)).

When the patterns for each cylinder are superimposed on the original trace, no large separation in the patterns should be seen. However if the cam or spindle is worn, the dwell angles will vary: this will appear as shown in Figure 5.81(b) and is called *dwell overlap*.

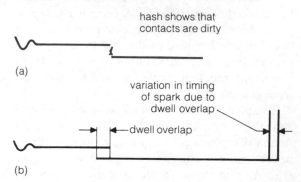

Figure 5.81   *Part of primary pattern; diagram shows two common faults*

The need for greater accuracy of dwell measurement has meant that most modern analysers now incorporate a digital meter to show the dwell angle. Furthermore, these modern units can measure the *individual dwell* given by each cam lobe. Having ascertained each dwell angle, the operator can subtract the smallest angle from the largest angle to find the dwell overlap.

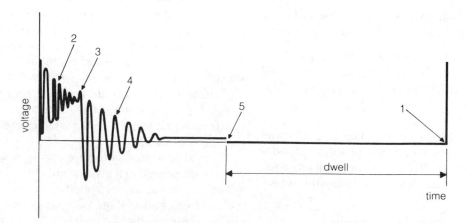

Figure 5.80   *C.R.O. pattern; primary circuit*

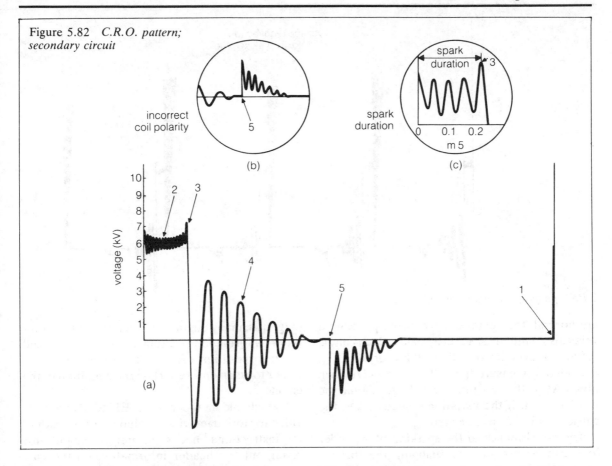

Figure 5.82 *C.R.O. pattern; secondary circuit*

Secondary patterns Figure 5.82(a) shows a normal pattern which represents the voltage variation in the secondary circuit. The opening of the contacts at (1) causes the voltage to rise until a spark is produced: the voltage during the sparking process is shown at (2). When sparking ceases, the voltage rises (point 3) and this is followed by the dissipation of energy in the secondary winding. As this occurs, the oscillating voltage will give four or five surges before dying away.

At point (5) the closure of the contacts and the start of the current flow in the primary winding mutually induces a small e.m.f. into the secondary winding.

If the pattern at point (5) is inverted (Figure 5.82(b)) it indicates that the coil polarity is incorrect. This is corrected by changing over the l.t. coil connections.

Faulty coil windings and poor h.t. leads give an unstable pattern and broken trace respectively.

Most modern analysers enable the length of the spark line to be measured in milliseconds (Figure 5.82(c)). Conventional systems have a spark line of length 0.75–1.5 ms. (Breakerless systems of the inductive discharge type give a 2 ms spark line, whereas capacity discharge systems have a very short line of length 0.15 ms or less.)

Secondary pattern, parade order Secondary output is shown by presenting each secondary trace, side by side; to ease the measurement the vertical trace lines are broadened (Figure 5.83) (overleaf). Normally the pattern is displayed in the order that the sparking plugs are fired.

This trace shows clearly the voltage needed initially to fire each sparking plug. If conditions

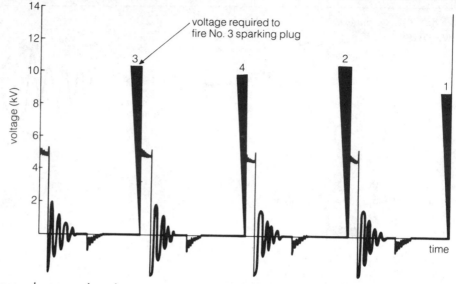

Figure 5.83
C.R.O. pattern; secondary, parade order

are normal, the height of each of the prominent verticals is similar. A vertical scale, which can be set at a maximum of either 20 kV or 40 kV is provided to measure the voltage applied to each plug. At 1000 rev/min the voltage should be 8–14 kV, but if the variation exceeds 3 kV, the cause should be investigated.

The performance of the sparking plugs under load can be assessed by snapping the throttle open. During this test, the voltage should increase to about 16 kV.

Shorting a plug to earth shows the voltage required to overcome the h.t. lead resistance and the rotor gap. Normally this voltage should be less than 5 kV.

Maximum voltage output of the coil is measured by using insulated pliers to remove the h.t. lead from the sparking plug and to hold it clear of the engine. During this test the 40 kV scale is selected, because most coils fitted to British vehicles since 1970 give a maximum output of at least 28 kV when supplied with a 14 V input. (*This test should not be performed on a breaker-less system unless a special lead is used to limit the voltage.*)

Engine diagnostic computer   To meet modern demands for accurate testing and efficient fault diagnosis, many engine analysers now use a computer. Information is displayed on a V.D.U. and the incorporation of a printer enables the customer to be given a report on the condition of the engine.

Detachable memory pods (EPROM) are provided to store data which include: test procedure for instructional needs, printer commands for laying-out the header information on the customer's report form, and the specifications for the engines to be tested.

Access to the engine data allows the computer to make a comparison between actual and specified values of a particular system so that it can pin-point any faults.

When set to an 'automatic mode' the computer guides the operator through the test program; this makes it a simple task to carry out a 'health-check' of the basic engine systems. Identification of the cause of a fault in a particular system can be made by a person having a higher skill level by using the computer in its 'manual mode'; in this setting the computer works to the instructions given by the operator.

Ignition patterns of the form associated with conventional C.R.Os are not displayed; instead values, such as maximum ignition voltages, are shown as a bar chart (Figure 5.84).

Maintenance of a breakerless ignition system

Extra care must be exercised when checking breakerless systems to avoid receiving an electric shock.

In some systems (e.g. C.D. systems) a high-voltage charge is still stored when the ignition is switched off.

Many different systems are in use, so the manufacturer's service manual should be consulted for the specific service and test information.

General maintenance Very little routine maintenance is necessary because checks on dwell angle and spark timing are not required. Also sparking plug service intervals are extended with these systems. When the distributor is timed initially by the manufacturer, a mark is stamped on the flange to indicate its correct position.

When a mechanical advance system is used, lubrication should be provided at the appropriate time. During this service operation, a timing light may be used to check the operation of the mechanism.

Ignition circuit testing

If an engine fails to start the following basic checks are made to most systems before conducting detailed tests on the control module and pulse generator.

1. *Visual check* Check all cables and connectors for security and ensure that the battery is in good condition.

2. *Coil output* Remove king lead from distributor cap, fit extension and hold with insulated pliers so that the end of the extension is about 6 mm from the engine block.

Switch-on ignition and crank the engine. A good spark indicates that the fault is beyond the coil h.t. lead: in this case proceed as with a conventional system.

IMPORTANT – *damage will occur to the control module if the engine is cranked and the king lead is set so that the gap is too large for a spark to jump to earth.*

The specific tests applied to a breakerless system depend on the type of system; the following descriptions are intended to show the basic principles.

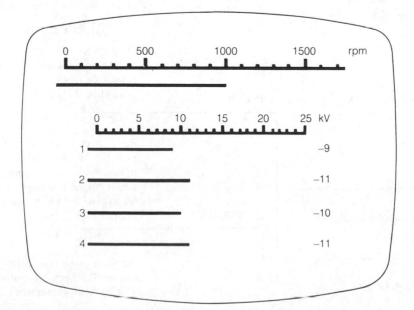

Figure 5.84 *Bar chart display*

Table XII Fault diagnosis chart: constant-energy ignition

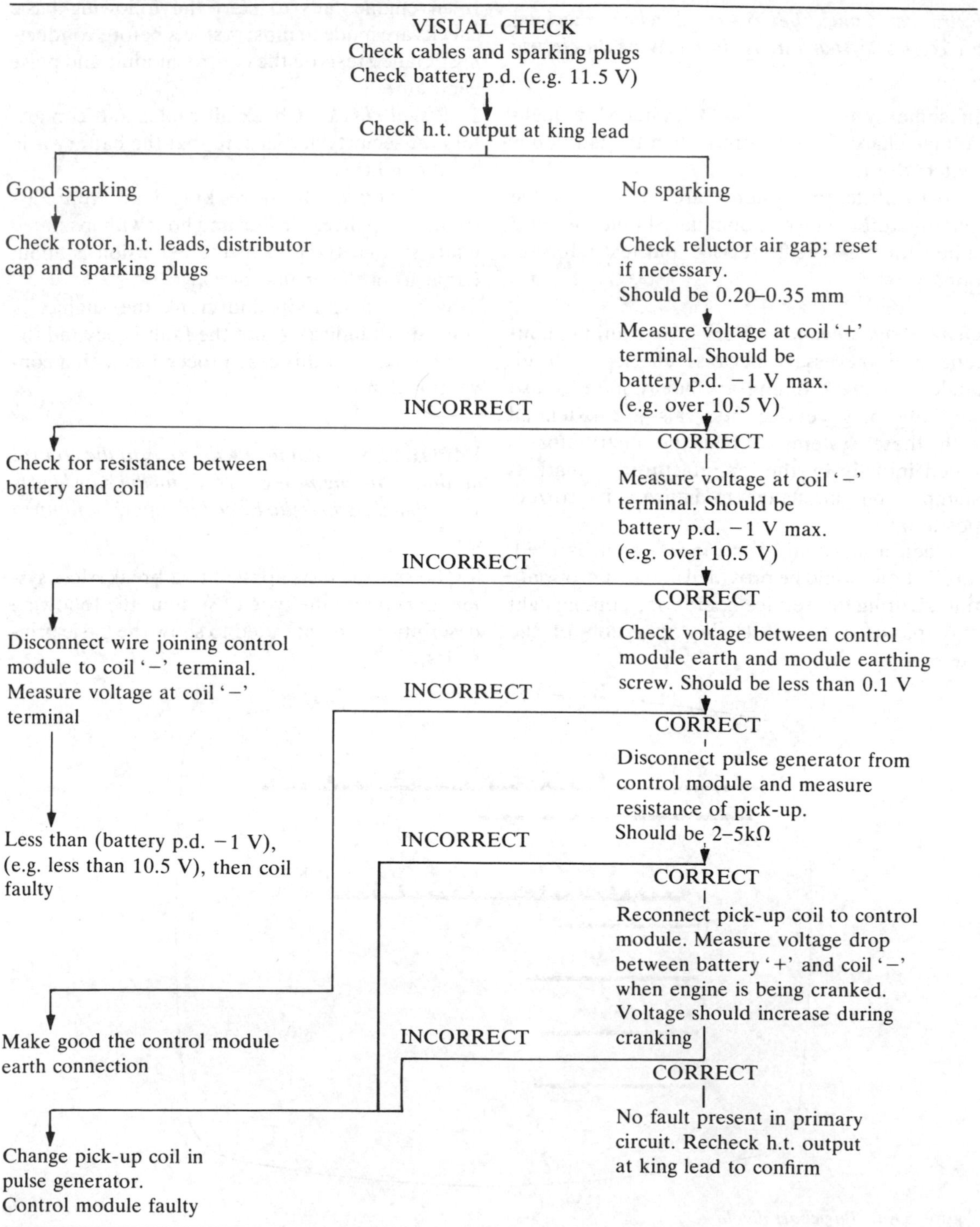

VISUAL CHECK
Check cables and sparking plugs
Check battery p.d. (e.g. 11.5 V)

Check h.t. output at king lead

Good sparking

Check rotor, h.t. leads, distributor
cap and sparking plugs

No sparking

Check reluctor air gap; reset
if necessary.
Should be 0.20–0.35 mm

Measure voltage at coil '+'
terminal. Should be
battery p.d. −1 V max.
(e.g. over 10.5 V)

INCORRECT

CORRECT

Check for resistance between
battery and coil

Measure voltage at coil '−'
terminal. Should be
battery p.d. −1 V max.
(e.g. over 10.5 V)

INCORRECT

CORRECT

Disconnect wire joining control
module to coil '−' terminal.
Measure voltage at coil '−'
terminal

INCORRECT

Check voltage between control
module earth and module earthing
screw. Should be less than 0.1 V

CORRECT

Disconnect pulse generator from
control module and measure
resistance of pick-up.
Should be 2–5kΩ

Less than (battery p.d. −1 V),
(e.g. less than 10.5 V), then coil
faulty

INCORRECT

CORRECT

Reconnect pick-up coil to control
module. Measure voltage drop
between battery '+' and coil '−'
when engine is being cranked.
Voltage should increase during
cranking

Make good the control module
earth connection

INCORRECT

CORRECT

No fault present in primary
circuit. Recheck h.t. output
at king lead to confirm

Change pick-up coil in
pulse generator.
Control module faulty

Constant-energy system (e.g. Lucas)
After performing the basic tests the following checks are made:

1. *Pulse generator gap* The gap depends on the type of engine but a typical gap is 0.2–0.4 mm (0.008–0.016 in). A plastics feeler gauge blade is used when checking the gap between the reluctor tooth and the pick-up limb (Figure 5.85).

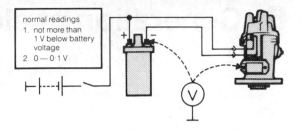

Figure 5.86 *Static test of control module*

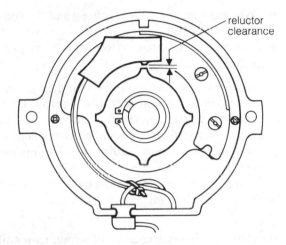

Figure 5.85 *Pulse generator gap*

2. *Control module* Two checks are possible:

Static check	A voltmeter check at the two points shown in Figure 5.86 shows if the voltage drop across the module is excessive or if the module has a poor earth.
Cranking test	The voltmeter is connected between the position battery terminal and the negative terminal of the coil. If the voltage does not increase when the engine is cranked, then Test No. 3 should be carried out.

3. *Pulse generator coil resistance (Pick-up resistance)* The pick-up leads from the pulse generator are disconnected at the harness connector and an ohmmeter is used to measure the resistance of the pick-up coil. The resistance value depends on the application (a typical value is 2–5kΩ).

An integral control module has to be removed from the distributor to carry out this test.

If the reading is incorrect then the pick-up is assumed to be faulty, but if the resistance is between the limits recommended, then the control module should be changed.

4. *Other components* The remaining parts of the system are checked in a manner similar to that used for a conventional ignition system.

A typical fault diagnosis chart is shown in Table XII.

Hall-effect generator Control module and wiring should be checked with a voltmeter similar to the method described previously. Meter readings at the various points, when compared with the specified values, show if the module and generator are serviceable.

Digital system The multi-plug connector is removed from the E.C.U. and voltmeter tests are made at the cable end of the connector to measure the voltage at the appropriate terminals. Voltage checks are also made to the pulse generator and the ignition coil as in other breakerless systems. Damage can occur to the Electronic Control Module (E.C.U.) if the current from an ohmmeter is fed to it. Spark advance given by the E.C.U. is checked in the normal manner with a strobe light.

6 Generation of electric energy

6.1 Charging-circuit principles

A large amount of electrical energy is required to drive the numerous electrical systems contained in a vehicle. The battery will supply this energy for a short time, but when the battery is exhausted the engine will come to rest and the electrical systems will cease to function.

To overcome this problem a charging system is provided to fulfil the needs of the various systems and maintain the battery in a charged state. This ensures that when the engine is stationary, the battery can meet reasonable electrical demands for a time which is governed by the size of the battery. Since a larger capacity battery means greater weight, the modern battery must be kept as small as possible; this places a larger emphasis on the charging system.

As time progresses, more and more electrically-operated devices are fitted to a vehicle, so the charging system now has to provide a much greater output. In addition to giving a high maximum output, the modern generator has to be made more efficient and lighter in weight. In the early 1960s these requirements forced the manufacturers to change from a dynamo to an alternator.

The dynamo is considered in this book at this stage only to show up the features which highlight the advantages of an alternator.

Charging systems
On page 17 it was shown how an e.m.f. can be generated by moving either the conductor or the field relative to each other. The parts that are moved, or fixed, give the main difference between the two types of generator, namely:

- Dynamo – magnetic field is fixed and the conductor is moved.
- Alternator – conductor is fixed and the magnetic field is moved.

The name 'alternator' implies that the dynamo generates direct current, but this is not so. Both machines generate alternating current, and in each case some form of rectifier is needed to produce the d.c. current required for charging the battery.

Figure 6.1 shows the main components for both types of charging system.

Dynamo
Basically the dynamo consists of a conductor coil which is rotated in a magnetic flux. The coil is wound around a soft iron armature and on the end of this is mounted a pulley which receives a drive from a vee-belt driven by the engine crankshaft. Although some dynamos use a permanent magnet to provide the magnetic flux, most vehicle units use electromagnets because the strength of these magnets can be easily varied to restrict the dynamo output.

Rectification in a dynamo system is performed by a commutator – a cylindrical part, made up of copper segments, that is fitted to the end of the armature. Two carbon brushes rub on the commutator to collect the armature output current; one brush is connected to the casing (earth) and the other brush is joined to the main output terminals (often marked 'D' in circuit diagrams).

Most dynamos, such as the one shown in Figure 6.2, have a small terminal (often marked 'F') situated adjacent to the main terminal. One end of the field coil, which is shunt wound in relation to the armature winding, is connected to this

Figure 6.1 *Vehicle charging systems*

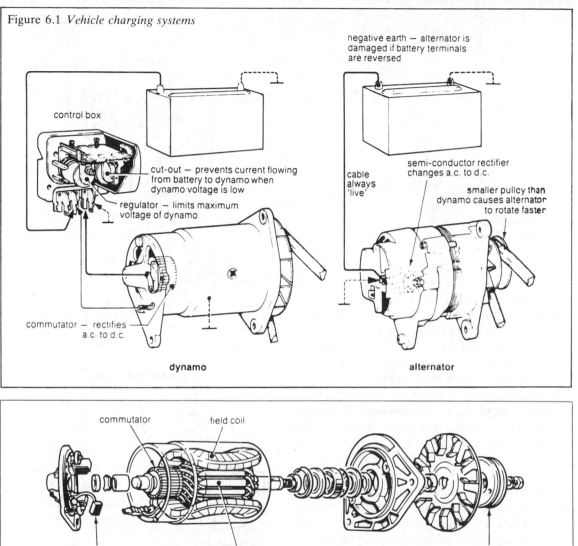

negative earth — alternator is
damaged if battery terminals
are reversed

control box

cut-out — prevents current flowing
from battery to dynamo when
dynamo voltage is low

cable
always
'live'

semi-conductor rectifier
changes a.c. to d.c.

regulator — limits maximum
voltage of dynamo

smaller pulley than
dynamo causes alternator
to rotate faster

commutator — rectifies
a.c. to d.c.

dynamo

alternator

commutator

field coil

brush

armature

driving pulley

Figure 6.2   *Dynamo*

small terminal; the other end of the coil is connected to the casing. The output of the dynamo depends on the current supplied to the field winding, i.e. the output depends on the strength of the magnetic flux.

The circuit diagram of a shunt wound dynamo is shown in Figure 6.3 (overleaf).

Control box for a dynamo   This box is mounted remote from the dynamo and houses the cut-out and regulator.

Cut-out   The cut-out is an electro-mechanical relay which allows current to flow from the dynamo to the battery, but not vice versa. This action prevents current from flowing from the battery to the dynamo when the dynamo is stationary or when the dynamo p.d. is less than the p.d. of the battery. A faulty cut-out causes the dynamo to act as a motor and if this occurs current conveyed by the cables soon causes the cables to become red-hot.

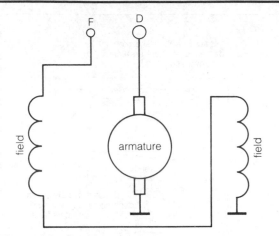

Figure 6.3   *Dynamo circuit*

Regulator   The regulator controls the dynamo output to suit the state of charge of the battery. It also prevents damage to the dynamo, especially at high rotational speeds, by limiting the output to a safe figure. When the battery is fully charged, its voltage is about 14 V, so by setting the regulator to limit the charge voltage to this figure, overcharging of the battery is prevented.

Once the engine is in operation the electrical energy is supplied by the dynamo, so voltage control of the dynamo is necessary if the vehicle's equipment, rated to operate at a given voltage such as 12 V, is not to be damaged.

Regulation is obtained by using electro-mechanically controlled contacts to interrupt the field current. These contacts vibrate at a near-constant rate and control the current by varying the closed/open period of one vibration cycle.

Often two regulators are fitted side-by-side; one controls the voltage, the other limits the current.

Alternator

The principle of an alternator is shown, in simplified form, in Figure 6.4(a). This shows a shaft and four-pole magnet fitted adjacent to a stator (stationary member) around which is wound a conductor coil. This winding is connected to form a simple circuit and in the diagram a galvanometer is included to show the output.

Rotation of the magnet generates an e.m.f. in the stator winding and since the North and South

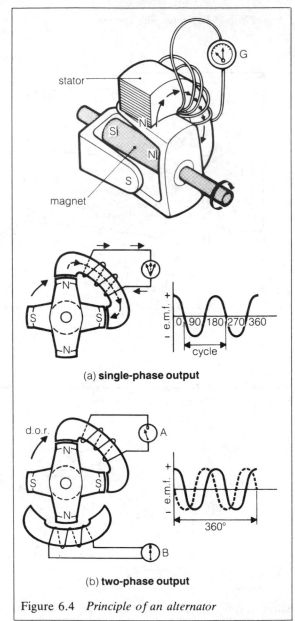

(a) **single-phase output**

(b) **two-phase output**

Figure 6.4   *Principle of an alternator*

poles present themselves to the stator in alternate order, the current produced will be a.c.

Output increases as the speed of rotation increases, but when the rate of change in current exceeds a certain value, self-inductance will retard the growth of current in relation to the increase in speed. This is fortunate because it causes the current to peak at a set speed and as a result gives protection to the machine against damage due to current overload.

Adding another stator winding in the position shown in Figure 6.4(b) gives two independent outputs as shown by the graph of e.m.f. Stator winding B gives an output which is 45° out-of-phase to winding A; this double-curve pattern is called *two-phase output*. Similarly if another stator winding is added and all three are spaced out around a multi-pole magnet, then a *three-phase output* is obtained (Figure 6.5). Due to the increase in the number of magnetic poles, each cycle will be shorter, therefore one revolution of the shaft will produce a large number of a.c. cycles (see page 22).

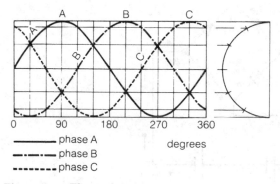

— phase A
—·—·— phase B
- - - - - phase C

degrees

Figure 6.5   *Three-phase output*

Rectification of the current is performed by semiconductor diodes. A single diode, as shown in Figure 6.6, gives *half-wave rectification* and since half the output is lost a more efficient means is needed. One method that is used for bench-type battery chargers (see page 76) is shown in Figure 6.7 (overleaf); this uses four diodes formed into a *bridge circuit*.

Arrangement of the diodes ensures that the current flow, in each direction, is channelled

Figure 6.6   *Half-wave rectification*

through the appropriate diodes to give a unidirectional flow through the battery, i.e. *full-wave rectification*.

Besides acting as rectifiers, the diodes also serve another purpose; they prevent current flow from the battery to the alternator when the battery p.d. is higher than the alternator p.d. This feature overcomes the need for a cut-out in the charging circuit.

Output voltage control is needed to prevent the voltage exceeding a given maximum; otherwise this would cause overcharging of the battery and also damage the electrical devices in other circuits. Output control is achieved by using an electromagnet instead of a permanent magnet for the field. Current to the field winding is governed by a regulator: some systems use an externally mounted 'bobbin type' regulator; this has vibrating contacts similar to that used with a dynamo. Today most alternators are fitted with microelectronic solid-state regulators; these control the output accurately to 14.2 ± 0.2 V.

Advantages of an alternator   Compared with a commutator-type generator (dynamo), the alternator has the following advantages:

Higher output	Rotating parts are more robust so a higher speed of rotation can be allowed; this is achieved by using a drive pulley of smaller diameter. Although extra output at high speed is needed, the improvement at low engine speed is more significant because of the time spent at this speed when the vehicle is operated in congested traffic situations (Figure 6.8) (overleaf).
Lower weight and more compact	The constructional features and improved efficiency allows the required output energy to be given by a smaller unit.
Less maintenance	Output current is not conducted through a commutator and brushes, so breakdown due to brush wear or surface contamination is eliminated.

More precise output control	The use of a solid-state regulator enables the maximum output limits to be reduced. This permits the use of maintenance-free batteries and other electronic systems which would otherwise be damaged by excessive voltage.
Requires no cut-out	Rectifier diodes also serve the same purpose as a cut-out.

6.2 The alternator

The many advantages of this type of generator has meant that since the late 1960s most vehicles have been fitted with an alternator. Although many different designs are in use, the basic principle of each is similar.

Construction

Figure 6.9 shows an exploded view of a typical alternator. This Lucas alternator is a 3-phase,

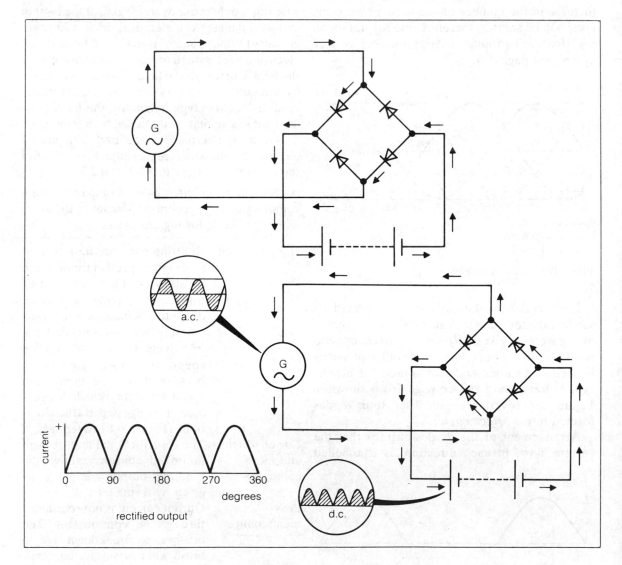

Figure 6.7   *Bridge circuit to give full-wave rectification*

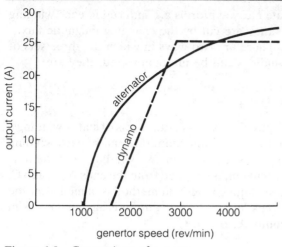

Figure 6.8 *Comparison of outputs*

12-pole machine which incorporates a rectifier and micro-electronic regulator.

The lightweight aluminium alloy casing of the alternator contains the following:

- Rotor to form the magnetic poles.
- Stator to carry the windings in which the current is generated.

- Rectifier pack to convert a.c. to d.c.
- Regulator to limit the output voltage.

Rotor This consists of a field winding that is wound around an iron core and pressed on to a shaft. At each end of the core is placed an iron claw to form 12 magnetic poles; one claw has 6 fingers to give N poles and the fingers on the other claw form S poles (Figure 6.10) (overleaf).

The magnet excitation winding is wound around the circumference of the soft-iron core and contact with the winding is made by two carbon brushes which rub on two copper slip rings. Two types of brush arrangement are in use, namely:

- Cylindrical or barrel type — Two slip rings placed side-by-side.
- Face type — The two brushes are fitted coaxially with the shaft.

The rotor is belt-driven from the crankshaft through a vee-pulley and Woodruff-type key. Since alternators are suitable for speeds of up to 15 000 rev/min, and because the belt tension must be high to prevent slip when a large current

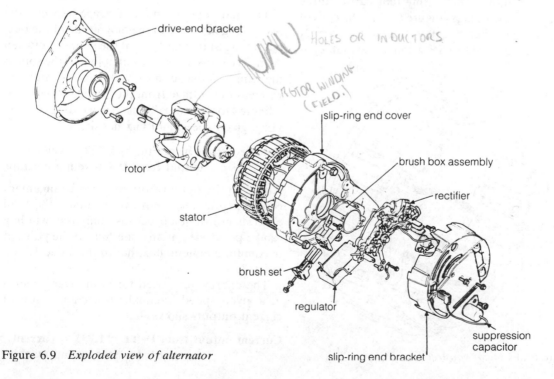

Figure 6.9 *Exploded view of alternator*

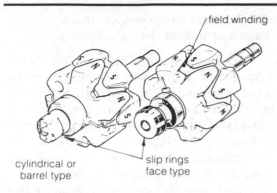

Figure 6.10   *Rotor construction*

output is being produced, ball bearings are needed to support the rotor. These bearings are packed with lubricant and sealed for life.

Forced ventilation of air through the machine is essential to cool the semiconductor devices and prevent the windings overheating. Air movement for this ventilation is achieved by a centrifugal fan fitted adjacent to the pulley.

Stator   This is a laminated soft-iron member attached rigidly to the casing that carries three sets of stator windings (Figure 6.11). The coils of comparatively heavy-gauge enamelled copper wire forming the stator are arranged so that sepa-

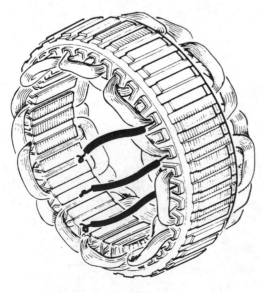

Figure 6.11   *Stator construction*

rate a.c. waveforms are induced in each winding as they are cut by the changing magnetic flux.

There are two ways in which the three sets of windings can be interconnected; they are:

- Star
- Delta

Figure 6.12 shows both forms of stator windings. In the Star connection, one end of each winding connects to the other two windings and the output current is supplied from the ends A, B, and C. The Delta connection method is named after the Greek letter 'Δ'; the output is again taken from points A, B and C.

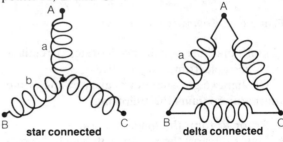

Figure 6.12   *Stator windings*

The main operational difference between the two arrangements is in the magnitude of the output. In the Star arrangement the voltage between A and B (or two other output points) is the sum of the e.m.f. induced into windings 'a' and 'b', whereas the voltage from the Delta arrangement is limited to the e.m.f. induced in winding 'a' only. For a given speed and flux density:

Voltage output from Star = 1.732 × voltage output from Delta-wound machine

Although the output from the Star arrangement is obtained mainly from two windings, the total e.m.f. is not doubled because only one winding can be positioned at any one time at the point of maximum magnetic flux, hence the value 1.732, i.e. $\sqrt{3}$.

The energy generated for both arrangements at a given speed is equal, so a comparison of current outputs shows that:

Current output from Delta = 1.732 × current output from Star-wound machine

The Star arrangement is used on the majority of alternators for light cars, but where higher current output is needed, the Delta-wound stator is preferred.

On some special designs of heavy-duty alternators, the operator can alter the stator windings from Star to Delta when a large output current is needed.

Rectifier Although some alternators use an external plate-type selenium rectifier, most machines use semiconductor diodes which are arranged to form a *bridge network* (see page 32). For a 3-phase output, 6 diodes are needed to give full-wave rectification and these are arranged as shown in Figure 6.13. The diodes act as 'one-way valves' so the current generated in any winding will always pass to the battery via the terminal marked 'B+'. Since a complete circuit is needed to give this d.c. current flow, the appropriate earth diode (the negative diode in this case) is fitted to pass current from 'earth' to the active windings.

The action of the diodes can be verified by inserting two arrows adjacent to any two of the stator winding shown in Figure 6.13. Irrespective of the position and direction of the arrows, it will always be possible to trace the circuit between earth and 'B+'.

Besides the rectifying function, the diodes also prevent flow of current from the battery when the alternator output p.d. is less than the battery p.d.

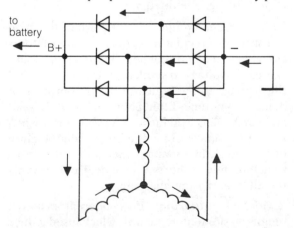

Figure 6.13 *Rectifier circuit*

Therefore the diodes overcome the need for a cut-out as is required in a dynamo charging system.

With the alternator stationary the connection to the alternator 'B+' is 'live'; this should be remembered by any person who has to remove an alternator from the engine. The battery earth terminal should be disconnected prior to starting work on the equipment.

Figure 6.14 (overleaf) shows alternative constructions used to mount the rectifier diodes. In all cases the semiconductors must be kept cool, so it is usual to mount the diodes in an aluminium alloy block or plate called a *heat sink*.

Field excitation Unlike the dynamo, there is insufficient residual magnetism present in the magnetic poles to start the charging process, so a battery is used to initially excite, or activate, the field magnets.

Early alternator systems used a field relay to connect the battery to the field when the ignition was switched-on. This *battery-excited* system has given way to a *self-excited* system, which uses three *field diodes* to supply the rotor field with a portion of the current generated by the alternator (Figure 6.15) (overleaf).

Although the self-excited machine supplies the field current when the alternator is charging, it is not able to provide the initial current to energize the field to start the charging process. This is achieved in a simple manner by utilizing the charge-warning lamp; in this way the warning lamp sub-circuit fulfils two duties; it provides a signal to warn the driver when the system is not functioning and also supplies the initial field current.

When the engine is to be started, the ignition is switched-on; this connects the lamp to the battery and makes a circuit through the field to earth. At this stage the lamp is illuminated and the field is excited to the extent controlled by the wattage of the lamp; a typical lamp size is 12 V, 2.2 W.

As the alternator speed is raised, the p.d. on the output side of the field diodes is increased. This gradually reduces the voltage applied to the

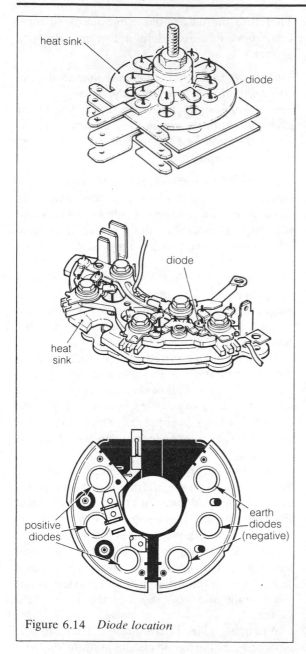

Figure 6.14 *Diode location*

lamp so the light slowly fades and eventually goes out when the output voltage of the alternator equals the battery voltage; i.e. when the alternator 'cuts-in' and starts to charge. When this happens the field diodes will be providing all the field current. The cutting-in speed, which is normally about 1000 rev/min, depends on the field

current so if an earlier cutting-in speed is desired, the wattage of the lamp should be increased.

In view of the dual role fulfilled by the warning lamp, it will be apparent that if the lamp filament is broken, the alternator will not charge.

Figure 6.15 shows the rectifier and field diode arrangement used in a Lucas ACR type alternator. The cable from the charge indicator light connects with the 'IND' terminal on the alternator which is, in turn, joined to the '+' side of the field.

Regulator   Output voltage from an alternator must be limited to prevent the battery from being overcharged and to protect the electrical equipment from excessive voltage. On a 12 V Lucas machine the regulator sets the alternator voltage to a maximum of 14.2 V.

Since this voltage corresponds to a fully-charged battery, the alternator must be made to vary its charging current to suit the state-of-charge of the battery.

Control of the field current is achieved by fitting a regulator on one side (earth for Lucas alternators) of the rotor field (Figure 6.16). The regulator uses a power transistor to act as a field-switching device; the current flow is controlled by the proportion of time that the switch is closed in relation to its open period. When the alternator is below 14.2 V the switch is closed, but at the maximum voltage the switch operates and keeps the output voltage at 14.2 V irrespective of the current being generated.

Surge protection diode   Breakdown of the main transistor in a regulator occurs if the alternator is charging and a poor connection, or similar fault, causes the voltage to suddenly increase. To avoid this damage to the regulator, a surge protection diode is sometimes fitted between the 'IND' lead and earth. This avalanche diode conducts when the surge voltage exceeds a given value. Failure of this diode, in a manner that causes it to continually conduct, shorts out the field and prevents the alternator from charging.

Regulator construction   Early alternators used a remotely situated regulator which used either vibrating contacts or solid-state switches to con-

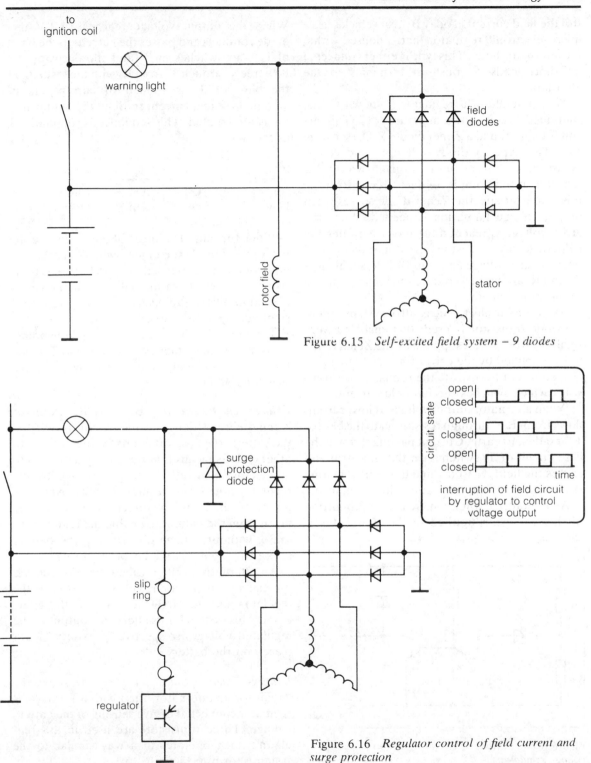

to
ignition coil

warning light

field
diodes

rotor field

stator

Figure 6.15   *Self-excited field system – 9 diodes*

surge
protection
diode

slip
ring

regulator

circuit state

open
closed

open
closed

open
closed

time

interruption of field circuit
by regulator to control
voltage output

Figure 6.16   *Regulator control of field current and
surge protection*

trol the field current. Today British vehicles use a micro-electronic regulator that is housed within the alternator body. This type is either connected by short leads or push-on terminals to the alternator.

The principle of a regulator is shown by the simplified circuit shown in Figure 6.17. This circuit is built around a Zener diode, Z.D. (see page 41). This type of diode will not conduct any appreciable current until a given voltage is reached; at this point it conducts freely. By using this characteristic, the Zener diode senses when output voltage limitation is needed. When the given voltage is reached the diode activates the field-switching transistor. The Zener diode used operates at a voltage less than 14.2 V, so resistors R_1 and R_2 are fitted to reduce the voltage applied to the Zener diode.

As in similar electronic control systems, more than one transistor is used; this enables a very small control current supplied by the Zener diode to be amplified by the driver transistors to a current sufficient to operate the robust power transistor which switches the full field current.

When alternator output voltage is low, current flows from 'B+' through resistor R_3 to the base of T_2 and then to earth. Current passing through the base circuit of T_2 switches-on the transistor and causes the field 'F' to be linked to earth. During this phase, a strong magnetic field is obtained.

As the output voltage increases, a proportion of this voltage is applied to the Zener diode.

When the output voltage reaches 14.2 V the diode conducts and passes the current to the base of T_1. This switches-on T_1 and allows current to flow freely through T_1 from R_3 with the result that the base of T_2 is robbed of current, T_2 is switched-off and current through the field winding is interrupted. This sequence is summarized as follows:

Z.D.	T_1	T_2	Field circuit
No flow	Off	On	Closed
Flow	On	Off	Open

When the output voltage falls below its operating value, the Zener diode switches back to a non-conductive state; this switches the transistors to re-establish the field circuit. This process continues in rapid succession to give a constant voltage output from the machine.

The diode D_1, fitted across the field winding, prevents a high voltage being applied to T_2 when the field is suddenly interrupted by the rapid switching of T_2.

Voltage-sensing circuits   Being as the alternator is remotely positioned in the circuit away from the battery, the take-off points for the supplies to other circuits means that the energy delivered to other equipment alters the p.d. sensed by the regulator situated in the alternator. To overcome this problem, a separate direct lead is sometimes taken from the battery to enable the regulator to sense, without voltage disturbance, the battery p.d.; this system is called *battery sensing*.

An alternative system called *machine sensing* uses an internally connected lead between the regulator and the 'IND' terminal of the alternator. This system limits alternator output to the regulated voltage irrespective of external loads placed on the battery.

Battery-sensed regulator   Figure 6.18 shows the circuit for an alternator fitted with a Lucas 8TR regulator connected to give sensing of the battery voltage. Three transistors are used in this regulator which operates in a way similar to the system shown in Figure 6.17.

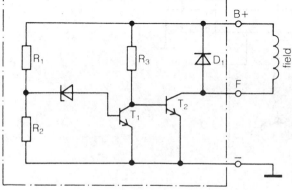

Figure 6.17 *Simplified circuit of a transistorized voltage regulator*

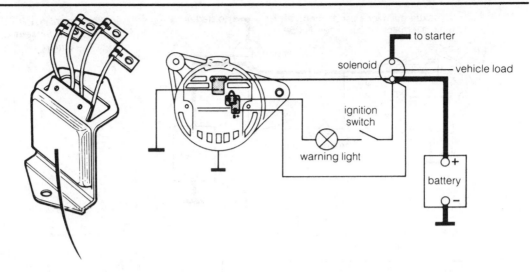

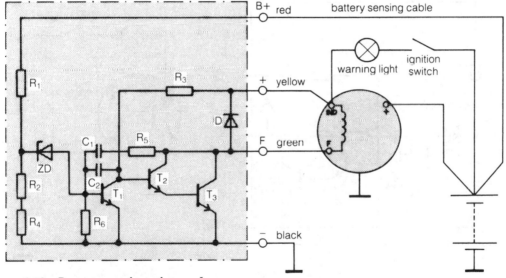

Figure 6.18 *Battery-sensed regulator – Lucas*

In this system the cable connected between the battery and the regulator terminal 'B+' acts as the sensing lead. Voltage applied to 'B+' signals the point at which the Zener diode starts to conduct.

Machine-sensed regulator When a Lucas 8TR was used for this layout, the regulator had a circuit similar to Figure 6.18, except the regulator lead 'B+' was internally connected to the '+' terminal. This arrangement sensed the voltage given at the 'IND' end of the field winding. The regulator had three leads; '+' (yellow), 'F' (green) and '−' (black).

Later designs of a regulator, e.g. Lucas 14TR, used a Darlington amplifier to perform the heavy-duty switching of the field winding. Figure 6.19 (overleaf) shows a typical machine-sensed circuit.

Two additional resistors R_1 and R_6 and two capacitors C_1 and C_2 are shown. This sub-circuit allows the regulator to oscillate at a frequency

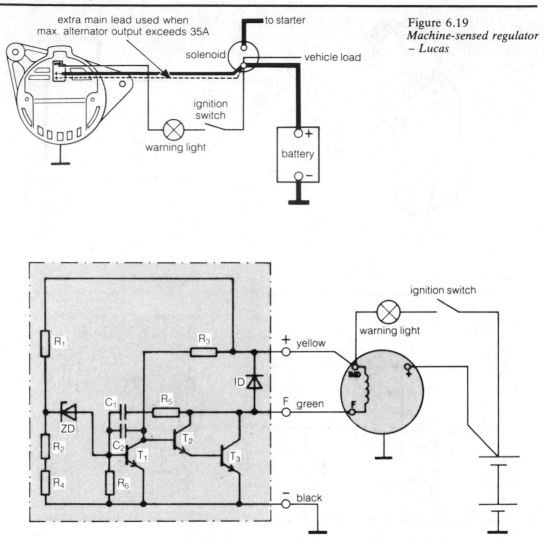

Figure 6.19
*Machine-sensed regulator
– Lucas*

controlled by the internal time-constant given by the charge–discharge action of the capacitors; this ensures that the transistor T_3 is rapidly switched-on and -off. Output-voltage control by the regulator is obtained by the modulation of the mark–space ratio, i.e. the ratio between the closed and open periods (Figure 6.16).

The adoption of spade connectors in Lucas type regulators 16TR–21TR improves reliability by eliminating interconnecting cables; this forms an 'integral' circuit with stator and field systems.

Lucas alternators The Lucas A-range alter-nator uses machine-voltage sensing and gives a regulated voltage of 14.2 ± 0.2 V.

An option makes provision for fault diagnosis by providing a positive indication of loss of output due to:

- open-circuit field/worn brushes
- open-circuit regulator
- open-circuit output cable
- broken drive belt
- overcharge through faulty regulator.

Output from an A127 type alternator is shown in Figure 6.20.

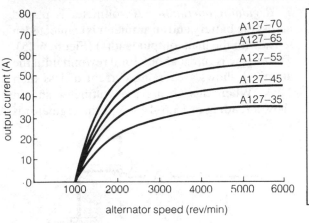

Figure 6.20   *Alternator – Lucas A127*

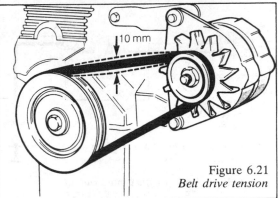

Figure 6.21
Belt drive tension

6.3 Maintenance of alternator charging systems

Very little maintenance is required on a modern alternator other than a check of the tension and condition of the driving belt at 10 000 km (6000 miles) and a brush check at 65 000 km (40 000 miles). At times when an under-bonnet check is being made for general security of all items, the alternator mounting and cable condition should be examined.

Fault diagnosis
Equipment needed for testing a charging system on the vehicle should include:

- d.c. moving-coil voltmeter, 0–20 V
- d.c. moving-coil ammeter, 5–0–100 A.

No-charge or low-charge
Many different types of charging system are used so the following is intended to outline the basic method for diagnosing a fault.

1. *Battery test*   The battery should be tested as described on page 78.

2. *Drive belt*   The condition and tension should be checked and adjusted as shown in Figure 6.21.

3. *Visual check*   All cables and connections should be checked for security.

4. *Cable continuity*   The connector is removed from the alternator, the ignition switched-on and the p.d. checked at each of the leads (Figure 6.22). No voltage at any one lead indicates an open circuit. In the case of the 'IND' lead, the charging light bulb may be defective.

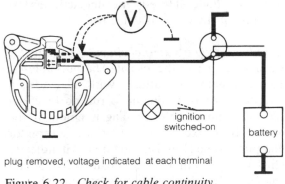

plug removed, voltage indicated at each terminal

Figure 6.22   *Check for cable continuity*

5. *Alternator output*   Maximum output will not be supplied if the battery is fully charged, so switch-on all loads (excluding wipers) for about one minute. With the ammeter securely connected in *series* with the output lead(s), the engine is run-up to about 3000 rev/min (Figure 6.23) (overleaf). The output should not be less than the manufacturer's specification, e.g. Lucas alternators:

15ACR – 25 A	20ACR – 60 A
16ACR – 30 A	23ACR – 50 A
17ACR – 33 A	25ACR – 60 A
18ACR – 40 A	

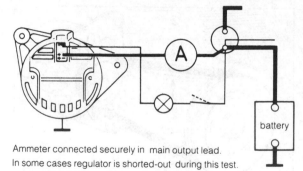

Ammeter connected securely in main output lead.
In some cases regulator is shorted-out during this test.

Figure 6.23 *Check for maximum output*

Lucas A-range alternators are classified by their output rating, e.g.
A127-35 – 35 A   A127-45 – 45 A

If the output is considerably less than the specified value, the alternator should be removed and examined. Prior to removing the unit, the output test should be repeated with the surge protection diode (if fitted) disconnected; this test will show if the diode is defective.

6. *Voltage drop of external circuit* A voltmeter is connected across the insulated output cables(s) from alternator to battery. With all loads on (except wipers) the engine is run-up to about 3000 rev/min and the reading is noted. If the voltage drop exceeds 0.5 V on any alternator charging system (English and foreign) it indicates a high resistance in that line (Figure 6.24). Sometimes a similar check is made on the earth line to ensure the voltage drop is 0–0.25 V.

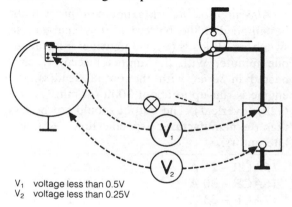

V_1 voltage less than 0.5V
V_2 voltage less than 0.25V

Figure 6.24 *Voltage drop of external circuit*

7. *Regulator operation* A voltmeter is placed across the battery and an ammeter is connected in *series* with the main output lead(s) (Figure 6.25). The engine is run at about 3000 rev/min until the ammeter shows a charging current of less than 10 A. When this occurs the voltmeter should show a reading of 13.6–14.4 V if the regulator is serviceable.

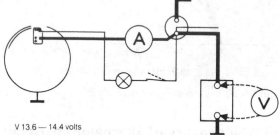

V 13.6 — 14.4 volts

Figure 6.25 *Regulator check*

Foreign alternators The service manual should be consulted when testing a charging system fitted to an imported vehicle. Where this is difficult to obtain, Table XIII may be used as a general guide in conjunction with the previously described tests.

C.R.O. tests on an alternator Faults in an alternator can be diagnosed by using a C.R.O. The patterns shown on the screen will enable the operator to pin-point the fault (Figure 6.26).

Figure 6.26 *C.R.O. alternator test*

Electrical tests on a dismantled alternator
Electrical tests are conducted on the:

- diode pack
- rotor field
- stator windings

The following description applies to alternators in general, but the specific values given apply to a type similar to Bosch type K1-35A.

Diode pack Prior to testing, the diode connections must be unsoldered from the stator wind-

Table XIII Foreign alternators[a]

	French	German	Italian	Japanese	Significance of reading
General features	9 diode Ext. regulator	6 or 9 diode Int. or ext. regulators	6 or 9 diode Ext. regulator	6 diode Int. or ext. regulators	
TEST 4 Cable continuity	Battery voltage at all leads (except earthing leads)				No reading indicates open circuit in that lead
TEST 5 Alternator output	'Exc' lead and output lead are shorted together during test	If low output is obtained with ext. reg., short 'B+' and 'DF' together to cut out regulator	Output '30' and field '67' are shorted together during test	Output 'B' and field 'F' are shorted together during test	Output current should be as stated. No output or low output indicates that alternator should be removed
TEST 6 Voltage drop	Maximum drop on insulated line should be less than 0.5 V				High reading indicates presence of resistance in that line
TEST 7 Regulator check	13.8–13.2 V	Int. reg.: 13.7–14.5 V Ext. reg.: 13.9–14.8V	13.9–14.5 V	Colt Honda 13.5–14.5 V Mazda Toyota 13.8–14.8 V Datsun 14.3–15.3 V	Suspect reg. if lower or higher

[a] Ext. = external; Int. = internal; reg. = regulator.

ings. Heat from the soldering iron must be prevented from passing to the diode by using a pair of pliers as a heat sink (Figure 6.27).

The leads of a 12 V, 5 W test lamp are connected to each diode in turn; when the current flows through the diode in one direction the lamp should light but when the leads are reversed the lamp should not light (Figure 6.28) (overleaf).

Rotor field The field is checked for:

- resistance
- insulation

To measure the resistance, the leads of an ohmmeter are applied to the slip rings. A typical resistance is 3.4 Ω (Figure 6.29) (overleaf).

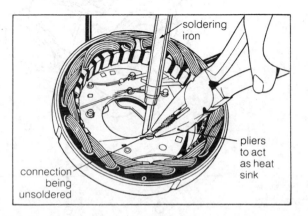

Figure 6.27 *Protection of diodes*

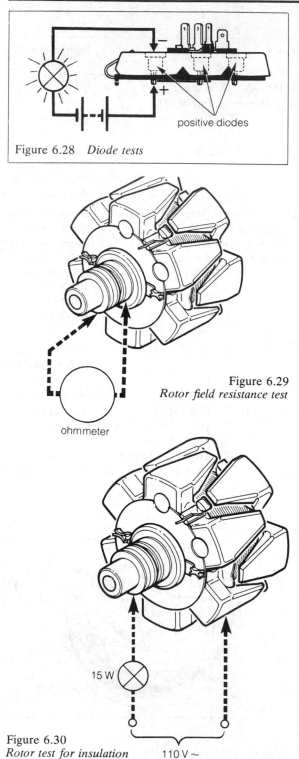

Figure 6.28   *Diode tests*

Insulation is checked by the use of a 110 V a.c. supply fitted with a 15 W lamp. The test prods are applied to the iron pole and one slip ring. Insulation is defective if the lamp illuminates (Figure 6.30).

Stator windings   The leads of an ohmmeter are applied to two of the three leads of the stator and the resistance is measured. One of the ohmmeter leads is then transferred to the remaining stator leads and the resistance is measured again. Both results should be equal and the resistance should be as specified by the manufactuers; a typical value is 0.09 Ω (Figure 6.31).

Insulation is checked in a way similar to the rotor; the tester should show good insulation between the winding and the iron laminations (Figure 6.32).

Figure 6.29
Rotor field resistance test

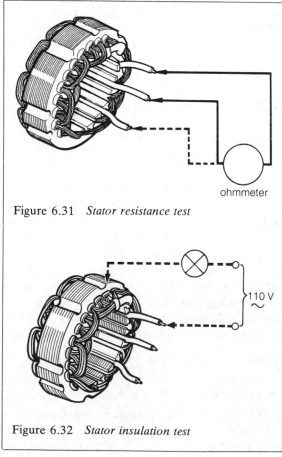

Figure 6.31   *Stator resistance test*

Figure 6.30
Rotor test for insulation

Figure 6.32   *Stator insulation test*

7 Vehicle lighting systems

7.1 Circuit layout

Lights are needed on a vehicle to allow the driver to see, and be seen, in conditions of darkness and poor visibility.

Statutory regulations dictate the number, position and specification of many of the external lights fitted to a vehicle. In addition to the *obligatory lights*, vehicle manufacturers and vehicle owners often fit other *supplementary lights* to fulfil other purposes.

Lamps are grouped in separate circuits; these include the folllowing:

- *Side and rear lamps* including lamps for the number plate, glove compartment and instrument panel illumination.
- *Main driving lamps* (headlamps) fitted with a dipping facility to prevent approaching drivers being dazzled.
- *Rear fog lamp(s)* for 'guarding' the rear of the vehicle in conditions of poor visibility.
- *Auxiliary driving lamps* including spot lamps for distance illumination and fog lamps that are positioned suitably and designed to reduce the reflected glare from fog.
- *Reversing lamps* to illuminate the road when the vehicle is moving backwards and warn other drivers of the movement.
- *Brake lights* to warn a following driver that the vehicle is slowing down.
- *Interior light* and courtesy lights on doors.
- Instrument panel lights for signalling either the correct operation of a unit, or the presence of a fault in a particular system.

In addition to these lights, directional indicators and hazard warning lights are fitted; these are covered at a later stage in this book.

Circuit arrangements

For maximum illumination the lamps are connected in parallel with each other (see page 9). This arrangement provides various circuit paths for the current so an open circuit in any branch will cause failure in that one branch only; the other lamps will still function normally.

Most vehicle lighting systems use an earth-return circuit; this requires less cable than an insulated return or two-wire system. When the vehicle body is used as an earth, a good clean connection must be made at suitable earthing points on the main body. This earthing lead is essential where the lamp is mounted in a plastics body panel.

Lighting circuit diagrams are drawn in either a *locational* or *compact* theoretical form. The former type shows each component positioned relative to its situation on the vehicle. Although this is useful in showing the location of the various connectors and component parts, it makes the diagram more difficult to trace out a particular circuit path. To minimize this problem, some manufacturers use extra diagrams to show separate parts of the circuit, e.g. the supply system is shown separately. Figure 7.1 (overleaf) shows a simple circuit drawn both ways.

This parallel circuit has the lamps controlled by three switches:

Switch 1 operates the side and rear lamps. It also supplies:

Switch 2 which operates the headlamps, and supplies:

Switch 3 to distribute the current to either the main beam or the dip-beam headlamp bulbs.

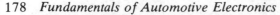
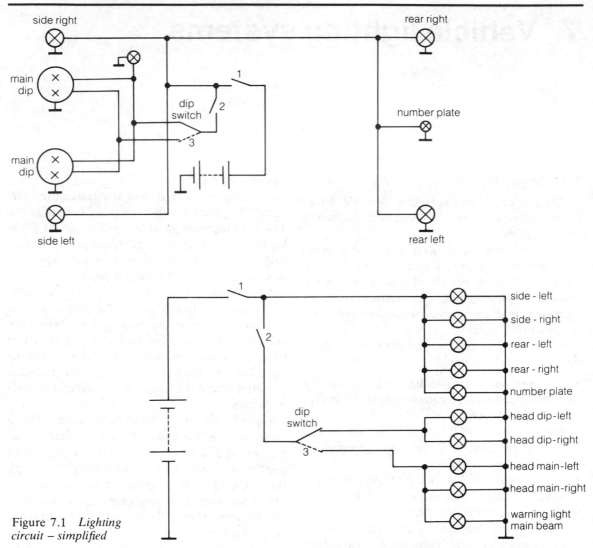

Figure 7.1   *Lighting circuit – simplified*

Circuit protection   A single fuse, mounted in the main supply cable, protects a circuit in the event of a severe short. This simple protection system cannot be used in the external lighting supply cable because all lights will go out when the fuse fails; a dangerous situation when the vehicle is travelling at speed along a dark road.

To avoid this danger manufacturers either fit separate fuses for each light system, or refrain from fusing the headlamp circuit altogether (Figure 7.2).

The circuit in Figure 7.2 incorporates extra features:

Headlamp flash switch   This switch enables the driver to signal to other drivers during daylight and avoids the use of the main light switches. The spring-loaded switch operates only when the lever is held in the 'on' position.

Ignition-controlled headlamps   Regulations insist that the headlamps should not be used when the engine and vehicle are stationary. This is achieved by using the ignition switch to control the feed to the headlamps. A relay is often used to reduce the current load on the switch.

Auxiliary driving lamps These long-range lamps (spot lamps) are used when the headlamps are set to main beam, but they must be extinguished when other vehicles are approaching. This is achieved by connecting the auxiliary lamps to the main beam branch of the circuit. As the power consumed by these lamps is considerable, the load on the lighting switches is reduced by using a relay to control these lamps.

Fog lamps (front) In fog the main headlamps cause glare so by using low-mounted fog lamps this problem is minimized. These twin lamps can be used instead of headlamps so the feed must be taken from the side lamp branch of the circuit.

Rear fog guard The high-intensity fog lamp(s) guard the rear of the vehicle; they must be used only in conditions of poor visibility. To prevent the driver using the lights illegally the feed is taken from either the dipped beam or the front fog lamps. A warning light must be fitted to indicate when the rear fog guard lamp(s) are in use.

Lamp failure indicator Many manufacturers now fit a warning system to inform the driver when a light is not functioning correctly. Often the lamp signal indicator on the instrument panel is a graphical map of the vehicle. On this display sections are illuminated either when the lights are operating normally, or as a signal to warn the driver that a light is 'out'.

In addition to the graphical display unit, a module (sometimes called a 'bulb outage module') is fitted to sense when a specific section of the circuit does not consume the appropriate cur-

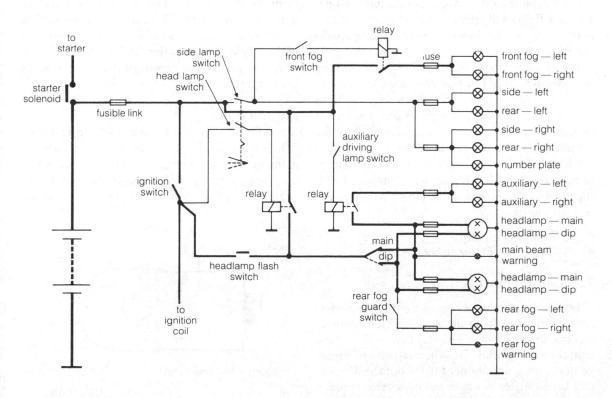

Figure 7.2 *Lighting circuit incorporating fuses and relays*

rent. When an open-circuit condition exists, the module triggers a light on the instrument panel to show the driver the actual lamp that is 'out'.

To enable the module to monitor the system, each branch of the circuit passes through the module. Unfortunately, this feature tends to complicate the circuit and also increases the weight and bulk of cable needed for the lighting system. However, its use is defended on grounds of safety.

Most graphical display systems are arranged to illuminate fully for a few seconds after switching-on the ignition; this tests the lights in the graphical display panel and shows that they are functioning correctly. For further details on lamp monitoring systems, see page 249.

Dim-dip lighting devices   British vehicles registered after 1 April 1987 must be fitted with dim-dip lighting. This regulation makes it impossible for the vehicle to be driven on side lights alone. The side lights will operate only when the ignition is switched 'off' so they may be regarded as parking lights.

Headlamps can be operated in two dip modes. A dim-dip light of low power is intended for use, without dazzling other road users, in conditions such as well-lit streets at night or dull weather at twilight. The dip beam of standard intensity is for normal night driving in out-of-town areas.

The regulation requires one pair of headlamps to incorporate a dim-dip device. Compared with the normal dipped beam, the dim-dip light intensity should be:

- 10% (halogen lamps)
- 15% (normal filament lamps)

7.2 Lamp construction

Illumination can be obtained from an incandescent filament or from the glow emitted when an electric current is passed through a glass tube containing a special gas. The majority of motor vehicle lamps are filament types but the alternative type is often used on public-service vehicles for interior lighting. These fluorescent lamps have the advantage of a light source that is spread over a large area, so passengers are not subjected to glare and eyestrain.

Light intensity
The intensity of light or luminous intensity is the power to radiate light and produce illumination at a distance. Luminous energy refers to the source of light and its intensity is measured in *candelas* (cd); in the past the unit 'candle power' (c.p.) was used. For practical purposes:

$$1 \text{ cd} = 1 \text{ c.p.}$$

The amount of light that falls on a surface is called the illumination and the intensity of illumination is measured in *lux* or *metre-candle*. A surface illumination of 1 metre-candle or 1 lux is obtained when a lamp of 1 candela is placed 1 metre from a vertical screen. When the distance is increased the intensity of illumination decreases; it varies inversely as the square of the distance from the light source. This means that if the distance is doubled, the illumination of the surface on which the light rays fall will be reduced to 1/4 of the original illumination; if the original brightness is required, the power of the lamp must be quadrupled.

Filament lamps
The main details of a lamp are shown in Figure 7.3. Enclosed in a glass container is a tungsten filament that is secured to two support wires; these are normally attached to contacts in a brass cap. Low wattage bulbs, such as those used for

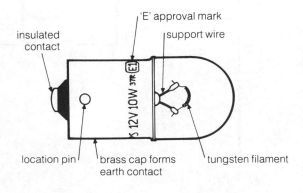

Figure 7.3   *Filament lamp*

side lamps, are normally of the vacuum type. Removal of the air prevents oxidation and vaporization of the filament, and reduces the heat loss. Oxygen in the air causes tungsten deposits to blacken the glass above the filament; also after a very short time the filament burns away.

When operated at the rated voltage, a filament temperature of about 2300°C is reached and a white light is produced. If the lamp is operated at a lower voltage both the temperature and light output will be low. Conversely the operation of a lamp at a higher voltage soon vaporizes the tungsten, blackens the glass and burns-out the filament.

Filaments of larger powered bulbs, such as those used for headlamps, can be made to operate at a higher temperature and give about 40% more light by filling the bulb to a slight pressure with an inert gas such as argon. Heat loss from the filament due to convection movement of the gas is reduced by winding the filament in the form of a helix.

Regulations state that all bulbs used on vehicles must be marked with the letter 'E' and a number that identifies the country where approval was given. This mark indicates that the bulb conforms to the E.E.C. Standard specified for a given application.

Tungsten–halogen type bulbs   During the life of a normal gas-filled bulb, evaporation of a tungsten filament causes the glass to become black. Although this can be minimized by spreading out the tungsten deposit over a larger glass bulb, the light intensity after a period of time is far from ideal.

The problem has recently been overcome with the introduction of the tungsten–halogen bulb; this type is also called quartz–halogen, quartz–iodine and tungsten–iodine. A much higher output is obtained from this new type; it also maintains its efficiency for a longer time.

Halogen refers to a group of chemical elements that includes iodine and bromine. When a halogen is added to the gas in a bulb a chemical action takes place which overcomes the evaporation problem. Evaporation of the tungsten still

occurs but as the tungsten moves from the hot filament towards the envelope it combines with the halogen and forms a new compound (tungsten halide). This new compound does not deposit itself on the glass envelope; instead the convection movement carries it back to the hot gas region around the filament. Here the tungsten halide splits up and causes the tungsten to redeposit itself back on the filament; the halogen particles released are returned to the gas. This regeneration process not only prevents discolouration of the bulb; it also keeps the filament in a good condition for a much longer time.

To produce this action, the bulb must be made to operate at a gas temperature higher than the 250°C needed to vaporize the halogen; this is achieved by using a small bulb of quartz. This material can withstand the heat and is sufficiently strong to allow the bulb to be gas filled to a pressure of several bars so as to give a brighter filament for a given life (Figure 7.4).

An added advantage is obtained from the smaller filament needed with this type; it allows more precise focusing than is achieved with the normal bulb.

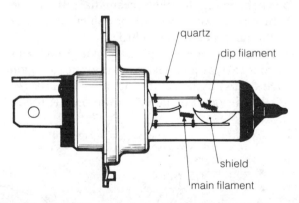

Figure 7.4   *Quartz-halogen lamp bulb*

Reflector
The function of a headlamp reflector is to redirect the light rays. An ideal reflector gives a beam of light that illuminates the road from far ahead to the region immediately in front of the vehicle.

A normal reflector is shaped in a paraboloid form, highly polished and then coated with a material such as aluminium to give a good reflective surface.

To give good illumination, the lamp filament must be accurately positioned at the *focal point* of the reflector; this allows it to reflect the light rays in the form of a parallel beam (Figure 7.5).

Other positions of the filament put the lamp out-of-focus; this reduces the illumination and, in the case of a diverging beam, may dazzle the drivers of oncoming vehicles. In the past, lamps incorporated a focus adjustment, but nowadays *pre-focus* bulbs have a fitting which sets the filament at the correct place.

Some bulbs are shielded on the lens side to ensure that all light rays are directed back to the reflector.

Lens

A glass lens, moulded to form several prismatic block sections, bends the rays and distributes the light to obtain the required illumination. The design of a headlamp lens pattern attempts to achieve good illumination for both main- and dip-beam positions. The main beam requirement is a long-range penetrating light whereas a dipped light needs a low-level beam that gives a wide light spread just in front of the vehicle and offset to the near-side of the vehicle.

Figure 7.6 shows a typical headlamp lens. This European-type lens design incorporates a region (marked A) to deflect the dip beam towards the left-hand side, if intended for use on UK roads.

For Continental touring this lens must be temporarily converted to 'dip to the right' either by fitting a pair of beam deflectors to the lens or by masking-out with tape the region marked 'A' in Figure 7.6.

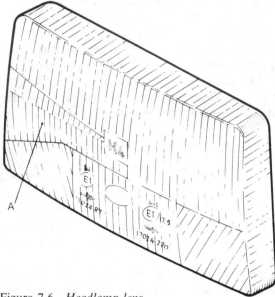

Figure 7.6 *Headlamp lens*

All lamp lenses must be 'E' marked and have an arrow moulded in the glass to show the dip direction. When two opposing arrows are shown, the lens is suitable for both dip directions; the actual direction is dictated by the position of the bulb.

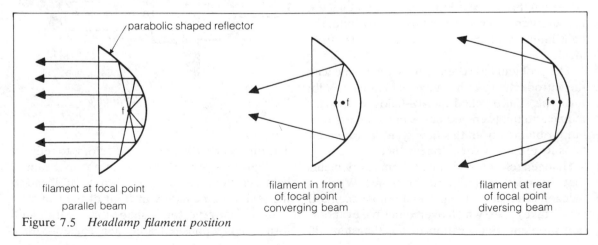

filament at focal point
parallel beam

filament in front
of focal point
converging beam

filament at rear
of focal point
diversing beam

Figure 7.5 *Headlamp filament position*

Dipping facility

The eye takes time to adjust when it moves from a brightly lit area to an area where the light is poor. For a few seconds that it takes the iris of the eye to open, vision is poor. Conversely, movement from a dim to a bright zone reverses the iris movement; the immediate effect is to squint the eye to restrict the sudden burst of light until the iris adjustment has been completed.

Applying this example to night driving shows that an arrangement is needed to prevent a dazzling light beam from entering the driver's eyes. This provision allows the driver's vision to be maintained during, and after, the time that the passing light is directed towards the driver.

Once the eye has adjusted itself to a high level of illumination it finds difficulty in seeing clearly the areas that are dimly lit. For this reason the headlamp design should graduate and distribute the light rays to even-out the illumination rather than concentrate the light in one small region. Also under this distributed pattern, dazzle is less pronounced.

Eye strain results if the eye has repeatedly to adjust to cope with varying illumination levels. This further emphasizes the need for efficient anti-dazzle headlamp arrangements.

The statutory requirements are detailed in the Road Vehicle Lighting Regulations. These state that the lighting system must be arranged, so that it is:

'incapable of dazzling any person standing on the same horizontal plane as the vehicle at a greater distance than 25 feet from the lamp whose eye level is not less than 3ft 6in above the plane.'

Nowadays the dip facility is achieved by redirecting the rays downwards and towards the near-side; this is obtained by using a *bifocal*, *twin-filament* bulb which has the dip filament either *offset* to the focal point of the reflector or *shielded* (Figure 7.7).

When these bulbs are combined with a modern lens as fitted to European-type lamps, the light pattern, as projected on to a vertical screen, gives an asymmetrical image with a sharp cut-off as shown in Figure 7.8. The dimensions shown indi-

cate the legal requirements, so to meet this regulation adjusters must be provided for horizontal and vertical alignment.

Sealed beam

In the past a headlamp was made up of separate parts and this made it difficult to locate the filament at the focal point. Also the efficiency deteriorated severely when dust and moisture settled on the reflector after entering the lamp through various cracks between the lens and reflector.

The designers of the sealed beam lamp unit overcame these problems by producing a one-piece sealed glass unit that incorporated the lens

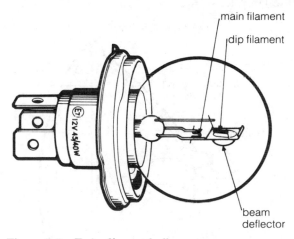

Figure 7.7 *Twin-filament bulb*

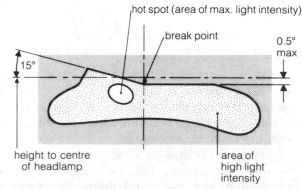

Figure 7.8 *Light pattern from European-type headlamp, dipped beam; left dip*

and aluminized reflector. Two tungsten filaments for the main and dip beams are precisely positioned at the correct points and the complete lamp is filled with an inert gas. Because the bulb has no independent glass envelope, tungsten deposits are spread over a very large area, so the light efficiency of this unit remains high for a long period of time (Figure 7.9).

Although this type of lamp is a great advance over earlier designs, the sealed beam unit has two disadvantages: it is more costly to replace when the filament fails; also sudden light failure occurs when the lens becomes cracked. In some countries a secondary glass screen is used to improve the aerodynamic line and give extra protection to the lamp lens.

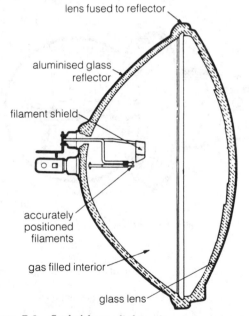

lens fused to reflector

aluminised glass reflector

filament shield

accurately positioned filaments

gas filled interior

glass lens

Figure 7.9   *Sealed beam light unit*

Four-headlamp system
Optically it is difficult to produce a single lens and reflector unit that gives an illumination to satisfy both main and dip conditions.

To improve this drawback some manufacturers use four headlamps: two for long-distance illumination and two for lighting the area immediately in front of the car. Each one of the outer lamps has two filaments; a dip filament

situated at the focal point to give good light distribution and a second filament positioned away from the focal point to provide near-illumination for main-beam lighting. When the lamps are dipped the inner lamps giving long distance illumination are switched-off.

For accommodation reasons the lamp of a four-headlamp arrangement is smaller than that used on a two-headlamp system.

British–American and European headlamps
Headlamps can be divided broadly into three main categories which can be identified by the lens marking and shape. Identification of the lamp type is necessary because the alignment method differs with each type.

British–American lamps having the number '1' or '1a' moulded on the lens, or in some cases no number whatsoever, are checked on main beam. These lamps are always circular in shape, are often of a sealed-beam construction, and have a symmetrical main-beam pattern.

Other British–American lamps having the number '2' moulded on the lens are checked on the dipped, or passing beam.

European headlamps may be circular, rectangular or trapezoidal in shape and incorporate a prismatic region as shown in Figure 7.6. These lamps normally have the number '2' moulded in the glass as well as the 'E' approval mark. All lamps of this type are checked on the dipped beam when carrying-out an alignment test.

This type of lamp has an integral lens and reflector assembly and the bulb is removable (Figure 7.10). This feature allows for variation in the bulb type and enables many different lens shapes to be developed to suit the body contour of the vehicle. The larger lens and reflector areas enable the illumination pattern given by the dipped beam to have a wide horizontal spread.

UK regulations require that four-wheeled vehicles must be fitted with a matched pair of headlamps, white or yellow in colour, and symmetrically positioned on the vehicle. Each filament should have a wattage of not less than 30 W.

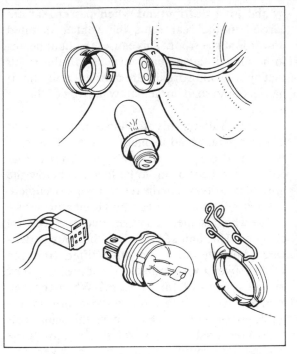

Figure 7.10 *Bulb types used in European headlamp*

Double-reflector headlamp

This highly efficient light source combines two bulbs and two reflectors in one headlamp unit. This makes it possible to meet, with precision, the optic requirements for main and dip beams as well as satisfying the European regulations covering bulb replacement and dazzle.

The Amplilux range of lamps, produced by SEV Marshal, gives out over twice the output of the average conventional headlamp, e.g. the 7 in quartz halogen unit produces 72 000 cd compared with 32 000 cd for an average non-iodine unit.

The circular or rectangular lead crystal-glass lens and reflector are bonded together to ensure maximum weather protection and constant performance over a long period of time.

Figure 7.11 shows the construction of this type of lamp. In front of the main reflector is a small inset reflector and bulb that provides the main beam. The back of the main-beam reflector is used to screen the lower half of the dip-beam reflector; this gives a sharp horizontal cut-off to the asymmetrical dipped-beam pattern and

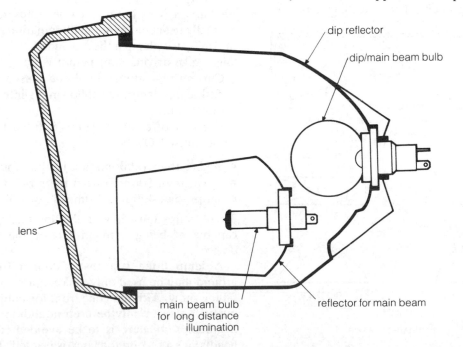

Figure 7.11 *Double-reflector headlamp*

avoids dazzle. On some lamps a screen (*occulteur*) is placed in front of the dip bulb to stop direct light from the bulb showing above the cut-off axis.

Homofocular headlamps

These headlamps use an advanced design of reflector divided into separate segments with different focal lengths, all sited inside the one light unit (Figure 7.12).

Use of this type of reflector is necessary where low bonnet lines limit the depth and height of the headlamps and where the lens has to be angled to blend with the body contour.

The homofocular reflector features parabolic segments of different focal lengths arranged about the same focal point. This 'stepped' reflector cannot be manufactured in sheet metal; instead the unit is moulded from a plastics material that combines a very smooth surface with resistance to heat that is radiated from halogen bulbs.

Headlamp range adjustment

Vertical alignment of the headlamps is affected

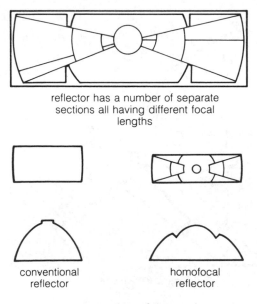

reflector has a number of separate sections all having different focal lengths

conventional reflector

homofocal reflector

comparison of size

Figure 7.12   *Homofocular headlamp*

by the load distribution; when passengers are carried in the rear seats the vehicle is tilted upwards at the front. This causes the light beams to be aimed higher than normal with the result that approaching drivers are dazzled; also illumination of the road immediately in front of the car is poor.

On the Audi car this problem is minimized by a headlamp range adjustment system. This provides the driver with a thumbwheel control that enables the light beam to be lowered below the normal position when the rear seats are occupied.

Each reflector is moved by an electric servo-motor which is operated by an electronic control element. This unit senses when the voltage delivered from the driver's control differs from the voltage given by a variable resistor positioned by the movement of the reflector. When the two voltage signals differ, the servo-motor moves until a position is reached where the signal voltages are equal. The control system moves the lamps to the setting selected by the driver.

Other external lamps

Auxiliary driving lamps   These lamps are fitted normally to supplement the illumination given by the obligatory lamps; they include fog lamps and long-range driving lamps (spot lamps).

Current regulations should be consulted prior to fitting any lamp; the following is intended to act as a guide.

If a pair of auxiliary lamps is fitted then the lenses must NOT be:

- higher than 1200 mm above the ground
- lower than 500 mm from the ground
- more than 400 mm from side of vehicle.

These lamps must have a dipping facility or be capable of being extinguished by the dipping device.

A lamp fitted less than 500 mm from the ground may be used only in conditions of fog or falling snow. Although one front fog lamp is permitted, this can only be used in addition to the headlamps. If glare is to be avoided from the headlamps a second fog lamp is needed. The two fog lamps must be placed:

- at the same height
- equidistant from the centre line of the vehicle
- so that the illuminated area is not more than 400 mm from the outermost part of the vehicle.

Fog lamps have a lens that gives a wide flat-topped beam with a sharp cut-off to illuminate the road immediately in front of the vehicle without causing glare in fog conditions (Figure 7.13).

Driving lamps incorporate a lens that projects a narrow spot beam of high intensity light to illuminate the road well ahead of the vehicle.

Sidelamps UK regulations require a vehicle to carry two white sidelamps each having a wattage of less than 7 W and be visible from a reasonable distance. Since it is now illegal in the UK to drive during the hours of darkness using only the sidelights, the role fulfilled by the sidelamps is now limited to marking the vehicle when it is parked.

On many cars the 'parking' light is incorporated in the headlamp; the bulb often used is a 5 W capless type as shown in Figure 7.14.

Rear lamps A car must carry two red 'E' marked rear lamps of a given size and of wattage not less than 5 W. They must be positioned:

- between 1500 mm and 350 mm from the ground
- spaced apart more than 500 mm
- set so that the distance between the edges of the vehicle and the illuminated area is not more than 400 mm.

The red lens must diffuse the light and be 'E' marked to show that it meets the specified standard.

In addition to the rear lamps, a car must be fitted with two red reflectors of approved design.

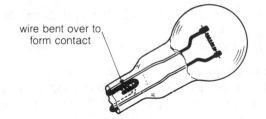

wire bent over to form contact

Figure 7.14 *Capless-type bulb*

Stop lamps Two stop lamps each of wattage between 15 and 36 must be fitted. These lamps must illuminate a red diffused, and 'E' marked, lens when the service (foot) brake is applied and be designed to be visible through a given angle.

The lamps must be positioned:

- between 1500 mm and 350 mm from the ground
- symmetrically at least 400 mm apart.

Often a single 6/21 W bulb with twin filaments is used to provide rear lamp and stop lamp functions; the bright light given by the 21 W filament is used for the stop lamp.

Number plate The rear number plate must be clearly illuminated by a white light, but the bulb must not be visible from behind the vehicle. The light is connected in parallel with the sidelights.

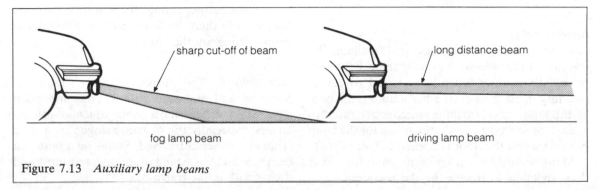

sharp cut-off of beam

long distance beam

fog lamp beam

driving lamp beam

Figure 7.13 *Auxiliary lamp beams*

Rear fog lamps   These high-intensity red lamps are used to improve safety in conditions of poor visibility; namely fog, falling snow, heavy rain or road spray. Either one or two lamps must be fitted by the manufacturer on current vehicles and these must be set:

- between 250 mm and 1000 mm from the ground
- more than 100 mm from any stop lamp.

If two lamps are used they must be set symmetrically, or in cases where only one lamp is to be used it must be fitted either on the offside or on the vehicle centre line.

Normally each lamp has a 21 W bulb and a lens of large area; both must be 'E' marked. The circuit must allow the rear fog lamp to be operated by an independent switch and a warning lamp must be fitted to signal to the driver when the rear fog lamp is in operation. Furthermore the circuit must function only when the headlamps, front fog lamps or sidelights are in use. In some cases a fog guard relay coil prevents operation of the rear fog lamps until the front fog lamps are switched-on.

Reversing lamps   When a reversing lamp is fitted it must conform with the statutory regulations. These state that not more than two lamps may be used and the total wattage per lamp must not exceed 24 W. The white light should be switched automatically by the gearbox and be subject to anti-dazzle requirements. Where automatic switching is not provided, a separate switch, together with a warning lamp, may be used.

Interior lights
One or more lights are used to illuminate the interior of the vehicle. Normally these lights can be switched independently or wired to act as courtesy lights to operate when a door is opened. In the latter case, earthing switches are fitted to each door so that an earthing contact for the lamp is made when the door is opened (Figure 7.15).

Many security car-alarm systems utilize these door switches to trigger the alarm system.

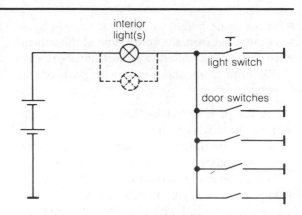

Figure 7.15   *Courtesy light system*

7.3 Maintenance of lighting systems

Other than the usual superficial checks for cable security and condition, most items of work occur only when a fault develops. On modern vehicles the graphical display on the instrument panel will warn the driver of a lighting fault.

Failure of a lamp
Initial warning to the driver of 'bulb outage' should be verified to ensure that the monitoring system is not giving a false signal. If an incorrect message is shown, then the circuit should be checked by using the method outlined in the Instrumentation section (Chapter 10) of this book. Most lighting faults are caused by the failure of a fuse or bulb.

Fuse   A 'blown' fuse should be replaced with a fuse of the correct rating. If the new fuse blows immediately then the fault must be pin-pointed before fitting another fuse.

Bulb failure   The suspected bulb should be removed and replaced with the recommended type. Glass surfaces must not be touched with the fingers, especially the quartz–halogen type, so a clean cloth should be used. Stains on a bulb can be removed by washing in methylated spirit and drying with a lint-free cloth.

Wiring faults If the initial check shows that the bulb and fuse are serviceable, and a visual check of the cables does not reveal the defect, then the circuit should be tested with a voltmeter. Figure 7.16 shows the principle as applied to a simple lighting circuit.

Tests

Test 1 A voltmeter (V_1) is connected across the battery to measure the voltage under lighting load.

Test 2 When the voltmeter (V_2) is placed across the lamp the voltage should be similar to the voltage at Test 1. If a resistance in the circuit causes the reading to differ by more than 10% of the battery voltage, the cause should be investigated by using Tests 3 and 4.

Test 3 Voltmeter V_3 shows the voltage drop on the insulated line. If an excessive drop is shown, the location of the fault can be detected by moving one voltmeter lead along the connection points in the circuit path until a stage is reached where the change in voltage is considerable.

Test 4 With the meter arranged as in V_4, the drop in the earth line is shown. By using a similar technique to Test 3, a high resistance can be located. The total voltage drop from Tests 3 and 4 should be less than 10% of the battery voltage, e.g. less than 1.2 V for a 12 V system.

A quick check for an open-circuit can be made by using a 12 V test lamp (Figure 7.17). A circuit break between points 1 and 4 is easily found. With one side of the lamp connected in good earth, the break can be located.

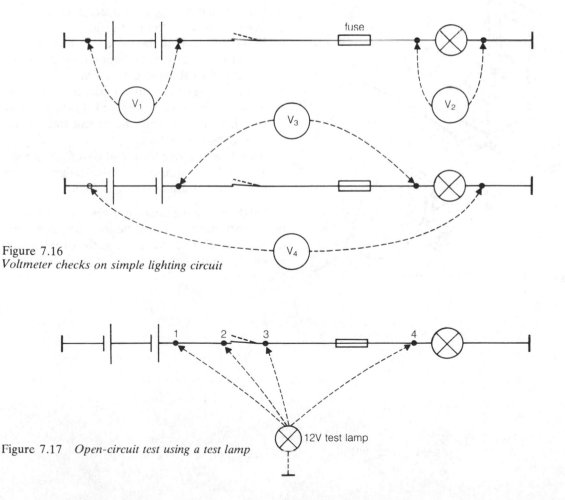

Figure 7.16
Voltmeter checks on simple lighting circuit

Figure 7.17 *Open-circuit test using a test lamp*

Headlamp alignment

The alignment of headlamps must be correct to meet the requirements of the Law in respect to dazzle and also to provide good illumination for the driver. Headlamp alignment, as well as lamp condition, forms a part of the annual M.O.T. Test.

Although the lamps may be checked by observing the illumination pattern on a vertical screen, most garages use special aligning equipment to achieve greater accuracy.

Special equipment   Figure 7.18 shows one type of optical beam setter. This equipment checks horizontal and vertical aim and enables the lamps to be set accurately. Adjusters are provided at each lamp to alter the setting.

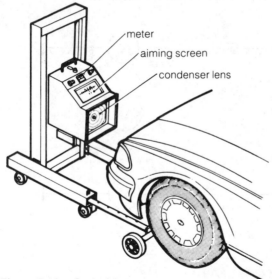

Figure 7.18   *Optical beam setter*

Initially the aligner is set level and positioned parallel with the front of the car. When the lamps are switched-on, light rays from the lamp pass through a condenser lens and are reflected by a mirror on to a small screen.

Most lamps, other than the British–American type having a symmetrical beam and identified by the number '1' or '1a' moulded on the lens, are set to dip beam when aligning the lamps.

Without special equipment   This method requires the vehicle to be positioned on level ground at a given distance in front of a vertical screen set parallel with the headlamps (Figure 7.19(a)). One method recommended for the Ford Escort is as follows:

1. Position car 10 metres (33 ft) from aiming board.
2. Ensure that tyre pressures are correct.
3. Bounce car to settle suspension.
4. Mark out aiming board as shown in Fig. 7.19(b). The distance x depends on the vehicle, e.g. Escort saloon, 130 mm.
5. Mark centres of front windscreen and rear window with wax crayon and position car so that it is aligned with the centre line of the aiming board.
6. Switch on dipped beam and cover one lamp.
7. Adjust horizontal and vertical alignment to give light pattern as shown in Figure 7.19(a).

Auxiliary driving lamps can also be aligned by using this method. The beams are deflected downwards a small amount, e.g. the distance x is about 180 mm when measured at 10 metres from the lamp.

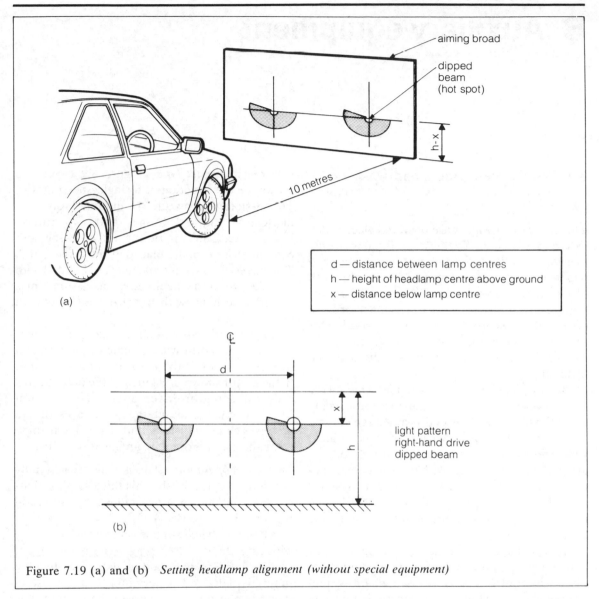

aiming broad

dipped beam (hot spot)

h-x

10 metres

(a)

d — distance between lamp centres
h — height of headlamp centre above ground
x — distance below lamp centre

d

x

h

light pattern right-hand drive dipped beam

(b)

Figure 7.19 (a) and (b) *Setting headlamp alignment (without special equipment)*

8 Auxiliary equipment

8.1 Windscreen wipers and washers

Wipers

The majority of wipers are operated electrically. Whereas in the past a vehicle was fitted with only one wiper, today it is common practice to use two wiper blades for the front windscreen with both blades driven from a single motor. The Law requires the wiper on the driver's side to operate effectively. Hatchback cars often have a wiper for the rear window and some cars in the more expensive range also have wipers fitted to the headlamps.

The force needed to drive a rubber wiper blade across a glass surface is considerable, especially when the blade has to sweep away a large volume of water or snow. Most modern vehicle windscreens have a double curvature, so long articulated wiper blades with the ability to flex to the contour of the glass are needed. Systems often have two wipe speeds to suit the driving conditions and in addition, an intermittent wipe facility is provided.

Modern requirements for a car wiper motor demand a high-powered quiet unit that operates on a current of 2–4 A. In the past, shunt-wound motors were used but the introduction of powerful magnetic materials during recent years has made the permanent-magnet motor the type commonly used today.

Figure 8.1 shows the layout of a typical wiper system. A worm on the armature drives a worm wheel which is connected to a crank to provide the reciprocating action needed to oscillate the wiper blades. The gearing gives the speed reduction and the torque increase needed to drive the wiper blades.

Permanent-magnet type Figure 8.2 (overleaf) shows the construction of a single-speed motor. The 8-slotted armature is mounted on self-lubricating sintered bushes and two carbon brushes, set 180° apart, rub on an 8-segment commutator normally placed at the driving end. Two strong permanent magnets are bonded with an adhesive to the steel yoke; this is sometimes coated externally with a non-ferrous metal to resist corrosion.

A steel worm formed on the end of the armature drives a worm wheel, made of plastics, at a speed of about 1/10th the speed of the armature. In the motor shown in Figure 8.2 the output drive is by a pinion gear driven directly by the worm wheel. Rubber seals at the joint faces of the motor exclude moisture and a polythene pipe vents the gases formed by arcing at the brushes.

Two-speed operation This is achieved generally by using an extra brush. This third brush is thinner than the main brushes and is set as shown in Figure 8.3(a) (overleaf).

When the switch supplies current to 'B', a low wipe rate of about 50 wiping cycles per minute is obtained; this is increased to about 70 when the supply is delivered to terminal 'C'. The rise in speed is due to an increase in the current flow through the motor.

When brushes 'A' and 'C' are used fewer armature windings are involved so the lower resistance gives a larger current flow and a higher rotational speed. As the speed is increased a rise in back-e.m.f. reduces the current flow. The shorter armature path between brushes 'A' and 'C' is shown in Figure 8.3(b). Figure 8.3(c) shows the interconnection of the coils of a 'lap wound' type of armature normally used for a wiper motor.

Figure 8.1 *Layout of a typical wiper system (simplified link-type drive)*

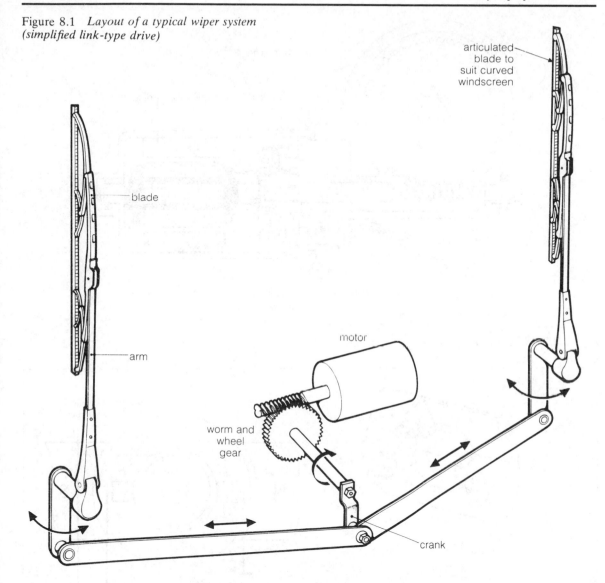

articulated blade to suit curved windscreen

blade

arm

motor

worm and wheel gear

crank

High speed should not be used when there is a heavy load on the wiper blade, e.g. in heavy snow or on a windscreen which has been swept clear of water and is dry.

Self-switching action When the wiper is not required, the blades should be set so that they are at the end of their wiping stroke. The driver finds it difficult to stop the blades in this position so a *limit switch* is fitted to achieve this requirement. This automatic switch is controlled by the gear-

box of the wiper motor and is arranged to open only when the wiper blades are at one end of their stroke.

Figure 8.4 (overleaf) shows the principle of the limit switch. If the driver switches-off the motor in any position other than that shown, the limit switch continues to supply current until the 'park' position is reached.

Even with this switch, the blades do not always come to rest at the correct place owing to the momentum of the moving parts. This problem is

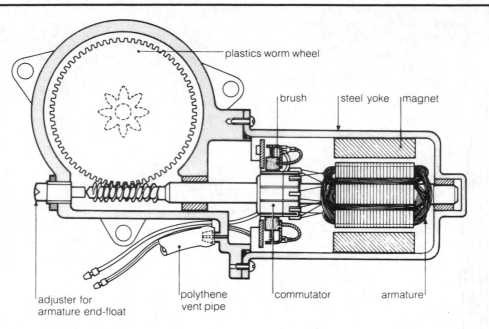

Figure 8.2   *Single-speed motor*

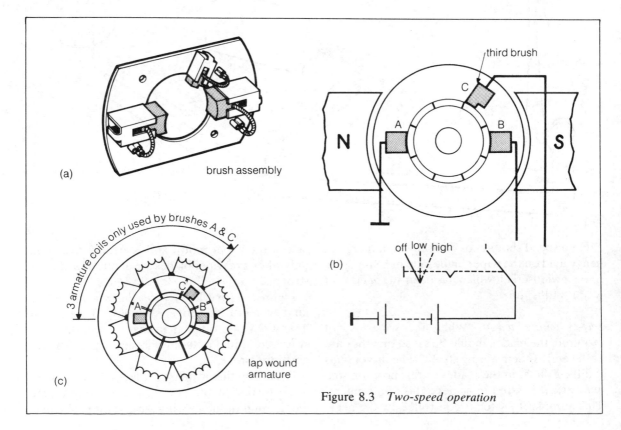

Figure 8.3   *Two-speed operation*

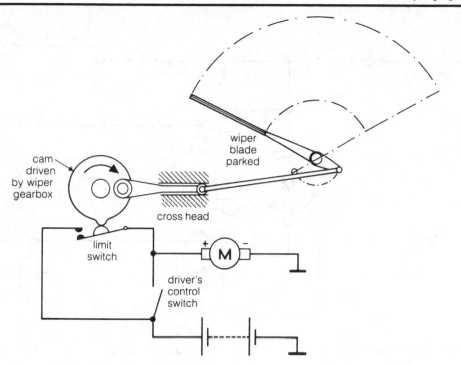

cam
driven
by wiper
gearbox

wiper
blade
parked

cross head

limit
switch

driver's
control
switch

Figure 8.4 *Limit switch to give self-switching action*

overcome by using an action called *regenerative braking*.

When the driver has switched-off the motor, another set of contacts on the limit switch is arranged to connect the two main brushes together (Figure 8.5). At this point the current generated by the moving armature creates a load on the armature which gives a braking action and quickly brings the motor to rest.

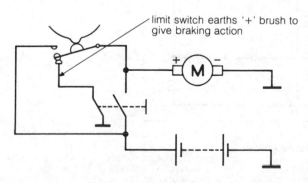

limit switch earths '+' brush to give braking action

Figure 8.5 *Regenerative braking*

Intermittent wipe Spray from passing vehicles and light drizzle conditions require the screen to be wiped every few seconds, so most vehicles have a switch position to provide this facility.

To overcome the regenerative braking provision on a permanent magnet-type motor, a current pulse of comparatively long duration is needed to rotate the armature sufficient to move the limit switch from its 'braked' position. Most vehicles use a semiconductor-controlled relay to provide this function; the time period between wipes is governed by the action of a capacitor. This time constant is governed by the resistance–capacitance (R-C) of a circuit, so by varying either 'R' or 'C' the interval can be varied to suit the requirement.

Figure 8.6(a) (overleaf) shows one electronic circuit layout which gives an intermittent wipe action.

The diagram shows the two main brushes interconnected through the two switches and relay contacts (1); regenerative braking takes place when the contacts are set in this position.

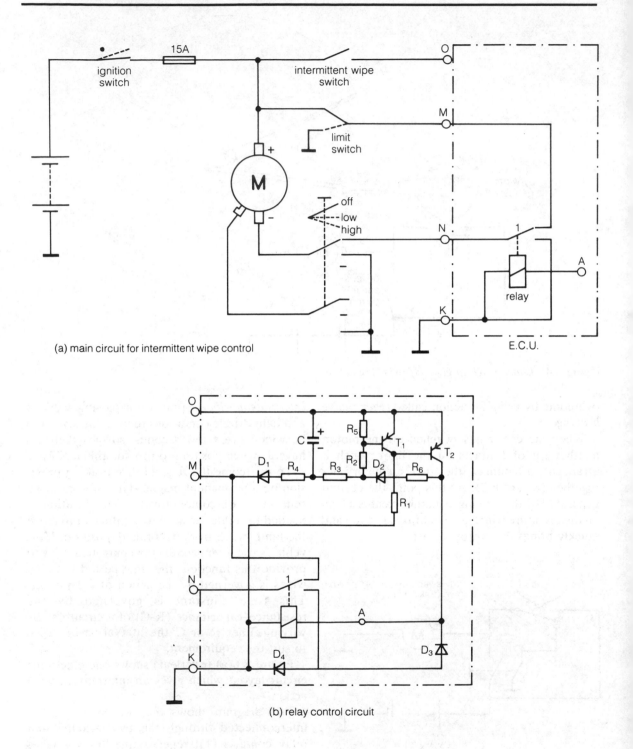

(a) main circuit for intermittent wipe control

(b) relay control circuit

Figure 8.6 *Intermittent wipe control*

When current is supplied to terminal (A), the relay is energized and the contacts are closed. This connects the negative brush to earth and causes the motor to operate, irrespective of the position of the limit switch.

If at this stage, the supply is disconnected from (A), the relay will open and the contacts (1) will close. Being as the limit switch is earthed, the motor will continue to operate until the earth contact at the limit switch is broken.

The control circuit for the relay is shown in Figure 8.6(b); the operating sequence is commenced from the point where the intermittent wipe switch is closed. The sequence is:

1. Current flows from the switch through the base of T_2 to earth via R_1. This switches-on T_2 to energize the relay and start the motor.
2. After a time the motor moves the limit switch to the earth position. Current from T_2 will now pass to the limit switch via R_6 and will cause the relay to de-activate; this closes the relay contacts '1' and provides an alternative path from the negative brush to earth. In consequence the motor continues to operate.
3. When the limit switch, makes its earth contact, current passes through the base of T_1; this switches-on T_1 and switches-off T_2. During this stage the p.d. across the capacitor causes it to charge-up.
4. Further rotation of the motor moves the limit switch to the stop position; this causes the motor to stop abruptly. Current flow through R_4 from T_1 now ceases but T_1 is prevented from switching-off by the discharge current from the capacitor. This current gives a flow in the sub-circuit incorporating the base of T_1 together with R_2 and R_3.
5. It takes about five seconds for the capacitor to release its charge so after that time T_1 will switch-off and T_2 will switch-on to repeat the cycle.

On some vehicles intervals can be varied to suit the conditions. This provision can be made by fitting a variable resistor control in the capacitor-discharge sub-circuit in place of resistor R_3.

Self-parking wipers On some vehicles the wiper blades are parked off the windscreen. This provision can be achieved by switching the circuit so that after the motor has stopped the current through the armature is reversed. When the brush polarity of permanent magnet motor is changed, the armature rotates in the opposite direction.

By arranging the gearbox linkage so that reverse motion extends the wiping stroke, the movement is made to park the wiper blades well away from the glass screen.

Wound motors These are seldom used today in view of the superiority of the permanent magnet type in respect of power, noise, efficiency, cost, reliability and current consumption.

Figure 8.7(a) (overleaf) illustrates the layout of a single-speed motor having a shunt wound field; Figure 8.7(b) shows the circuit when a limit switch is fitted to provide a self-switching facility.

Two-speed operation Although this could be achieved by switching-in a resistor in the battery feed line, the loss of efficiency due to heat loss at the resistor makes it unsuitable.

Figure 8.8 (overleaf) shows one arrangement for obtaining a two-speed operation. When 'low-speed' is selected, the current passing through the field winding is divided between the armature and the field.

Moving the switch to 'high-speed' inserts a resistor in the shunt field. This causes a larger current to flow through the armature and this results in an increase in the motor speed.

Overload protection Under snow or ice conditions the load on a motor becomes excessive; this causes it to slow down or in extreme conditions it stops. A decrease in armature speed reduces back-e.m.f.; this lower opposition allows a large flow of current in the order of 11 A through the motor and leads to overheating and possible damage to the motor.

Protection is normally given by incorporating a thermal switch in series with the supply lead. The switch is controlled by a bi-metallic strip; when

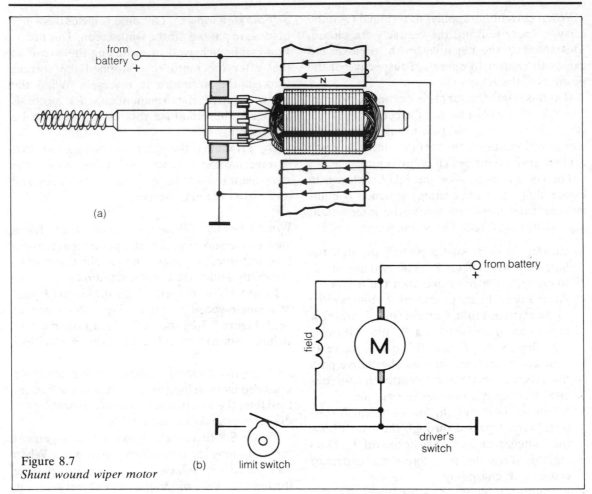

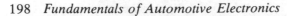

(a)

Figure 8.7
Shunt wound wiper motor

(b)

limit switch

driver's switch

from battery

field

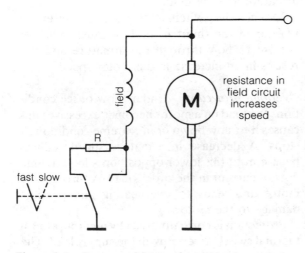

Figure 8.8   *Two-speed wiper motor circuit*

resistance in field circuit increases speed

R

fast slow

the strip is heated by a higher-than-normal current, the contacts are opened.

Mechanical-drive systems

For accommodation reasons the motor is situated remote from the wiper blades. This means that a mechanical drive must be used to transfer the motion to the blades. The two main systems used are:

- Link
- Flexible rack.

Link system   Figure 8.1 shows the layout of a link system. This efficient system uses a crank on the output shaft of the motor to reciprocate a transverse link. This drives the levers and partially rotates the shafts on to which the wiper arms are connected. The relative lengths of the levers control the angle of sweep of the wipers. Self-lubricating bushes are normally fitted at each connection.

Flexible rack This system is more compact and quieter than the link system. Also it allows the motor to be situated in an accessible place, normally under the bonnet.

Figure 8.9 shows a crank pin on a worm wheel driving a rod which connects with, and reciprocates, a flexible rack contained in a rigid tube. The rack is similar to a speedometer cable except that it is wrapped with a wire to form a 'thread'. Drive from the rack to the wiper is by means of a pinion which engages with the rack teeth. Each pinion is held in a wheelbox (or gearbox), the casing of which is screwed to the rigid tube.

Washers

Statutory regulations require that a screen washer must be fitted to clean the driver's side of the windscreen. Today most vehicles fit an electrically-operated pump to supply water or cleaning fluid to two or more jets that spray the windscreen. On some vehicles an extra pump is fitted to supply a headlamp wash system; some of these vehicles are also fitted with headlamp wipers.

The small centrifugal pump is either fitted directly on to the water reservoir or mounted in

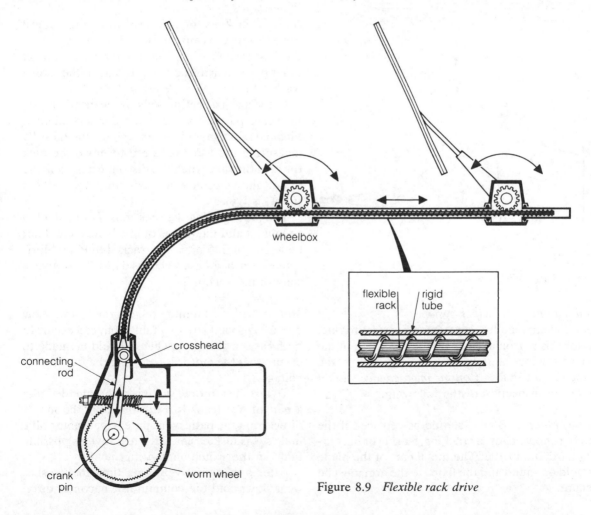

Figure 8.9 *Flexible rack drive*

the hydraulic line. The pump is driven by a permanent-magnet motor controlled by a switch that is often operated from the wiper switch stalk on the steering column (Figure 8.10). The pump is self-priming and is protected by a filter at the inlet. Polythene tubing is used to supply the jets. A typical motor consumes about 3 A and supplies about 0.75 litre/min at a pressure of 0.67 bar (10 lbf/in²).

In the winter, a small quantity of methylated spirit added to the water lowers the freezing temperature.

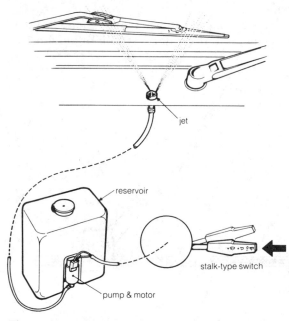

Figure 8.10   *Windscreen washer*

Maintenance of a wiper system
Good clear visibility is essential for safe driving, so for this reason a check on the operation of the driver's wiper and washer is included in the annual M.O.T. test. *Routine maintenance* should cover an inspection of the following:

Wiper blades   Blades should be replaced if the rubber shows signs of cracking, tearing or becoming hard and brittle. The metal part of the blade should be sound and the fixing to the arm must be secure.

A wiper motor must not be operated when the screen is dry; this overloads the motor and also severely scratches the surface of the glass.

Screen   Traffic film can be removed from the screen and wiper blades by using methylated spirit. Polishes containing silicone and wax should not be allowed to contaminate the screen or blade surface.

Wiper arms   These should be checked to ensure that the spring is serviceable and is applying sufficient force (generally about 350 gram) to the blade. The arm should not be bent because this can cause the blade to chatter on one stroke.

Wiper faults
Noise   Slackness or tightness of the mechanical drive system can cause noise. Also noise occurs in the link system when the moving parts contact other parts such as the metal tubing of the screen washer tube.

If a visual inspection fails to locate the noise, then each part should be checked independently. Flexible racks should be checked for tightness by measuring the force required to move the rack through the tube when it is disconnected from the motor and wheelbox; a maximum force of 27 N (6 lbf) is typical.

The tube holding the rack must not be dented or kinked and the radius of any bend should not be less than 230 mm. The rack should be lubricated with a grease such as H.M.P. to give a smooth movement.

Motor faults   'Failure to operate' and 'low operating speed' are two faults that can occur. In both cases a voltmeter check should be made to ensure that the motor is receiving the full battery voltage.

To test the motor *in situ* it is recommended that a pair of test leads is used to supply the motor direct from the battery; a spare wiper motor plug makes this task easier. This test reveals possible faults in the switch and wiring.

After a comparatively long time the brushes wear down and the commutator becomes dirty.

On many models brush replacement is necessary when the main brushes are worn to a length of less than 5 mm, or the stepped part of the third brush has worn away. New brushes, complete with springs and plastic mounting plate can normally be obtained.

The commutator should be cleaned with a petrol-moistened rag or a strip of glass-paper if the surface is badly blackened.

Some motors have a screw for adjustment of the armature end-float; a typical setting is 0.2 mm (0.008 in).

8.2 Fan and heater motors

Motors used to operate the engine-cooling fan and the heating ventilation system are normally of similar construction, in many cases the same model.

The motor normally used is a 2-pole permanent-magnet type having two brushes set at 180°. This type is also used for window operation, seat adjustment and many other general applications. Since this motor must operate in both directions the brushes are positioned at the mean magnetic axis.

Brush position
Poor efficiency and excessive sparking between the brushes and commutator will arise if the angle between the brushes and magnet poles is incorrect. This sparking occurs at the point of commutation, i.e. when the brush contact changes from one commutator segment to the next.

The simplified diagrams (Figure 8.11(a)) shows how the brush position is affected by the armature current.

In Figure 8.11(a) no armature current is flowing and in the diagram the brushes are positioned at 90° to the magnetic axis; this position is called the *geometric neutral axis* (g.n.a.).

When current is supplied to the armature, a secondary field is set up around the conductors and this distorts the main field (Figure 8.11(b)). The distortion moves the axis of the main field away from some of the armature conductors and this lowers the efficiency of the motor. To over-

come this problem the brushes are moved through an angle to a position called the *magnetic neutral axis* (m.n.a.). The position shown in Figure 8.11(c) depends on the magnitude of the current but because the current varies with the speed, the final position is a compromise.

Although the brush position BB is suitable for the direction as shown, a reversal of the armature current to give an opposite direction of rotation

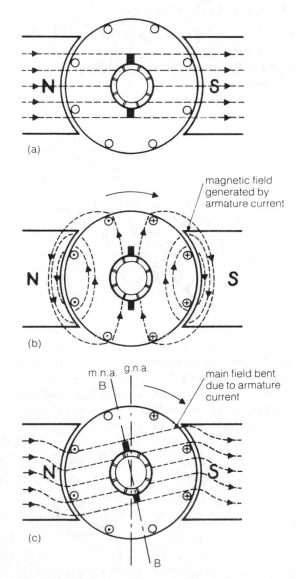

Figure 8.11 *Brush position*

requires the brush axis to be moved through an angle in a clockwise direction.

Due to this problem, a reversible motor requires the brushes to be set at the mean of the two m.n.a. positions. Consequently the efficiency of a reversible motor is lower than a unidirectional unit.

Cooling fan

A cooling fan driven by an electric motor instead of a belt gives the advantages:

- Energy saving – fan can be switched-off when not needed
- Easy to accommodate, especially for transverse engines
- Close control over engine operating temperature.

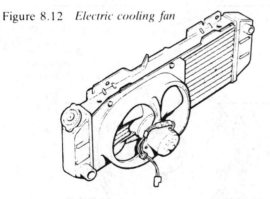

Figure 8.12   *Electric cooling fan*

Figure 8.12 shows a typical installation. A plastics fan is fitted to the armature shaft of the motor and the assembly is situated to provide the appropriate air movement through the radiator.

The control circuit (Figure 8.13) uses a bimetal type thermal switch located at the radiator side of the thermostat housing. This switch operates the fan when the coolant temperature reaches about 90°C and switches it off when the temperature drops to about 85°C. In the layout shown, the current for the motor is supplied from an ignition switch relay. This relay energizes when the ignition is switched-on.

Heating and ventilation fan

Most systems use a centrifugal-type fan to boost the air flow into the interior of the vehicle (Figure 8.14). Variable motor speeds are generally required and this is achieved by changing the voltage applied to the motor either by a variable resistor or by a resistor network.

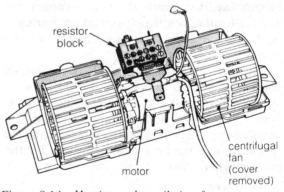

Figure 8.14   *Heating and ventilation fan*

Figure 8.15 shows a circuit for a 3-speed operation. Moving the switch through the three positions – low, high and boost – shorts-out a resistor at each stage and steps-up the applied voltage.

Maintenance

Each system is fused so this is the first check to make when the system fails to operate. If this is

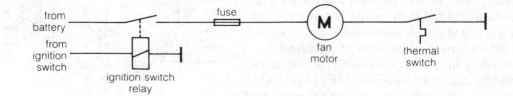

Figure 8.13   *Control circuit for engine cooling fan*

cle.

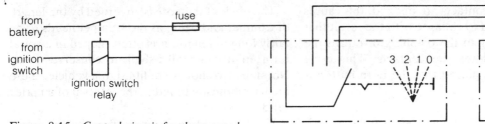

Figure 8.15 *Control circuit for three-speed operation of heating and ventilating fan*

heater blower switch   heater blower motor

not the cause then a voltmeter check should be made to ensure that battery voltage is applied to the motor when the switch is set to give maximum fan speed.

In the case of a fan motor a quick check of the thermal switch can be made by disconnecting the leads at the switch and earthing the lead from the motor.

8.3 Signalling equipment

Horns
The Law requires a motor vehicle to be fitted with an audible warning device which emits a continuous note that is not too loud or harsh in sound.

Most horns are operated electrically and the note is obtained either by magnetically vibrating a diaphragm or by pumping air past a diaphragm into a trumpet.

The note produced should be musical rather than a noise likely to cause annoyance to the public. To give a pleasing and penetrating note two horns are often fitted: one emits a high-pitch note, to overcome traffic noise, and the other a low pitch, to carry over a distance. The pitch of a note is its frequency, i.e. the number of oscillations made per second. The unit of frequency is the *hertz* (Hz); one oscillation per second is one hertz.

There are three types of electric horn; they are:

- high frequency
- windtone
- air horn

High-frequency horn (HF) This comparatively cheap type of horn has been in use for many

years. An electro-mechanical action vibrates a steel diaphragm at about 300 Hz which causes a tone disc to oscillate at about 2000 Hz. The combined sound produces a highly penetrating note that meets the normal requirements of a horn.

Figure 8.16 shows how the oscillatory movement of the diaphragm is produced by the action of a magnetic field winding on an iron core.

When the horn button is pressed, the closed contacts allow a current of about 4 A to flow around the field coil; this generates a magnetic flux which attracts the armature. After the armature has moved a given distance as set by the

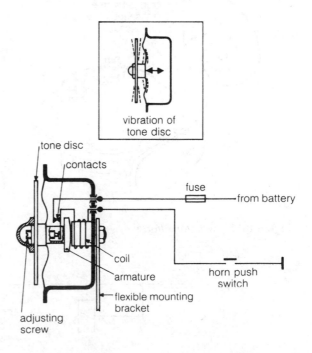

Figure 8.16 *High-frequency horn and circuit*

adjustment, the contacts are opened; this causes the magnetic field to collapse with the result that the natural spring of the diaphragm returns the armature and closes the contacts. This cycle continues for the period that the horn button is pressed.

Windtone horns   This type uses a similar electrical operating system to the HF horn for moving the diaphragm, but instead of oscillating a tone disc, a windtone horn vibrates, or resonates, a column of air contained in a trumpet (Figure 8.17). The sound is produced in the trumpet in a manner similar to that used in a wind instrument.

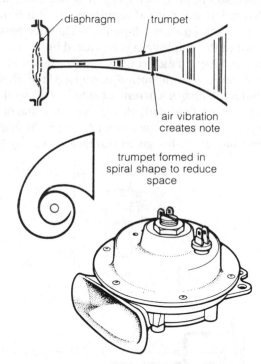

Figure 8.17   *Windtone horn*

The pitch of the note is governed by the length of trumpet and being as the trumpet needs to be fairly long in length, it is often shaped in a spiral form, or like a snail's shell, to conserve space. Normally two horns are fitted to a vehicle; these are harmonically tuned to the interval of a major third.

Horns should be mounted flexibly; this limits the transmission of external shocks that would otherwise affect the quality of the horn note. The majority of horns, especially the windtone types, consume a total current in excess of 10 A, so to prolong the life of the horn switch a relay is generally fitted (Figure 8.18).

Air horns   An air horn consists of a trumpet through which air is forced by means of an electrically-driven air pump. Vibration of the air column in the trumpet is initiated by a diaphragm valve positioned at the end of the trumpet.

Operation of the horn switch causes the motor-driven air-compressor pump to discharge air into the pressure chamber in the horn. The air pressure deflects the centre of the diaphragm and this allows some air to escape into the trumpet. As this occurs the slight pressure drop in the chamber closes the diaphragm valve and the cycle is repeated. The rate of vibration of the diaphragm combined with the trumpet length governs the pitch and quality of the note emitted.

Maintenance   Horns are often positioned in an exposed place behind the radiator grill, so special attention is needed to avoid the ingress of moisture.

Most faults are due to cable and connector problems, so when a fault is present a visual check should be followed by a voltmeter test to determine that the p.d. applied to the horn is correct.

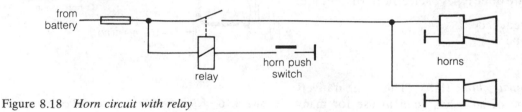

Figure 8.18   *Horn circuit with relay*

Adjustment   After a long period of time the horn contacts may require adjustment, so an adjusting screw is provided to alter the contact setting. The position of the screw is varied until the horn consumes the current specified for the particular type; a typical current consumption for a HF horn is 4 A.

Directional indicators and hazard warning systems

Pre-warning that the driver intends to turn or overtake is signalled to other road users by the flashing of amber-coloured lights. The law requires vehicles made after 1986 to have three lamps on each side that flash simultaneously at the rate of 60–120 flashes per minute. In addition a tell-tale indicator for the driver must be fitted.

Each directional indicator light must be sited so that it is visible through a given angle, the light unit must carry the appropriate 'E' marking for its position and the bulb must be rated at 15–36 W.

Figure 8.19 shows the layout of a typical directional indicator circuit. When the switch is moved to the left or right, current is supplied to the appropriate lamps. Regular interruption of the current to give a flashing light is performed by

the flasher unit: this is situated on the battery side of the switch.

If the vehicle breaks-down on the highway the driver should be able to warn other drivers by arranging all the directional indicator lamps to flash simultaneously. This hazard-warning feature is activated by a separate switch by the driver (Figure 8.20) (overleaf).

There are three types of flasher in common use:

- thermal
- capacitor
- electronic

Thermal-type flasher   This type uses the heating effect of an electric current to bend or extend a metal strip. One type uses two bi-metallic strips; each strip is wound with a heating coil and fitted with a contact. When the switch is operated, the strips bend to open and close the contacts.

Another type of thermal flasher is the vane type. The Lucas 8FL uses a vane construction and this gives a compact, reliable and cheap unit (Figure 8.21) (overleaf). It consists of a rectangular, snap-action, spring-steel vane supported at a point midway along the longer side. A thin metal ribbon, diagonally connected to the corners of

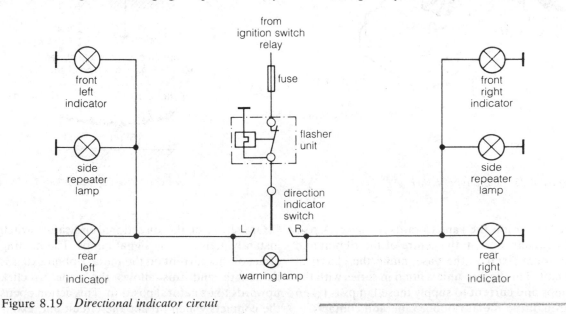

Figure 8.19 *Directional indicator circuit*

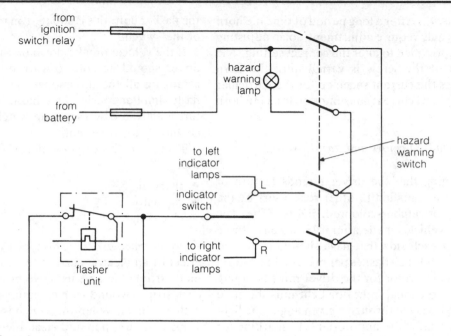

Figure 8.20 *Hazard warning circuit*

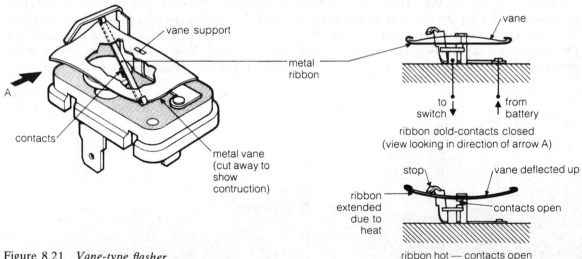

Figure 8.21 *Vane-type flasher*

the vane, pulls the vane towards the base. A pair of contacts, one on the centre of the ribbon and the other fixed to the base, make the electrical circuit. This flasher unit is fitted in series with the lamps and current to supply these lamps is taken through the metal ribbon, vane and contacts.

Operation of the directional indicator switch instantly activates the signal lamps. The heating effect of this current on the metal ribbon extends its length and this allows the vane to click upwards to its natural position. This action opens the contacts which breaks the circuit and extinguishes the signal lights. A short time after this action the lack of heating current causes the ribbon to cool and contract; this clicks the vane downwards and once again closes the contacts to repeat the cycle of events.

The time taken to heat the vane depends on the current, so the flash frequency is governed by the lamp load. Failure of one lamp reduces the electrical load so this prevents the vane from operating; as a result the remaining lamps stay on continuously.

Capacitor type Some systems use the timer action of a capacitor while it is performing its charge–discharge cycle to trigger a relay which in turn controls the switching and flashing of the signal lamps. The improved performance of electronic control has resulted in the development of the capacitor-type into a new design called an electronic type.

Electronic type Turn-signal flasher units of the electronic type are more efficient than the thermal pulse generator types such as the vane flasher unit. The electronic type meets international standards which include the provision of an audible and visual warning system to signal when a bulb has failed.

Bulb failure in a system fitted with an electronic flasher is indicated by arranging the lamps to flash at twice the normal rate or by using an extra warning lamp.

The type shown in Figure 8.22 can handle a directional indicator signal load of up to 98 W without altering the flash frequency; in addition it provides a hazard warning signal for many hours of continuous use.

Electronic flasher units normally use an electromagnetic relay to control the current to the signal lamps. This method is preferred to transistor switching because in addition to giving an audible signal it is not affected by the high-voltage spikes that are generated during the switching operation. Also the relay contacts give very little voltage drop. Compared with the transistor, the drop across the relay contacts is about one-tenth of that of a transistor. Although tran-

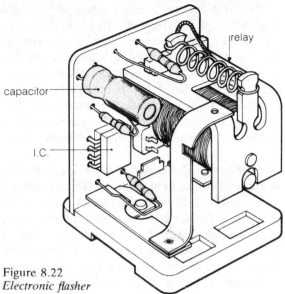

capacitor

I.C.

relay

Figure 8.22
*Electronic flasher
unit; Lucas FL19*

sistor switching is ideal for high-speed applications, the relay is still preferred when slow-speed, heavy-current switching is required.

The basic construction of a typical flasher unit such a Lucas FL19 consists of a printed circuit board which carries an I.C., capacitor, relay and three resistors (Figure 8.23) (overleaf).

The I.C. chip has three main sections; an oscillator, relay driver and lamp-failure detector. A Zener diode in the I.C. regulates the operating voltage of the chip to ensure that the flash frequency remains constant over a supply voltage range of 10–15 V.

Timer control for the oscillator is achieved by using the charge–discharge action of the capacitor C; this operates in conjunction with the resistor R_1 to give a R-C time constant for a flash frequency of 90 per minute with a 50–50 off–on signal time (see page 36).

Pulses from the oscillator are passed to the relay driver which is a Darlington amplifier; this provides the current pulses to energize the relay coil. Transient protection of the output transistor is achieved by fitting a diode across the collector–emitter of the power transistor; this allows the self-induced charges in the relay coil to bypass the transistor.

The lamp-failure detector senses the voltage drop across the resistor R_2. This resistor senses the current passing to the signal lamps via the relay contacts. In the event of a lamp failure, the lower current flow will cause the voltage drop (IR drop) across the resistor to decrease. This reduced voltage will cause the detector to alter the resistance of the R-C time element and as a result the frequency of flashing will be doubled.

Maintenance   The common faults are bulb failure and loss of earth at the lamp. These faults are diagnosed by observing the rate of flashing, but the type of flasher governs whether the rate will be slower or faster. Earth problems are often caused by a bad connection of the earth cable and this can normally be detected by a visual check.

Assuming the bulbs are serviceable, a system that does not flash is due to an open-circuit in the supply line; this can be pin-pointed by using a voltmeter.

A flasher unit that fails to operate when full voltage is applied to it is identified as being faulty when a substitute unit fitted in the circuit shows that the remainder of the circuit is serviceable.

8.4 Window winding and central door locking

Central door locking and 'electric' windows are two features which, in the past, were restricted to luxury cars but nowadays are becoming quite common across the range of cars.

Electrically-operated windows
Most electric-window applications use a d.c. permanent magnet motor for each window; this is operated via a three-position rocker switch to enable the polarity to be changed to give up-and-down motion of the window. The driver's panel has four main window switches, one for each window, and an isolation switch to disconnect the supply to the rear windows.

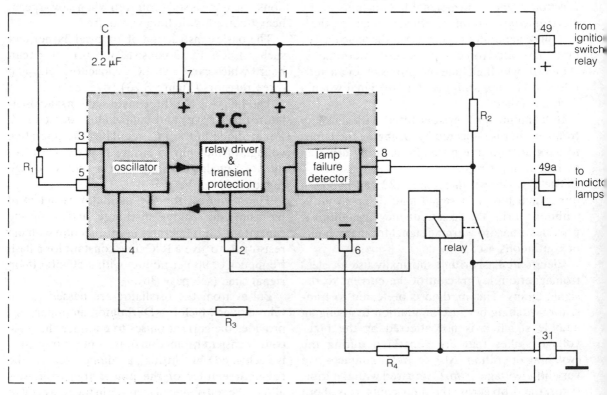

Figure 8.23   *Electronic flasher unit circuit*

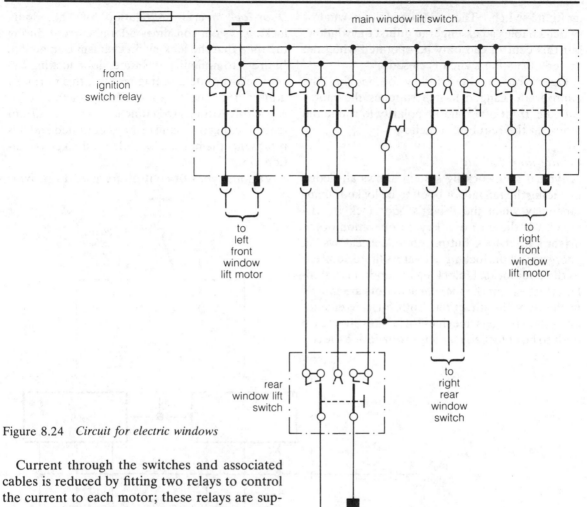

from
ignition
switch relay

main window lift switch

to
left
front
window
lift motor

to
right
front
window
lift motor

rear
window lift
switch

to
right
rear
window
switch

Figure 8.24 *Circuit for electric windows*

M

left rear window
lift motor

Current through the switches and associated cables is reduced by fitting two relays to control the current to each motor; these relays are supplied by a common feed. Since the signal current for a relay is low, the size of cable for this current is considerably smaller than that needed for a direct supply.

Drive between the motor and the window glass is by means of a gearbox; this amplifies the torque sufficient to raise the window which is more difficult to undertake than the downward motion. The output gear of the gearbox drives either a flexible rack or acts directly on to the window winding mechanism similar to a manual system.

One or more thermal cut-out switches are fitted in the circuit, sometimes in the motor, to limit the current in the event of overload. The cut-out is opened if the operating switch is held closed when the window reaches its limit of movement or in a case where ice prevents free movement of the glass. When a main overload switch is fitted it is often a type that requires resetting after the circuit has been exposed to an overload situation.

Figure 8.24 shows a circuit for electric operation of a rear-passenger window; the remainder of the circuit has been omitted for simplicity. The motor in this layout is supplied directly via the ignition switch. The additional rear-window switch enables a passenger to adjust the window, but this control can only be operated when the driver's isolation switch is closed.

Operation of the window by moving the appropriate ganged switch supplies the motor with a current of a suitable polarity to rotate the motor in the required direction.

Central door locking

A typical door-locking system allows all doors, including the tailgate or boot, to be locked simultaneously when the driver's door lock is activated. On the turn of a key, or operation of the driver's door-lock button, the electrical system energizes all the locking activators fitted adjacent to the door locks. Unlocking the door has a similar effect except the locking actuators are moved in the opposite direction. Both for convenience and safety reasons, mechanical latches allow each door to be unlocked manually from inside the car.

Door-lock relays Activation of the door-locking system consumes a large current during the time that the locking mechanism is in action. In order to minimize this time, door-locking circuits incorporate a timing feature; this normally utilizes the charge or discharge action of a capacitor. After a given time interval, the current to the locking mechanism is discontinued and this represents the normal locked or unlocked condition.

Various types of control are used to provide

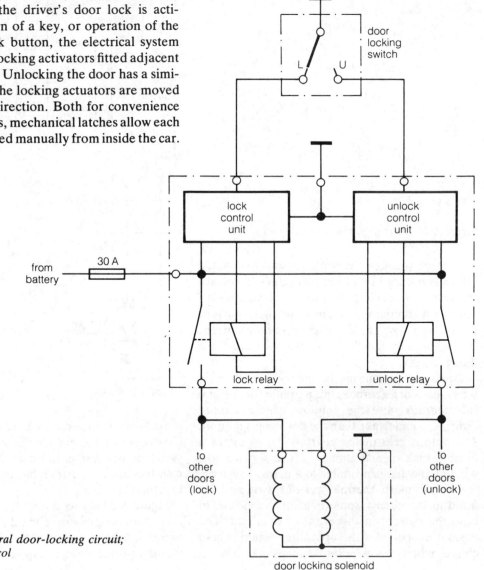

Figure 8.25 *Central door-locking circuit; transistorized control*

the locking/unlocking pulse to operate the actuators. One arrangement uses two relays: one for locking and the other for unlocking the doors. These relays are controlled by a transistorized switching circuit, which is timed by the charge–discharge action of a capacitor to give a current pulse length sufficient to activate the locks (Figure 8.25).

An alternative arrangement uses the discharge current from a previously charged capacitor to energize the relay. When the door key is turned and the appropriate switch is closed by the key movement, the capacitor releases its charge through the relay coil. After the capacitor has been fully discharged the relay opens and the system goes out of action (Figure 8.26).

Door-lock actuators The electrical methods used for actuating the actual door-lock mechan- ism are electromagnet solenoids, linear motors or permanent magnet rotary motors. In each case the reversing action is achieved by changing the polarity.

Figure 8.27 (overleaf) shows one arrangement for operating a door lock.

External circuit The layout of the complete cir- cuit is shown in Figure 8.28 (overleaf). In this case the driver's door switch operates the locks on four other doors.

Maintenance
Faults in actuator circuits should be diagnosed by using a systematic approach. A system such as a central door-locking arrangement, should be ini- tially checked to isolate the fault to a particular area before the trim is removed to expose the

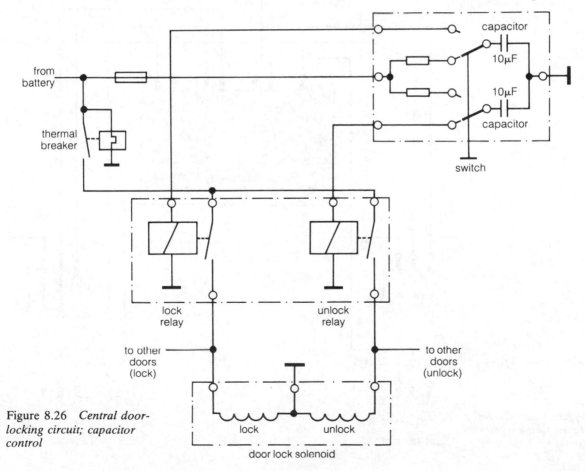

Figure 8.26 *Central door-locking circuit; capacitor control*

Figure 8.27
Door-locking actuator

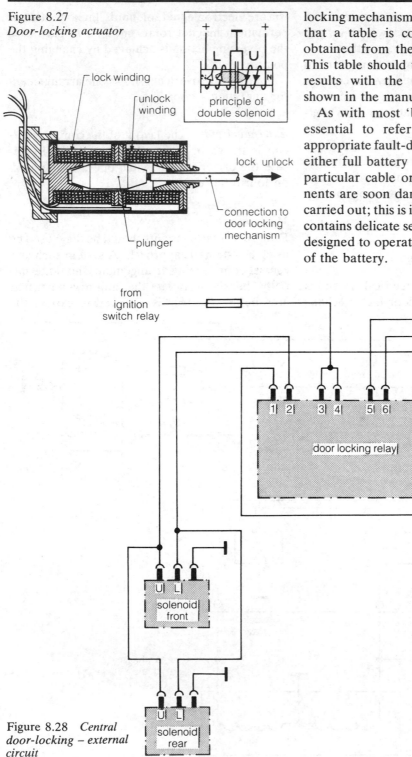

lock winding

unlock
winding

principle of
double solenoid

lock unlock

connection to
door locking
mechanism

plunger

locking mechanism. Many manufacturers suggest that a table is compiled to show the results obtained from the door-lock switching action. This table should then be used to compare the results with the fault-finding chart normally shown in the manufacturer's service manual.

As with most 'black-box' components, it is essential to refer to the circuit diagram or appropriate fault-diagnosis chart before applying either full battery voltage or an ohmmeter to a particular cable or test pin. Expensive components are soon damaged if haphazard testing is carried out; this is important especially if the unit contains delicate semiconductors or I.Cs that are designed to operate at a voltage well below that of the battery.

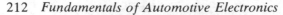

from
ignition
switch relay

door locking relay

solenoid
front

solenoid
rear

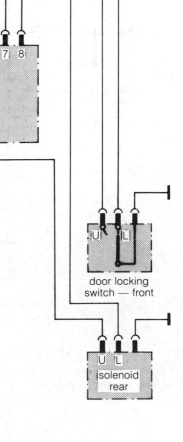

door locking
switch — front

solenoid
rear

Figure 8.28 *Central
door-locking – external
circuit*

An example of a check-out sequence as applied to a system similar to that shown in Figure 8.28 is as follows:

All locks fail to operate This suggests that the fault is due to:

- No power to the relay unit
- Defective switch
- Faulty relay unit

Each one of these possible faults should be tested to narrow-down the actual cause. The following method can be used:

1. After checking the circuit breaker, the input voltage applied to the relay unit pins 3 and 4 should be checked.
2. Assuming the supply is good, then a jumper lead should be used to connect pin 5 to earth to reproduce the switch action; this will prove if the switch section of the circuit is faulty.
3. Having ascertained that the input and switch are satisfactory, the output voltage, at one of the pins 1, 2, 7 or 8, should be measured when the switch is operated. If battery voltage is not shown at the instant the switch is closed, then the relay is defective and should be changed.

This description can be reduced considerably by using a fault diagnosis chart as shown in Table XIV. Since the path followed through the chart is controlled by the result given by a particular test, then the need to read other test data is avoided. For this reason the use of charts has become commonplace for fault diagnosis of electronic equipment.

Table XIV Fault-diagnosis chart – central door-locking system (V = battery voltage)

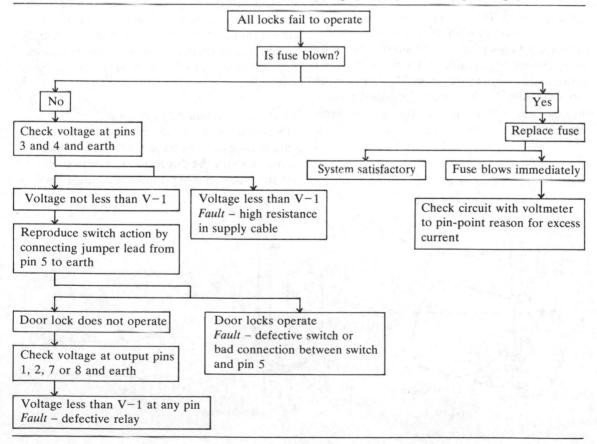

8.5 Other auxiliary equipment

Each year extra features are introduced on motor vehicles to simplify the tasks undertaken by the driver. An item classified as a luxury one year becomes a normal fitting in later years. Many of these 'extras' are controlled or driven by electricity; some of the more common items are considered here.

Electric radio aerial

Extension or retraction of the aerial is achieved by using a permanent-magnet rotary motor. Often this reversible motor acts through a reduction gear to drive a cable by means of two crimped plates (Figure 8.29). This cable, normally made of a plastics material to avoid radio interference, is joined to the top section of the aerial. When the aerial is retracted, provision is made to enable the spare cable to be wound in a coil adjacent to the driving plates.

Up-and-down motion of the aerial is obtained by using a two-way switch to supply the motor with current of suitable polarity to drive the motor in the required direction. This aerial switch can be mounted remote from the radio or can be incorporated in the radio to operate the aerial automatically.

Electrically-operated door mirrors

Adjustment of a door mirror is difficult for the driver, especially the setting of a mirror on the passenger-side door. This task is made easier by using an electrical control system.

Figure 8.30 shows a system for the control of two mirrors. Each mirror is electrically adjusted by two reversible permanent-magnet motors fitted behind the mirror. One motor controls the vertical tilt of the mirror and the other the horizontal tilt.

Each mirror is set by means of a single switch; this has a stalk with a universal movement to enable single or combined operation of the two motors.

When the switch is depressed to vertically tilt the mirror, the switch cage (A) is moved downwards and the two contacts (B) and (C) make a circuit with the '−' and '+' surfaces, (D) and (E).

Conversely, when the switch stalk is moved upwards, (B) contacts the positive surface and (C) contacts the negative or earth surface. In this switch position, the potential applied to the vertical tilt motor is opposite to that given when the switch was in the previous position. This causes the motor and mirror to move in the opposite direction.

Electrically-operated sun roof

Operation of a switch, instead of winding a handle to open/close a sun roof, is safer and more convenient for the driver, so electricity is utilized for this duty on many top-of-the-range cars.

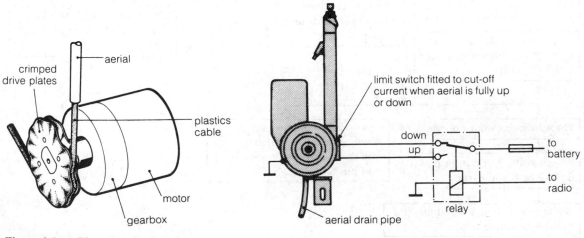

Figure 8.29 *Electric radio aerial*

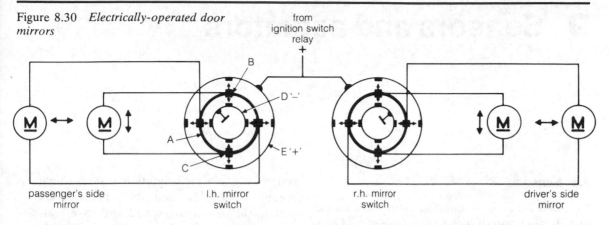

Figure 8.30 *Electrically-operated door mirrors*

Movement of the roof panel is achieved by a rack that is driven by a pinion gear. This is rotated by a reversible permanent-magnet motor after passing the drive through a gearbox to amplify the torque.

A three-position switch giving open–stop–close positions acts as a 'polarity changer' to give the required rotation of the motor and a relay adjacent to the motor reduces the current carried by the switch.

Often the gearbox incorporates an extra rack to control the switching mechanism. This uses a switch to limit the travel of the roof panel and another switch to insert a resistance in series with the motor to reduce the speed of movement as the roof panel approaches the limits of its travel.

9 Sensors and actuators

9.1 Sensors

An electrical control unit is in many ways similar to a human brain. It receives messages from various sources and after processing the information, it either instructs an actuator to perform some physical action or it stores the data away in its memory for use at some time in the future.

Electronic sensors perform the information-gathering role in this system. Each sensor feeds the Electronic Control Module (E.C.U.) with information that relates to some particular mechanical action or thermal effect (Figure 9.1).

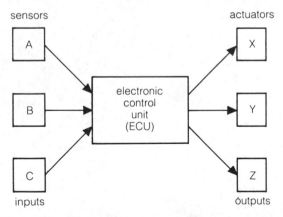

Figure 9.1   *Electronic system*

A sensor, or *transducer* as it is called when it gives an output signal proportional to the physical quantity it is measuring, converts the physical actions it notices into either an analogue or digital electrical signal. A switch is a simple type of sensor; it opens or closes to signal when a specific action has taken place. The number of sensors required to control a system efficiently depends on the factors that affect the operation of the

system, e.g. a simple ignition system shown in Figure 9.2 has three sensors to measure crankshaft position, engine speed and manifold depression; these provide signals to the E.C.U. to enable it to time the spark to occur at the correct instant. Each sensor measures one item only. When greater accuracy of spark timing is required other variables must be measured; these often include the engine temperature and the degree of knock produced during the combustion process. Extra sensors are needed to signal the changes that take place in these areas.

Sensors can be separated into two main classes; these are:

- active or self-generating
- passive or modulating

The passive type requires an external energy source to drive it and the sensor acts only as an energy controller.

Cost normally influences the accuracy and reliability of a sensor, so a typical general-purpose sensor used for a vehicle system has an

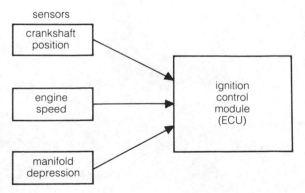

Figure 9.2   *Sensors for ignition system*

accuracy of only about 2–5% and a moderate life cycle. In cases where this error is unacceptable such as pollution control, a more accurate sensor must be used. Being as cost is associated with volume, the manufacturers of vehicles are often able to use a higher-quality sensor than is possible in cases where the demand is limited.

Signal output to the E.C.U. should relate closely to the physical quantity the sensor is intended to measure. Once the signal has been transmitted, no amount of signal processing can improve the original data accuracy, so if control precision is expected, the sensor quality must be high. One method of achieving an accurate signal is to use an 'intelligent' or 'smart' type sensor; this type incorporates a microcomputer to correct for systematic errors. A type in which the microcomputer deals only with random errors is named a 'soft' sensor.

Normally the sensor used to perform a specific task has to fulfil the requirements of cost, size, power consumption and compatibility with other electronic circuits.

In this book there are many cases where electronic sensing devices are used. At this stage it is necessary to summarize some of the main types. Further descriptions of the sensor applications are given where the full system is covered.

Sensors are used on vehicles to measure various things; these include:

- pressure
- position
- flow
- temperature

Pressure sensing

Engine oil pressure   The engine oil-pressure switch was one of the first sensors to be used on motor vehicles. A pressure switch-type sensor signals when a certain pressure is reached or it initiates a warning message when the pressure drops below a given point. Using a more costly transducer for oil-pressure indication gives an output signal, of digital or analogue form, that is proportional to the pressure. This signal can be processed to indicate to the driver the actual pressure reading.

Engine load   The extent of the load on the engine at a given instant is required by the management systems responsible for ignition timing, fuel metering, emission control and automatic transmission operation. In a spark-ignition engine the induction manifold pressure changes with the load and for many years this pressure variation has been used to control the vacuum unit incorporated in the ignition distributor. When the engine is lightly loaded the *manifold absolute pressure* (MAP) is low, i.e. a high depression exists in the manifold. Opening the throttle to maintain a given speed and provide sufficient output power to overcome the load acting against the engine causes the MAP to rise. This pressure change is both reliable and convenient to use, so modern vehicles utilize this feature to signal the degree of engine loading to the systems that require the information.

Manifold depression is created by the pumping action of the pistons. Since the induction stroke in one cylinder is restricted to one stroke in four, the pressure in the manifold is far from cons[...] Although the pressure pulsations of the p[...] strokes reduce as the number of cylin[...] engine speed are increased, the pre[...] variation is still large and must b[...] filtered before it is allowed to a[...] sensor. Generally this is achiev[...] orifice between the manifol[...] into which the sensor is f[...] overleaf).

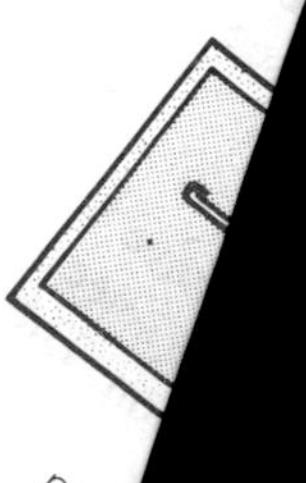

Types of pressure s[...] large range of se[...] the type that s[...] range includ[...]

- variab[...]
- vari[...]
- v[...]
- [...]

respon[...]
tronic sen[...]

Figure 9.4   Aneroid[...]

Figure 9.3 *Use of an orifice to filter pressure waves*

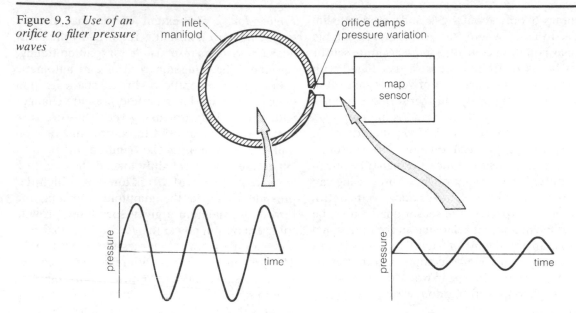

Aneroid pressure capsule An aneroid is a flexible metal box exhausted of air. It is designed to contract and expand as the external air pressure is varied. Figure 9.4 shows an aneroid capsule formed by placing two thin metal diaphragms together and joining them under vacuum conditions.

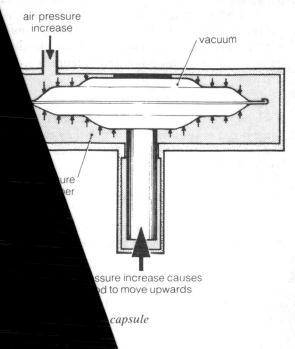

When the capsule is subjected to a positive pressure such as that given by the atmosphere, the walls of the aneroid are forced inwards. This principle is used in an aneroid-type barometer; in this case a mechanical linkage from the centre of the capsule to a pointer registers the atmospheric pressure.

MAP can be measured by connecting the pressure chamber around the aneroid to the inlet manifold of the engine. As manifold pressure changes, the aneroid moves the operating rod inwards or outwards to relate an increase or decrease in pressure respectively. By careful shaping of the aneroid, the relationship between rod movement and manifold pressure can be made linear (Figure 9.5).

Conversion from mechanical movement of the aneroid to an electrical output signal is carried out in various ways. These include:

- variable resistor/potentiometer
- variable inductance
- variable differential transformer

Variable resistor/potentiometer sensor When a wiper contact blade is rubbed over a resistance material, it acts as a divider of the voltage that is applied to the potentiometer resistor. In the cir-

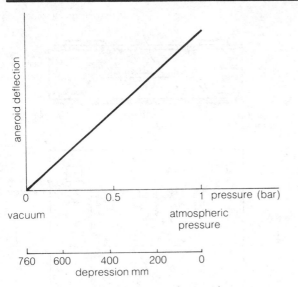

Figure 9.5 *Linear movement of aneroid*

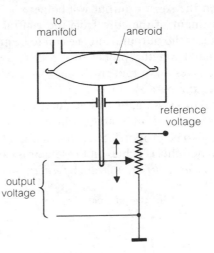

Figure 9.6 *Variable resistor pressure sensor*

cuit shown in Figure 9.6 a drop in air pressure causes the wiper blade to move towards the earthed end of the resistor; this reduces the output voltage signal. In order to obtain reasonable sensitivity, a comparatively large movement of the wiper blade must be provided.

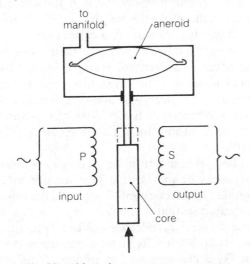

Figure 9.7 *Variable inductance sensor*

Variable inductance sensor When an a.c. current from an oscillator is passed through a primary coil as shown in Figure 9.7, mutual induction

causes an output voltage to be given at the secondary coil (see page 19).

The extent of this output voltage depends on the concentration of magnetic flux, so when an iron core is moved towards the centre of the two coil windings, the output signal is increased.

In this sensor the position of the iron core relative to the windings is controlled by an aneroid. As the pressure is decreased, the expansion of the aneroid moves the iron core towards the centre of the coil windings; this causes the output signal to increase.

A smooth analogue signal is normally required from the sensor so the a.c. output from the secondary is converted and amplified to meet this need.

Variable differential transformer The basic construction of this sensor is similar to the variable inductance type except two output windings are used instead of one (Figure 9.8, see overleaf).

As before, an a.c. current of the order of 10 kHZ is applied to the primary winding and this induces a voltage in both secondary windings. These windings are positioned so that they give an equal voltage output when the core is centrally situated. By winding the two output coils in opposite directions the two coil outputs will cancel each other out when the core is in the central position; in this position the sensor output will be zero.

Movement of the core from the central position causes the output from one coil to be greater than that from the other, so the difference in voltage gives an output signal appropriate to the distance the core is moved.

Signal processing, by a demodulator and filter, gives a d.c. output voltage which is normally arranged to be proportional to the manifold pressure. When this is achieved the sensor is called a *linear variable differential transformer*.

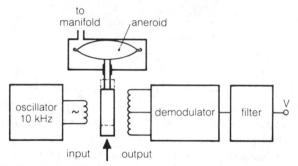

Figure 9.8 *Variable differential transformer*

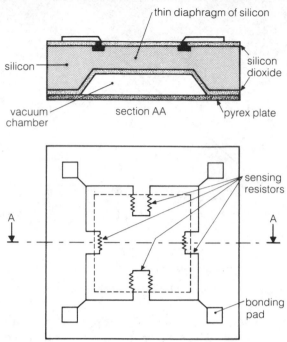

Figure 9.9 *Strain gauge pressure sensor*

Strain-gauge pressure sensor When a material is strained, a change in length occurs which gives a change in resistance. Early designs of strain gauge had a fine wire filament, but nowadays solid-state semiconductors are used to provide a low-cost and compact sensor.

Figure 9.9 shows a strain-gauge pressure sensor. This is built around a silicon chip about 3mm square which is formed into a thin diaphragm of thickness about 250 μm (0.25 mm) at the outer edges and about 25 μm, at the centre. The chip is sandwiched between two silicon dioxide layers into which is formed four sensing resistors positioned along the edges of the silicon diaphragm. Connection to the resistors is by metal-bonding pads formed at each corner of the sensor.

Air-pressure sensing requires a vacuum chamber and in this case, the bonding of a Pyrex plate to one face of the chip under vacuum conditions forms the chamber.

The strain gauge is set in a container which is normally connected by a rubber pipe to a region where pressure measurement is to be made.

When the air pressure is varied, the silicon chip deflects; this alters the length of each resistor. By arranging the resistors in a certain manner, two resistors are made to increase their value while the other two decrease their resistance by an equal amount. This resistance change due to pressure, or piezo-resistivity, is harnessed by using a Wheatstone Bridge circuit to give an output signal proportional to the pressure (Figure 9.10).

The four resistors in the strain gauge form the four arms of the bridge. This resistor array is supplied with a constant voltage and the bridge is calibrated so that it is balanced when the strain gauges are undeflected. When pressure is increased, R_1 and R_4 increase resistance and R_2 and R_3 decrease a similar amount. This unbalances the bridge and gives a difference in potential at AB which provides an output signal which is proportional to the pressure.

Using resistors in this form compensates for temperature change. Any increase in resistance

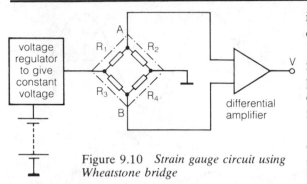

Figure 9.10 *Strain gauge circuit using Wheatstone bridge*

due to heat affects all resistors equally, so the bridge balance is maintained over a wide temperature range.

Capacitor-capsule MAP sensor A capacitor consists of two plates separated by a dielectric. Capacitance is varied by altering the distance between the plates, so this feature can be utilized in a pressure sensor.

The basic construction of a capacitor-capsule sensor is shown in Figure 9.11. It consists of two aluminium oxide plates which are coated on the

inner surfaces with a film electrode and a lead is connected to each electrode.

The two plates are held apart by an insulation material shaped in the form of a flat washer to provide an aneroid chamber at the centre. The capsule is placed in a container which is connected by a pipe to the pressure source.

Operation of the unit is achieved by using the change in pressure that is communicated to the sensor; this pressure change deflects the plates and alters the distance between the two electrodes.

Signal processing for capacitor-capsule MAP sensor Change in capacitance caused by alteration in the MAP is made to generate an output voltage signal by using various circuit arrangements. One method of signal processing is to use a series resonant circuit (Figure 9.12, see overleaf). In this arrangement a change in capacitance is made to alter the phase of the frequency produced by an oscillator.

The main circuit consists of an inductor, resistor and sensor capacitor; these are supplied with an a.c. current from an oscillator.

The output from the oscillator gives a normal a.c. waveform, therefore the voltage and current peaks occur at the same time. When the current is supplied to either an inductor or capacitor, the wave patterns are changed. Compared with the voltage peak, an inductor retards the current peak by 1/4 cycle and a capacitor advances the current peak by 1/4 cycle. These phase alterations are caused by self-inductance by the inductor and charge–discharge action by the capacitor.

At one particular frequency called the *resonance frequency*, the discharge time of the inductor balances the time required for the capacitor to charge. At this frequency the oscillation of the current between the inductor and capacitor is at a maximum, so the voltage of the circuit is high. This feature has been used for many years to amplify radio waves received by an aerial; in this case a variable capacitor is used to 'tune' the circuit to the resonance frequency.

Figure 9.13 shows two voltage pick-up points in an LCR series circuit. Voltage V_1 is the supply

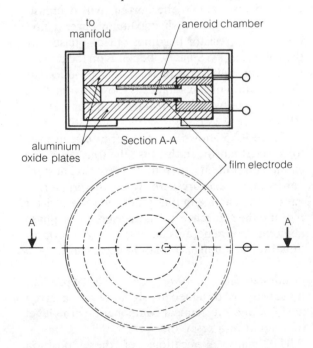

Figure 9.11 *Capacitor capsule pressure sensor*

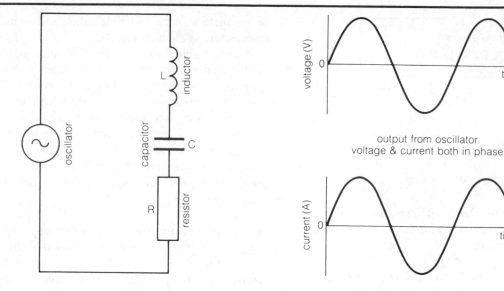

Figure 9.12   *Series resonant circuit*

or reference voltage and V_2 registers the p.d. across the resistor R. Voltage at V_2 is proportional to the current in the circuit.

At the resonance frequency the peak voltage at CD occurs at the same instant as the voltage at BD, so the voltage across the resistor is in-phase with the reference voltage. At this frequency the

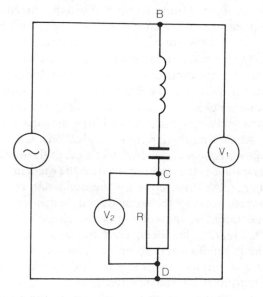

Figure 9.13   *L.C.R. series circuit*

phases coincide because the 1/4 cycle retard in current by the inductor equals the 1/4 cycle advance given by the capacitor.

When the capacitance is altered, the circuit will cease to resonate so the current will decrease. Also the time at which maximum current flow through the resistor becomes out-of-phase with the voltage at V_1. Since V_2 depends on the current passing through R, the phase change produced by the alteration in capacitance is shown by the time difference between the voltage peaks at CD and BD (Figure 9.14).

Figure 9.15 shows an LCR series circuit used for a MAP sensor. In this case the frequency and circuit components are tuned to resonate at standard atmospheric pressure. When the pressure is varied, the change in phase between the resistor p.d. and the reference p.d. is measured by a phase detector; this generates an output signal proportional to the change in manifold pressure.

Position and level sensing   This group of proximity sensors is designed to signal either a given position of a mechanical component or the level of a liquid in a reservoir.

The many applications of these position sensors include:

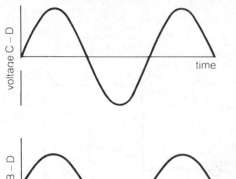

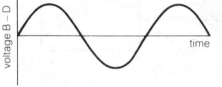

voltage at C – D in phase with B – D
(resonant frequency)

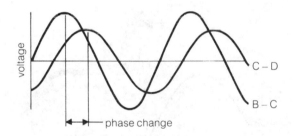

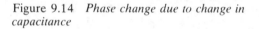

Figure 9.14 *Phase change due to change in capacitance*

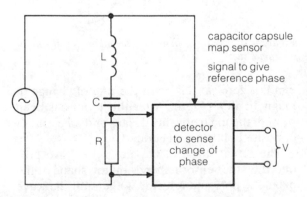

Figure 9.15 *LCR circuit for capacitor type MAP sensor*

- Crankshaft angular position for timing of the ignition and injection systems
- Crankshaft movement for computation of engine speed for engine management and tachometer operation
- Throttle position for fuel injection and automatic transmission systems
- Gearbox output shaft movement for speedometer, odometer and trip computer operation
- Axle to body position for indication of axle loading
- Maximum wear of brake friction material
- Door position for 'door ajar' indication
- Road wheel movement for anti-skid systems.

Sensing of liquid levels is needed for the monitoring of the vehicle condition. The levels checked by these sensors include: engine oil, coolant, fuel, brake and window washer reservoirs.

Types of position sensor   The types of sensor used on vehicles are:

- Magnetic – variable reluctance
- Magnetic – d.c.-excited inductive
- Magnetic – Hall effect
- Magnetic – reed switch
- Optical and fibre-optics
- Capacitance

Magnetic – variable reluctance   This type of position sensor is very robust and can be applied to a crankshaft, camshaft or distributor drive shaft to signal a given position of a shaft for ignition and fuel timing purposes or for measurement of engine speed.

The sensor in Figure 9.16 (overleaf) consists of a permanent magnet and a sensing coil winding. A steel disc, having a series of cut-away portions, is attached to the driving shaft and is set so that the disc passes between the poles of the magnet.

Reluctance is the term used in magnetic 'circuits' to indicate the circuit's resistance to the 'passage' of a magnetic flux. Placing a ferrous metal between the poles of a magnet makes it easier for the flux to link the two poles; this low

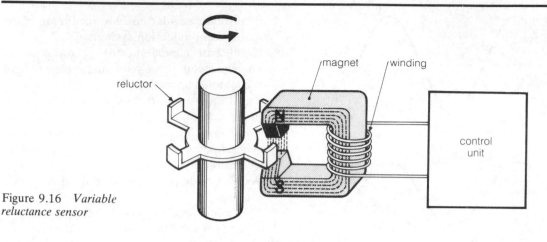

Figure 9.16   *Variable reluctance sensor*

reluctance path gives a high magnetic-field intensity.

Turning the disc to bring the cut-away portion of the discs between the magnetic poles considerably increases the reluctance of the magnetic circuit. As the permeability of air is much greater than that of the steel, the difficult flux path through the air gives a field of low intensity.

In simple terms the reluctor disc is like a water tap – it controls the 'flow' of flux around the magnetic circuit. When the disc protruding tab is between the magnetic poles, the tap is open, but when the cut-away portion is in place, the tap is closed.

The sensing signal is generated by the *changes* which occur in the magnetic flux intensity. When the magnetic flux either increases or decreases, an e.m.f. is inducted into the coil winding. Since the magnitude of the e.m.f. is proportional to the rate of change of the magnetic flux, the faster the change, the greater is the e.m.f. Conversely, no change of flux results in no e.m.f., i.e. when the reluctor disc is stationary, no output is obtained. This means that when the variable reluctor unit is used as an ignition sensor the engine timing cannot be set statically.

Figure 9.17 shows the waveform produced as the reluctor tab is passed through the magnetic pole. As the reluctor approaches the pole the e.m.f. builds-up to a maximum. When the reluctor tab reaches a point where it is aligned with the pole such that the flux is at its maximum, the

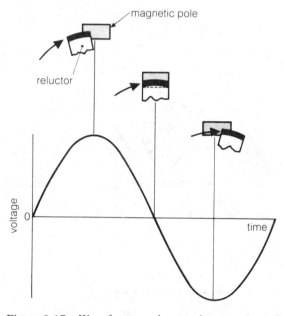

Figure 9.17   *Waveform as reluctor tab moves through magnetic flux*

e.m.f. is zero; at this point the rate of change of magnetic flux is zero. After this point has been passed, the magnetic flux decays and an e.m.f. of opposite polarity is generated.

The actual shape of the a.c. pulse wave produced by the sensor depends on the shape of the magnetic pole and reluctor. In general the wave resembles that shown in Figure 9.17.

Two crankshaft position sensors are shown in Figure 9.18. Flywheel applications use either a

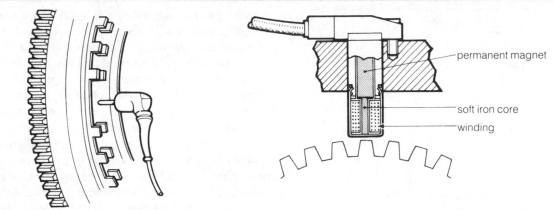

Figure 9.18 *Crankshaft position sensors*

ferro-magnetic pin or a partially machined-away tooth to trigger the sensor pulse. Typical peak-to-peak outputs for this type varies from zero when stationary to about 200 V when the peripheral speed of the flywheel is 60 m/second.

A notch cut in the pulley is another way of varying the reluctance. On both position sensors shown the lobe centre line corresponds to the point where the positive pulse half-wave changes to the negative half-wave.

This type of sensor has many advantages: it is robust; can easily be adapted to use the engine flywheel teeth; needs no exciting voltage or amplifier; operates over a wide temperature range and has a long life. These advantages show why the variable reluctance sensor is often used.

Magnetic – d.c.-excited inductive sensors The disadvantage of low output as given from a variable reluctor sensor is overcome by using a d.c.-excited inductive sensor. This passive type uses a d.c. current to excite a field magnet to give a suitable minimum output voltage such as 2 V peak-to-peak irrespective of the operating frequency.

Figure 9.19 (overleaf) shows the principle of this type of sensor. It consists of a W-shaped iron core with a field coil wound around the centre leg of the core. The spacing of the other legs is arranged to bridge the teeth of the rotating ferro-magnetic member. In the example shown, the flywheel teeth provide the paths for the flux.

Unless precautions are taken, the generation of eddy currents causes the pulse peaks to alter in respect to the position of the rotating member. This 'phase shift' problem is minimized in this design by using a ferrite material for the magnetic core. A typical phase shift is given as 0.3° over a speed range of 0–6000 rev/min.

Operation of the sensor is achieved by exciting the field magnet with a d.c. current from a control unit. This unit varies the current to give a constant-voltage pulse. At low speeds the exciting current is comparatively large, but as the speed is increased the current is reduced.

As the flywheel teeth pass the legs of the iron core, the change in the reluctor path varies the magnetic field intensity and as a result it produces an a.c. pulse in the field winding. This pulse is detected by the control unit which processes it to provide the output signal.

Sensors of this type are often used in conjunction with diagnostic test equipment for measurement of ignition and fuel-injection settings.

Magnetic – Hall effect This passive sensor is often used in an ignition distributor as an alternative to the contact points of a Kettering system. For its operation it needs a supply current and this feature enables it to detect zero movement. For this reason it is also used for many other position- and speed-sensing applications.

The Hall generator, as it is often called, utilizes the principle whereby a voltage is generated

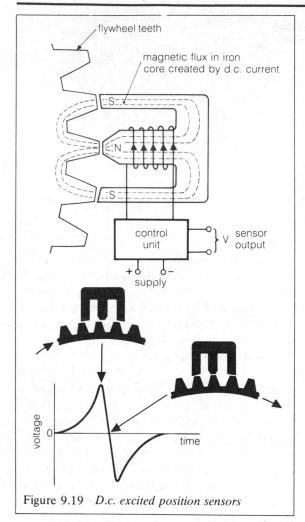

Figure 9.19   *D.c. excited position sensors*

Although the Hall voltage is limited to a few millivolts, the incorporation of the semiconductor wafer into an integrated circuit (Hall I.C.) enables the voltage to be amplified sufficiently to give a comparatively strong output signal.

The sensing feature is achieved by using mechanical means to vary the field strength. One method is to use a rotating vane to act as a switch for controlling the magnetic flux (Figure 9.20(b)). When the vane is in the air gap between the magnet and the Hall I.C., the magnetic field is diverted away from the wafer. This 'switches-off' the Hall voltage and signals the position of the shaft controlling the vane.

When used as an ignition timing sensor, ignition occurs when the Hall I.C. is switched-on, i.e. at the point when the vane leaves the air gap and the Hall voltage is re-established.

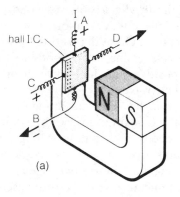

(a)

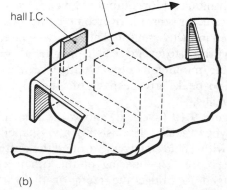

(b)

Figure 9.20   *Hall-effect sensor*

across a plate carrying an electrical current when the plate is exposed to a magnetic flux.

The Hall effect is shown in Figure 9.20(a). This shows a wafer of semiconductor material placed in a magnetic field and supplied with constant current I across the wafer from A to B. Under the influence of a magnetic flux, the electron flow of the current I is deflected as it passes through the semiconductor; this causes a potential difference across the plate in the direction CD. This difference in voltage across the plate is called the *Hall voltage*.

Alteration of the magnetic field strength varies the Hall voltage; the stronger the field, the higher the voltage.

Magnetic-reed switch   A reed switch consists of two or more contacts mounted in a glass vial to exclude contaminates. The vial is evacuated of air or filled with an inert gas to reduce damage by arcing.

When used as a position or proximity sensor the switch is operated by a permanent magnet. Exposing the reeds of the switch to a magnetic flux causes each reed to take-up the polarity of the magnetic pole nearest to it. Being as the reed polarities are opposite, they will be attracted together (Figure 9.21(a)). This will close the switch and allow the flow of current to create the sensor pulse.

Although it is possible to operate the reed by using a magnet attached to a rotating shaft, it is more common to keep the magnet stationary and use a mechanical means such as a toothed wheel to divert the flux away from the reed when the switch is to be opened (Figure 9.21(b)).

The reed switch is used for many applications; it can sense shaft position for fuel injection, road wheel movement for speedometer operation and it provides a simple means of indicating a liquid level.

Optical   Opto-electronic sensors can be used to signal a shaft position. When used in conjunction with fibre-optics, the numerous advantages suggest that this system will be used extensively in the future.

The principle is shown in Figure 9.22 (overleaf). A LED light source is positioned opposite a phototransistor; a chopper plate, in the form of a disc, is attached to the moving shaft. A slot or hole in the disc allows the light to pass to the

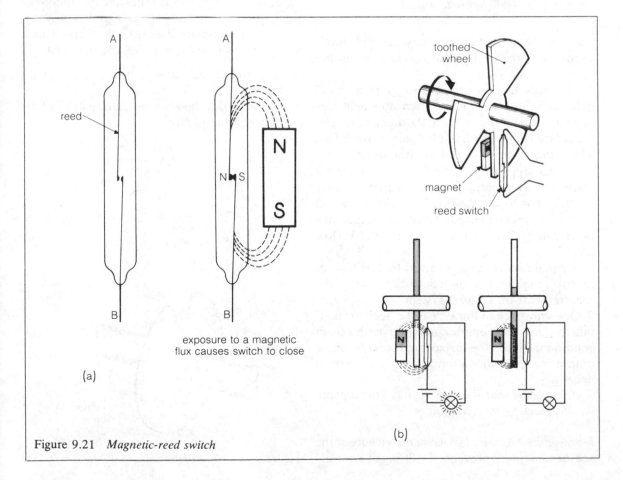

exposure to a magnetic
flux causes switch to close

(a)

(b)

Figure 9.21   *Magnetic-reed switch*

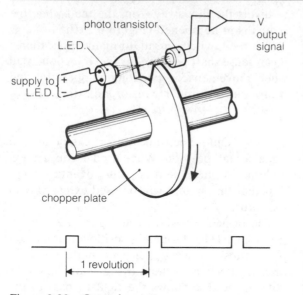

Figure 9.22   *Opto-electronic sensor*

phototransistor at the instant when the slot is aligned with the LED: at this point it signals the position of the shaft.

Although this description suggests a visual light radiation, the system often uses light frequencies outside the vision of the human eye. The LED fitted generally has a frequency which falls within the range infra-red to ultra-violet.

Voltage applied to the LED is set to give a phototransistor output which is sufficient after amplification to suit the E.C.U. In a case where a TTL logic circuit is used, the typical output voltages will be 2.4 V (high level) and 0.2 V (low level).

A square-wave pulse is given by this type of sensor; this makes it compatible with the requirements of a digital system.

One drawback of this type is the need to maintain the lens of the emitter and receiver in a clean condition. Dirt on the surfaces reduces the sensor output so periodic attention to this task is necessary.

Being as the sensor is passive, it has the capability of detecting zero motion.

Fibre-optics   Limited space in the vicinity of the sensor sometimes prevents the fitting of a normal type of optical sensor. This drawback can be overcome by using two fibre-optic strands to transmit the light signal from the LED to the phototransistor (Figure 9.23).

This light-transmission system offers many advantages for motor vehicles; these include:

- space saving
- rate of data transmission is very high
- extra safety because fibre-optic cables do not carry an electrical current
- can be used in hostile regions
- do not suffer 'noise' problems (i.e. electrical charges are not induced from surrounding electrical equipment).

Capacitance   This type of sensor uses a toothed wheel to form either a plate or the dielectric of a capacitor. Its operation relies on the following principle:

If a parallel-plate capacitor is charged to a given value the voltage across the plate will:

- increase as the distance between the plates increases
- decrease if a dielectric is used instead of air between the plates.

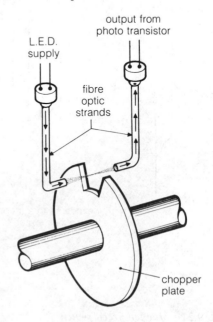

Figure 9.23   *Fibre-optics sensor*

Capacitive sensors can be designed to utilize one or other of these two features; they can be excited by a.c. or d.c. means.

Figure 9.24(a) shows a sensor which produces a signal by varying the distance between the plates; this is achieved by rotating a toothed wheel. Since the plate and the toothed wheel are electrically charged, the movement of the wheel has the effect of changing the distance between the plates. In the case of a d.c.-excited sensor, this change in voltage provides a pulse which is processed to give the required output.

The alternative construction uses a dielectric wheel (Figure 9.24(b)). Rotation of the wheel between the two charged plates varies the voltage across the plates to give the pulse.

This arrangement can be used for a high rotational speed but when used at a lower speed, and for detecting zero motion, the difficulty of maintaining the plates in a clean state and at a constant electrical charge makes this type of sensor less attractive than many other types.

When the sensor is a.c. excited, the capacitor forms part of an oscillator circuit. Changes in capacitance due to movement of the toothed wheel results in a change in the frequency (Figure 9.24(c)). To detect high-speed movement, a high frequency must be used but this often causes interference with radios and in-car telephones unless the sensor system is adequately screened. Because the screening affects the capacitance of the sensor system, the associated circuitry of the sensor must be situated as close as possible to the sensor. Often lack of space makes this difficult. (See also page 221 for a typical application of another type of capacitor sensor.)

Flow sensing

Measurement of fluid flow in a motor vehicle is needed in two cases:

- air flow for fuel metering as in fuel-injection systems
- fuel flow for computation of fuel consumption and associated information.

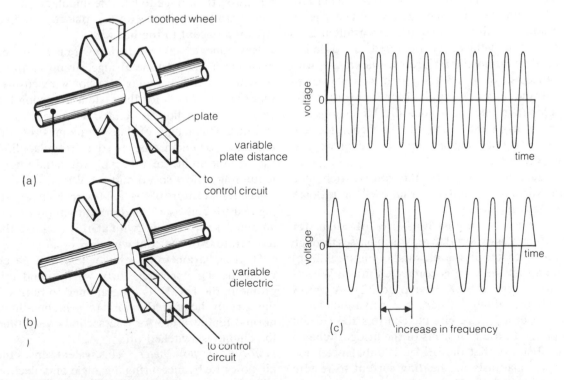

Figure 9.24   *Parallel-plate capacitor*

Air flow sensors   Full electronic systems for petrol injection use air-measurement sensors such as:

- hot wire
- flap type
- aneroid MAP

Hot-wire measurement of air mass   A hot-wire air-mass meter relies on the cooling effect of air as it passes over a heated wire. If this wire is heated by passing a constant current through it, then the temperature of the wire will fall as the air flow is increased.

Similarly if a hot wire is kept at a constant temperature, then the amount of current required to maintain this temperature will be governed by the air flow; the larger the air flow, the greater the current.

Both the constant-current and constant-temperature methods use electronic means to measure the temperature. Generally this is achieved by utilizing the change in resistance which occurs when the temperature is changed.

Hot-wire systems take into account changes in air density. This is particularly important in cases where the vehicle is operated at different altitudes. Atmospheric pressure decreases with altitude, so in an area situated well above sea level, the air mass supplied for a given throttle opening is reduced considerably. Unless this feature is taken into account, the richer mixture received by the engine would cause high exhaust gas pollution.

Figure 9.25(a) shows the construction of a hot-wire meter similar to that used on a Bosch LH-Jetronic system.

A thin platinum hot-wire of diameter 0.070 mm, is exposed to air which passes through a tube situated in the air intake. Figure 9.25(b) shows that the wire is connected into a Wheatstone Bridge circuit, (see page 32). A power amplifier, situated where the galvanometer is placed in a Bridge circuit, controls the current supplied to the four arms of the bridge. When a signal shows that the bridge is unbalanced, the amplifier adjusts the heating current to restore the bridge to a balanced state.

Operation of the sensor is based on the constant-temperature principle. When air is passing at a constant rate, the supply current holds the hot wire at a given temperature, consequently the bridge is maintained in a balanced state. Any increase in air flow cools the hot wire and causes its resistance to decrease. This unbalances the bridge and as a result causes the amplifier to increase the heating current until the original temperature, and resistance, is restored. This increase in the heating current causes a higher voltage drop across R_1, so by measuring this drop across a precision resistor placed at R_1 a sensor output signal is obtained. The signal shows the heating current but it also indicates the mass of air flowing through the meter.

By relating the hot-wire sensor signal to values stored in a 'look-up' table in the E.C.U's memory, the computer can determine the amount of fuel that needs to be injected to give the required air–fuel ratio.

Any alteration in the temperature of the intake air causes the bridge to become unbalanced, so a compensating resistor wire r_b is placed in the air stream adjacent to the hot wire.

Resistance change of the hot wire is measured by using a Wheatstone Bridge circuit. Careful selection of the diameter of a hot wire ensures that the time taken for the system to respond to changes in air flow is limited to a few milliseconds. This length of time overcomes the air-pulsation problem due to the irregular flow through the air intake, especially when the engine is running at low speed under full load.

The temperature of the 'cold-wire' compensating resistor acts as a 'standard'. In operation the amplifier keeps the hot wire at 100°C above the temperature of the cold wire.

Heat radiation from a hot wire is decreased when the wire becomes dirty, so to avoid this problem, the E.C.U. is programmed to burn-off the dirt by heating the wire to a higher-than-normal temperature for one second every time the engine is switched-off.

Flap-type air-flow meter   This system senses the air-flow rate by measuring the angle of deflection of a flap or vane. Mechanical movement of the

Figure 9.25 *Hot wire air-mass meter*

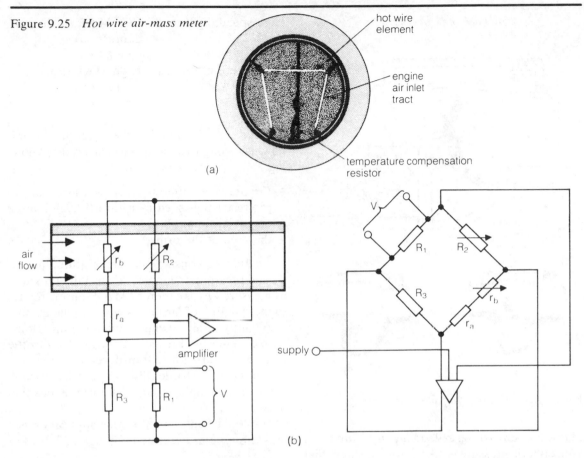

flap is translated to an electrical analogue signal by a potentiometer (Figure 9.26, see overleaf).

The sensing flap, placed in the air tract, is spring loaded to oppose the force given by the air flow. Flap pulsations, caused by the irregular air flow, are damped by a small air chamber adjacent to the flap.

The flap moves a slider of a thick-film type potentiometer which provides a voltage signal; this increases as the flap is opened. Using a thick-film resistor minimizes the effects of temperature changes and by varying the resistance of each resistor segment of the potentiometer, a linear relationship can be obtained between the sensor's output voltage and the fuel to be injected.

When the throttle is opened and the air flow is increased, the greater force given by the air opens the flap. This allows air to spill past the flap until a balance is reached between the air force and the flap spring force. As the flap moves to this position, the higher voltage output signal to the E.C.U. results in the injection of a larger quantity of fuel.

Manifold absolute pressure (MAP) sensors
Zero on the absolute scale is where pressure ceases to exist; at this point a complete vacuum is formed. Standard pressure of the atmosphere is 1 bar or 101.3 kilopascals when expressed as an absolute pressure.

During the induction stroke of an engine, the pressure in the induction manifold falls below atmospheric pressure; a typical value for an engine operating under light load at 1000 rev/min is 0.4 bar absolute. Relating this value to atmos-

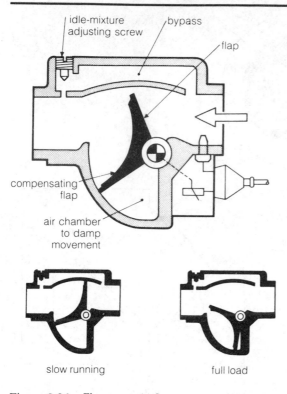

Figure 9.26 *Flap-type air flow meter*

pheric pressure (often called *gauge pressure* when atmospheric pressure is taken as zero) shows that the 'suction' effect on the air by the pumping action of the engine piston has lowered the pressure by 0.6 bar.

A pressure lower than atmospheric is called a *depression*; this term is used in preference to 'vacuum' because a vacuum indicates that no pressure exists whatsoever.

MAP is affected by a number of things; the two main variables are engine speed and throttle opening. Under ideal conditions where the manifold and piston has no air leaks and the throttle is fully closed, a very high depression approaching a vacuum is formed. In practice this is not possible and the maximum obtained is only about 0.17 bar absolute. Even this low pressure can be achieved only when the throttle is closed as the engine is decelerated from a high speed. During normal loading of the engine, the MAP varies between the following limits:

High speed, light load	Low MAP (high depression), e.g. 0.4 bar
Low speed, heavy load	High MAP (low depression), e.g. 0 bar

Aneroid MAP sensor The quantity of air flowing through a duct can be measured by knowing:

- the pressure difference across a given orifice (or restriction such as a throttle valve in an engine air intake)
- the area of the orifice

Assuming a constant air temperature, humidity and pressure, the manifold absolute pressure, as measured by an aneroid MAP sensor, relates to the quantity of air that is entering the engine.

Quantity of air flowing due to pressure difference depends on throttle opening, so the throttle must be fitted with a transducer to signal the extent that the throttle valve is opened. Normally a potentiometer is used as a throttle position sensor.

The MAP sensor used in this application is a pressure type as described on page 217.

Throttle position sensor Often this type of sensor is a potentiometer (pot); this is a means for altering electrical potential. One of the simplest arrangements is a variable resistor as shown in Figure 9.27. In this case the throttle is connected to a contact blade which wipes across a resistor coil. As the throttle is opened the number of

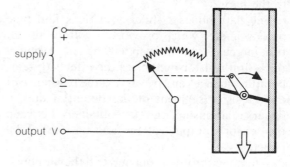

Figure 9.27 *Throttle position sensor*

resistor coils in the circuit is reduced. This alters the voltage and potential of the output in relation to 'earth'.

An improved accuracy and a longer life is obtained by using alternative resistor materials to wire. Figure 9.28 shows a typical sensor with two pairs of contact blades; one pair acts as the main potentiometer, the other pair acts as a micro-switch to signal the throttle-closed position. A constant voltage of 5 V is applied to the sensor. As the contact blades slide along the resistor in accordance with the opening of the throttle valve, the output voltage increases proportionally. This linear signal is transmitted to the E.C.U.

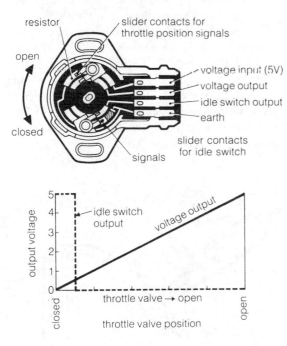

Figure 9.28 *Throttle position potentiometer*

Fuel flow sensor A trip computer, as fitted to many modern vehicles, has the facility to calculate fuel consumption. This feature requires an input signal that indicates the quantity of fuel used by the engine in a given time.

One type of fuel line transducer has a fuel-driven turbine which rotates a chopper plate of an optical sensor unit.

Temperature sensing A number of sensors are used on a vehicle to measure temperature. These include:

- Engine temperature for ignition, fuel metering and instrumentation
- Air-intake temperature for fuel metering and vaporization control
- Ambient conditions for driving safety
- Exhaust temperature for fuel metering

The majority of temperature sensors use a thermistor, but there are occasions where a thermocouple is used.

Thermistor sensor This type normally consists of a brass bulb, which is in contact with the substance it is sensing. The bulb contains a capsule called a thermistor.

Resistance of common metals increases with temperature and thermistors which have a sensing capsule that responds in this way are said to have a *positive temperature coefficient* (p.t.c.).

Conversely, a capsule made of a semiconductor material has a resistance which decreases with temperature. These materials, of which silicon is the most common, have a *negative temperature coefficient* (n.t.c.).

Figure 9.29 (overleaf) shows the construction and resistance variation with temperature of a typical sensor as fitted into an engine block to measure the coolant temperature. In addition to temperature measurement the thermistor is often used in an electronics circuit to safeguard semiconductor devices when the circuit components are cold. It compensates for temperature to keep the circuit 'in-tune'.

Thermocouple The thermistor is excellent for measuring temperature up to about 200°C, but above this temperature a thermocouple is normally used.

The principle of a thermocouple is shown in Figure 9.30 (overleaf). It consists of two wires of dissimilar material joined together and connected to a galvanometer. When the hot junction is heated, an e.m.f. is generated which is registered by the galvanometer. Up to a given temper-

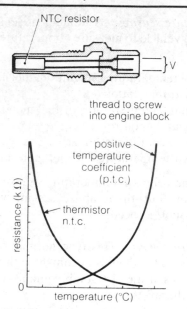

Figure 9.29   *Thermistor sensor*

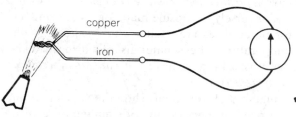

Figure 9.30   *Thermocouple*

ature, which depends on the metals used (250°C for copper–iron), the current increases with an increase in the temperature difference between the hot and cold ends of the wires. Indication of temperature is achieved by scaling the galvanometer accordingly.

This effect was discovered by Seebeck in 1822. He showed that *thermo-electric currents* are obtained from a pair of metals when their junctions are maintained at different temperatures.

A thermocouple can be made by using two metals from: antimony, iron, zinc, lead, copper and platinum. Current will flow from the higher to the lower in this list across the cold junction.

Nowadays other metals and alloys are used for

a thermocouple, e.g. nickel–chromium/nickel–aluminium alloys are used for the wires of a common type of thermocouple. This type is suitable for a temperature range 0–1100°C such as exists in an engine exhaust system.

Measurement of exhaust gas temperature is necessary when an oxygen sensor is used.

Other sensors

Oxygen sensor   The detection of oxygen in the gas exhausted from an engine provides a useful means for controlling the air–fuel mixture. Oxygen in the exhaust indicates that the air–fuel ratio is weak, whereas the absence of oxygen shows that the mixture is rich and exhaust gas pollution of the atmosphere is taking place.

A sensor designed to signal when combustion in an engine cylinder completely burns the fuel (i.e. that the air–fuel mixture delivered to the engine is chemically correct) enables the engine management system to operate effectively irrespective of the mechanical state of the engine. The gas detector used to provide this signal is called a *Lambda sensor* (Figure 9.31).

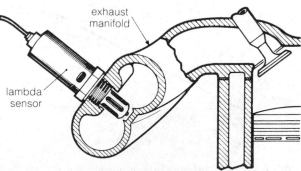

Figure 9.31   *Lambda sensor*

Lambda sensor   This type of exhaust gas oxygen (EGO) sensor normally uses zirconium oxide (ZrO_2) for its active material. Figure 9.32(a) shows the basic construction of an EGO sensor.

It consists of a thimble-shaped portion of ZrO_2 covered with two thin and porous platinum electrodes. The internal electrode is in contact with

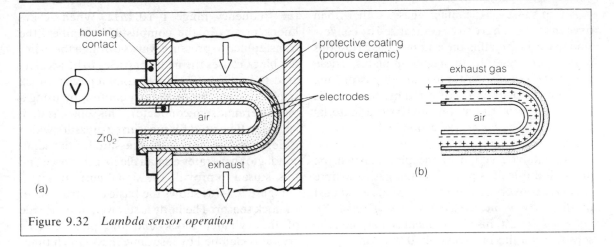

Figure 9.32   *Lambda sensor operation*

air in the centre of the dome and the outer electrode is placed so that it is in contact with exhaust gas. Extra protection against gas erosion is given by covering the outer electrode with a porous alumina ceramic coating through which the gas can penetrate.

The operation of the sensor is similar to a galvanic battery cell; in the case of the sensor the ZrO_2 acts as the electrolyte. At high temperatures this electrolyte becomes conductive so if the two plates are in contact with different amounts of oxygen, then a small voltage will be generated across the two plates. This action is produced because oxygen atoms carry two free electrons so this means that the atom carries a negative charge. The ZrO_2 attracts oxygen ions with the result that negative charges build up on the surface of the ZrO_2 adjacent to the platinum electrode (Figure 9.32(b)).

When the sensor is exposed to exhaust gas, a greater concentration of oxygen on the air side of the ZrO_2 causes this side to have a greater number of negative charges. In consequence a potential difference is built up across the plates which will depend on the difference in oxygen levels. It will be about 1 V when the engine is operated on an enrichened air–fuel mixture but this will drop abruptly at the point where the air–fuel ratio is chemically correct, i.e. at the stoichiometric point. In the weak zone beyond this point, the p.d. across the sensor will remain near constant at about 50 mV (Figure 9.33).

The time taken for the sensor to respond decreases as the temperature is increased. Common types in use become operational above about 300°C and have a response time of less than 200 ms. The point on the air–fuel ratio scale at which the abrupt voltage change occurs is not affected by the temperature, but temperature produces a small change in the voltage; a drop of about 200 mV occurs when the temperature is altered from 500 to 1000°C.

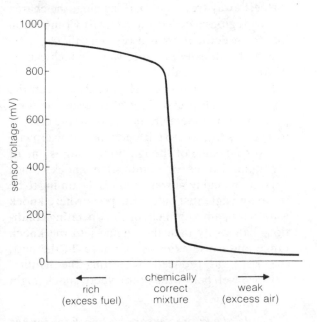

Figure 9.33   *Lambda sensor output voltage*

An exhaust gas contains gases other than oxygen so these have to be neutralized by conversion to avoid affecting the action of the ZrO_2; this duty is performed by the platinum plates. These plates are porous, so the unwanted gas passing through the platinum is oxidized by the catalytic action of the platinum. (A catalyst is a material which produces a chemical action without undergoing any change itself.)

The catalytic action of the platinum will be prevented if lead is present in the gas, so where catalytic converters are used to oxidize exhaust products *the vehicle must only be operated on unleaded fuel*. If this requirement is met, then the sensor has a life in excess of 50 000 km.

Knock sensor

The main purpose of a knock sensor is to detect combustion knock (detonation) in an engine combustion chamber.

Combustion knock   This fault occurs when the gas charge in one part of the combustion chamber ignites spontaneously instead of burning gradually. The region of the chamber where this explosive action takes place is normally the zone farthest away from the sparking plug; the charge in this region is called the *end gas*. Combustion knock, or detonation as it is often called, generates high-intensity pressure waves which can be heard as a noise called *pinking*.

Severe combustion knock not only subjects the engine to high pressure; it also raises the temperature of the metal in the region of the end gas to a point sufficient to part melt the piston crown.

Over-advance of the ignition timing is one of the main causes of combustion knock. Since maximum engine power requires the timing to be set to an angle just below the point where knock is initiated, any slight change in operating conditions can easily take the engine into the knock range and cause extensive damage. In the past, this risk was reduced by setting the ignition advance well below the point where knock might occur.

Principle of a knock sensor   A knock sensor has to detect vibrations from combustion knock in the frequency range 1–10 kHz. When severe knock occurs in the combustion chamber, the transference of pressure waves through the cylinder block causes the metal particles to be accelerated to and fro. The accelerometer-type knock sensor detects this oscillatory motion by using a piezo-ceramic semiconductor. This sensor is similar to the type used for measuring pressure which is described on page 220. Pressure on the semiconductor generates a small electrical charge and this is used to provide the signal current.

Figure 9.34(a) shows the basic construction of a knock sensor. The body is screwed into the side of the cylinder block and the piezo-electric crystal is clamped by a seismic mass which tunes the sensor to the required frequency range.

When an oscillation of the form shown in Figure 9.34(b) is applied to the sensor body, the sound waves vary the compression of the crystal; this causes a small e.m.f. of the order of 20 mV/g to be generated.

The small analogue signal current produced by the sensor when the engine is knocking is transmitted to the E.C.U. After filtering to remove the unwanted waves, the signal is then averaged and converted to a digital form to represent the knock/no-knock conditions. Whenever a digital pulse is sensed by a logic circuit, the system responds by reducing the spark timing advance.

9.2 Actuators

Actuators communicate motion; they produce mechanical motion when commanded by an electrical signal. Two types of electric actuator are:

- Linear
- Rotary

Most automotive components require a linear operating force, i.e. a force that moves the device in a straight line. This motion can be produced by a linear:

- Solenoid
- Motor

Linear solenoid

A simple solenoid consists of a bobbin that holds

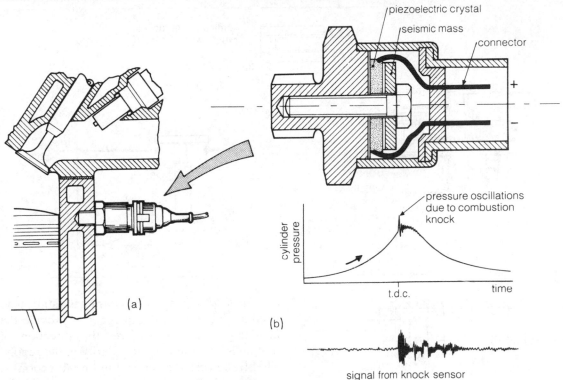

Figure 9.34 *Accelerometer-type knock sensor*

a coil of thin copper wire which is enamelled to insulate the coils from each other (Figure 9.35(a), see overleaf). A soft iron armature or plunger, of diameter sufficient to permit axial movement, slides into the bobbin when the coil is energized (see page 17). A spring is normally used to return the plunger when the current is switched off.

When the solenoid is energized for long periods, the current consumption is reduced by using two coils; a closing coil and a holding coil (Figure 9.35(b), see overleaf). Closing the switch supplies current to both coils until the plunger nears the end of its travel. At this point a pair of contacts is opened to disconnect from the circuit the powerful closing coil; this leaves the holding coil to retain the plunger in position.

Linear movement in both directions can be produced by using two coils, A and B, placed end-to-end (Figure 9.36, see overleaf). When B

is energized the plunger moves to the right and when coil A is energized the plunger is returned. Double-coil solenoids are used on some central door-locking systems.

A solenoid can produce large forces and give a rapid operation but has the disadvantage that the stroke is limited to about 8 mm. This limitation arises because the force on the plunger is proportional to the square of the distance between the plunger and the pole piece. As a result of this, the force decreases considerably when the air gap is increased. Often the solenoid plunger is connected to an extension arm or lever to make it suitable for the application.

Linear motor
At first sight this appears to be similar to a solenoid; the difference is that the linear d.c. motor uses a powerful permanent magnet to increase the magnetic action. In view of this, a

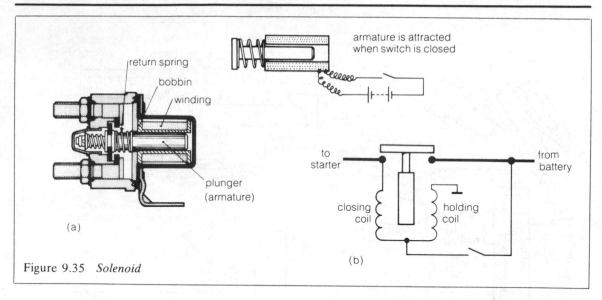

Figure 9.35   *Solenoid*

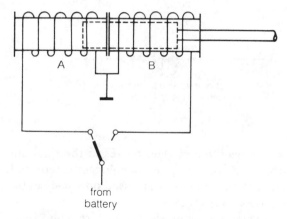

Figure 9.36   *Double-acting solenoid*

near-constant force over a longer stroke is achieved.

Two main types of linear motor, shown in Figure 9.37 are:

- Moving winding
- Moving magnet (moving field)

Moving winding   This has a fixed magnet around which is fitted a hollow armature and coil winding. When a d.c. current is supplied to the coil, the armature is either pushed outwards or pulled inwards, depending on the direction of the current.

Moving field   This type of motor has a static field winding and a moveable magnet to provide the actuating force. As before, the direction of motion is governed by the polarity of the supply. The stroke (distance moved in one direction) is limited to half the length of the magnet; to be effective the width of the coil winding should equal the stroke.

An alternative design uses two coils wound in opposite directions. In this case one coil moves the magnet one way and the other moves the magnet the other way. This double-coil system overcomes the need for changing the polarity.

Rotary actuators

The conventional permanent-magnet d.c. motor, as used originally for windscreen wipers, is still the most common type used to actuate vehicle systems such as washers, fuel pumps, windows, seats, sunshine roofs and radio antennae.

Since the motor is compact, a high speed is needed for it to generate sufficient power. This tends to reduce reliability and means that the motor requires a gearbox to reduce speed and increase torque. Cost and weight of the gearbox are important, so a plastics material is generally used for the gears even though a plastics gear is weaker than metal. Friction is a problem and this is not helped where the motion has to be changed

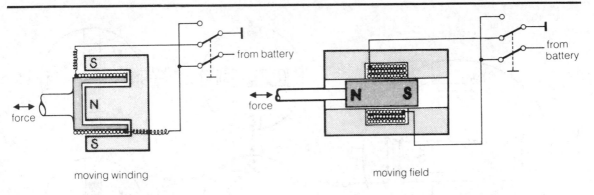

Figure 9.37 *Linear motors*

from rotary to reciprocating by means of a flexible rack. Nevertheless, the rotary motion together with its mechanical linkage gives a long-stroke action demanded by many automotive components.

Stepper motor   The introduction of digital electronic units in automotive control systems has been accompanied by the use of special actuators called stepper motors; these respond to electrical pulse signals. The motor is used to move a mechanical control unit in accordance to the messages it receives from the electronic 'brain'.

Figure 9.38 shows a control system which is based on three main stages: the actuator is part of the final stage. For precise control, the system requires a motor that moves through set angles in either direction. The type of motor used governs the smallest step angle through which it can move; typical angles used are: 1.8°, 2.5°, 3.75°, 7.5°, 15° and 30°.

Stepper motors are made in three versions:

- Permanent magnet (p.m.)
- Variable reluctance (v.r.)
- Hybrid

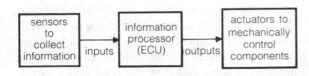

Figure 9.38 *Electronic control system*

Permanent-magnet stepper motor   The principle of this type is shown in Figure 9.39(a) (overleaf); in this case the active rotor is a two-pole permanent magnet. The stator has two pairs of independent windings AA_1 and BB_1 through which current may be passed in either direction to make the rotor turn through 90° steps.

When current is passed to phase winding BB_1, the magnetic laws of attraction and repulsion align the rotor with the active poles of the stator (Figure 9.39(b), see overleaf). Complete rotation is obtained by applying to the motor four electrical pulses of suitable polarity. Direction of rotation depends on the polarity of the stator during the first pulse, e.g. if the current direction in Figure 9.39(b) is reversed, the rotor will move in a clockwise direction.

Altering the frequency of the pulses applied to the stator varies the speed of rotation; also by controlling the number of pulses a given angular displacement can be obtained.

By increasing the number of rotor and stator poles, the step angle (°) can be reduced; this is calculated so:

$$\text{step angle} = \frac{360}{\text{number of step positions}}$$

Each winding has two possible current-flow directions so the number of step positions will always be an even number. Permanent-magnet motors are generally available with basic step angles between 7.5° and 120°.

One advantage of this type of stepper motor is that the magnet holds the rotor in place when the

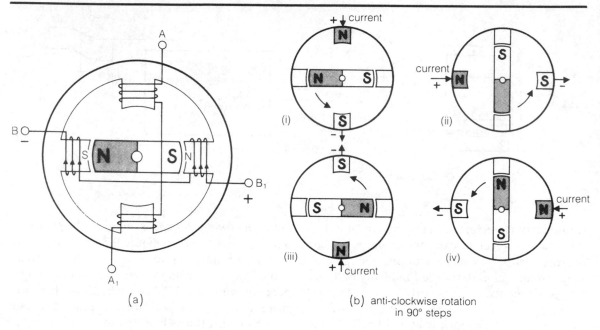

(a)

(b) anti-clockwise rotation in 90° steps

Figure 9.39　*Permanent-magnet stepper motor*

stator is not energized; this is called *detent torque* and not all stepper motors have this feature. Disadvantages include high inertia and the fall-off of performance due to changes in magnetic strength.

Variable-reluctance stepper motor　This type has a soft-iron rotor with radial teeth and a wound stator having more poles than the rotor. Figure 9.40(a) shows a simplified layout of a three-phase, 15° step-angle motor; this has eight rotor teeth and twelve stator poles around which the current flows in one direction only.

The number of step positions (N) is calculated so:

$$N = \frac{SR}{S - R}$$

where S = number of slots in stator and
R = number of slots in rotor.

In this case

$$N = \frac{8 \times 12}{12 - 8} = \frac{96}{4} = 24$$

So:

$$\text{step angle} = \frac{360}{24} = 15°$$

Figure 9.40(b) shows the winding arrangement for phase 1 and Figure 9.40(c) shows the way in which the other phases are connected.

When a current flows through one phase of the stator windings, the rotor aligns itself to give the shortest magnetic path, i.e. the path of minimum reluctance. In each step position, the rotor aligns with four stator poles so this gives the motor greater power.

An angular movement of one step from the position shown in Figure 9.40(a) is obtained by energizing either phase 2 or phase 3 depending on the required direction. For a clockwise motion the phases would be energized in the order: 3, 2, 1, 3, 2, 1. The angle turned by the rotor by these six current pulses is 90° and the time for the total movement is governed by the time taken by the control circuit to energize the windings sufficient to move the rotor to the next step.

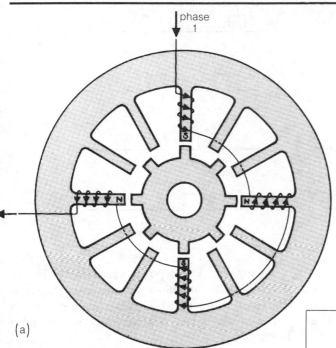

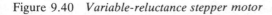

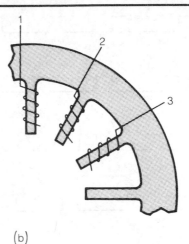

(b)

(a)

Figure 9.40 *Variable-reluctance stepper motor*

This type of motor is obtainable with step angles between 1.8° and 15°. It has a fast response because of its low rotor inertia and has a fast stepping rate. As it has no detent torque it is prone to oscillate and resonate unless it is damped externally.

Hybrid stepper motor As the name suggests this type is a combination of the p.m. and v.r. types. Figure 9.41 shows that in this type the rotor is constructed in a manner similar to that used in an alternator. A permanent magnet, with its poles coaxial with the shaft, is sandwiched between two iron claws having teeth which form two sets of poles.

The stator has eight main poles which are cut to form small teeth on the surface adjacent to the rotor.

The operation is similar to the p.m. type; the rotor aligns itself so that the magnetic reluctance is lowest.

This type of motor has stepping angles as low as 0.9°, a high torque and the ability to operate at

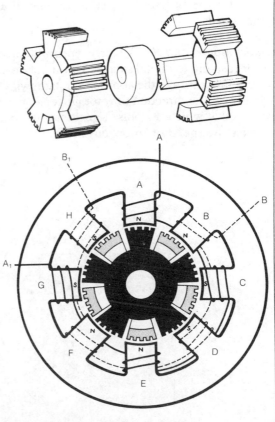

Figure 9.41 *Hybrid stepper motor*

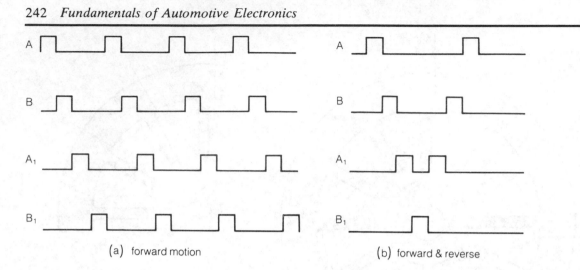

(a) forward motion   (b) forward & reverse

Figure 9.42   *Pulse signals to control stepper motor*

high stepping rates. Its disadvantages include high rotor inertia and the risk of resonance at some speeds.

Stepper motor control   All three types of stepper motor respond to digital signals. The direction of current flow through the appropriate stator winding governs the direction of rotor movement and the speed at which the pulse signals are supplied controls the speed of rotor movement.

Taking the p.m. type as an example, Figure 9.42(a) shows the pulses that are applied to turn the rotor. Note that the pulses do not overlap and that the speed is controlled by the pulse frequency.

Figure 9.42(b) shows the pulse pattern needed to move the rotor forward through three steps (270°) and then reverse it to its original position.

The input to the drive circuit, normally from a logic signal source, has a low power so this has to be amplified to a relatively high power to drive the motor.

10 Instrumentation

10.1 Basic instrumentation systems

Only a few years ago the instrument panel consisted of:

- Speedometer incorporating an odometer to register mileage covered
- Fuel contents gauge
- Engine temperature gauge
- Engine oil-pressure gauge or warning light to signal low pressure
- Signal lights to indicate battery charge (ignition warning light), directional indicator operation and headlamp main beam

These instruments formed the basic panel and even today they still form the minimum equipment used on 'low-line' models. With the exception of the speedometer, other instruments were electrified at a comparatively early stage to simplify the drive arrangement, save on cost and reduce weight. Today, very few instruments are mechanically operated; even the speedometer is often electrically operated. In addition, numerous other displays and signals have been introduced to provide the driver with information relating to normal operation or malfunction of a vehicle system.

Before considering a sophisticated modern system of instrumentation, the electrical equipment forming a basic panel is covered. The main items of equipment are shown in Figure 10.1.

Fuel contents gauge
In the past the moving-iron cross-coil gauge was commonly used. The gauge unit of this system is mounted in the panel and is similar to a simple ammeter; the tank unit or transmitter is a variable resistor, the sliding arm of which is operated by a float.

Although this type operated immediately the ignition is switched-on, surging of the fuel in the tank causes the gauge needle to swing about; as a

Figure 10.1   *Basic instrument panel*

result the gauge is difficult to read. It was mainly due to this that the thermal, or bi-metal, type was introduced.

Thermal type   This system, shown in Figure 10.2, has a gauge unit operated by a bi-metal strip and a variable-resistor sensor fitted in the fuel tank. Since voltage affects the current flow through the circuit, to obtain an accurate reading the supply voltage has to be kept constant; this duty is performed by a voltage stabilizer.

The stabilizer shown in Figure 10.2 is a bi-metal type. A heating coil is wound around a bi-metal strip, which consists of two metals: brass and steel. When the strip is heated, the strip bends towards the side of the steel due to the brass expanding nearly twice as much as the steel. In the stabilizer, the bending of the strip opens a pair of contacts and interrupts the current flow in the circuit. After cooling for a fraction of a second, the circuit is re-made; this cycle is repeated and a pulse current is provided. The frequency of the digital pulse output is preset to give a heating

effect on another bi-metal strip similar to that produced if a constant voltage is supplied to the circuit. Stabilizers in use generally provide a mean voltage of either 10 V or 7 V.

Instead of a bi-metal stabilizer some vehicles have a solid-state voltage regulator; these use a Zener or avalanche diode in a control circuit to provide the required voltage.

In addition to providing current for the fuel gauge, a stabilizer also supplies the other thermal instruments such as the engine temperature gauge.

The sensor, or transducer, fitted in the fuel tank is operated by a metal or plastics material float. Movement of the float arm causes a contact blade to wipe over a wire-wound rheostat. Raising the fuel-level moves the blade and decreases the resistance. Later designs of fuel level sensor make more use of plastics and have a laser-trimmed thick-film resistor set on to a ceramic substrate to give greater accuracy and reliability. A typical resistance range is from 19 Ω (full) to 250 Ω (empty).

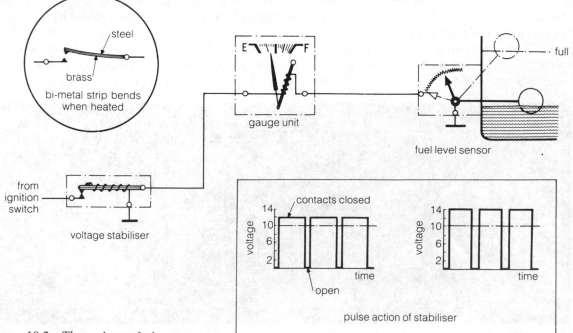

Figure 10.2   *Thermal-type fuel contents*

The gauge unit, or indicator, has a bi-metal strip around which is wound a heater coil. One end of the strip is anchored and the other end is attached to the gauge needle.

When the ignition is switched-on, the intermittent current causes the bi-metal strip to heat up to a temperature dictated by the resistance of the sensor in the fuel tank. Under full-tank conditions the low resistance of the tank sensor allows each pulse supplied by the stabilizer to give a large current flow so the bi-metal strip in the gauge bends a large amount.

After switching-on it takes about two minutes before an accurate reading is obtained; this sluggish action is one advantage of the thermal-type unit.

Care must be taken to avoid an explosion when testing the operation of the system with the sensor unit removed and positioned close to the tank. Sparks generated in a fuel-saturated tank area is safe, but it is highly dangerous to produce sparks where both oxygen and fuel vapour are present.

Engine temperature

The conventional thermal-type gauge used for this instrument is similar to that used to indicate the fuel contents. Gauge reading is affected by supply voltage, so the system is fed via a voltage stabilizer, the same unit as that used to supply the other thermally-operated gauges.

The engine temperature sensor is generally situated in the coolant system on the engine side of the thermostat. Normally the brass bulb in contact with the coolant contains a sensing capsule called a *thermistor*; this semiconductor resistor pellet is thermally sensitive, but varies its resistance in the opposite way to most metals: when the temperature is increased, the resistance is decreased. The thermistor's high *negative temperature coefficient* (n.t.c.) of resistance causes the resistance to vary from about 220 Ω at 50°C to 20 Ω at 115°C.

The circuit for a bi-metal indicator gauge and thermistor is shown in Figure 10.3. In operation, a rise in engine temperature lowers the thermistor resistance and increases the current passing through the gauge unit; this causes the bi-metal strip to bend more than before.

Damage will occur if the thermistor is connected directly to the battery.

Engine-oil pressure indication

Nowadays the majority of engines use a signal

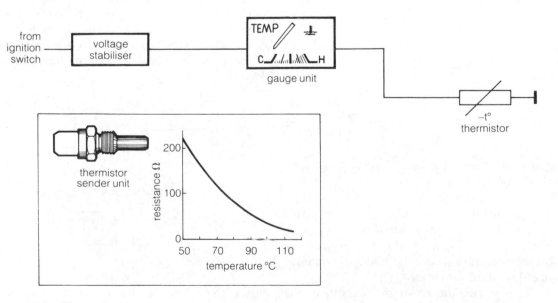

Figure 10.3   *Thermal-type engine temperature gauge*

lamp, or electronic display, to warn the driver of low oil pressure. The types of transducer used to sense engine oil pressure are:

- Spring-controlled diaphragm
- Thermal
- Piezo-resistor

Spring-controlled diaphragm Figure 10.4 shows the principle of this type. It consists of an oil-pressure sensor switch which is controlled by a spring-loaded diaphragm. Under low-pressure conditions the switch is closed and the warning light is 'on', but when the oil pressure is sufficient to move the diaphragm and overcome the spring, the contacts are opened and the light is switched-off.

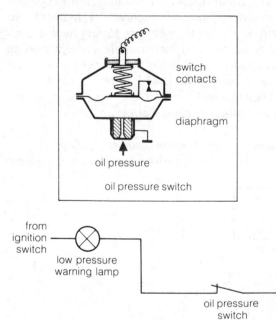

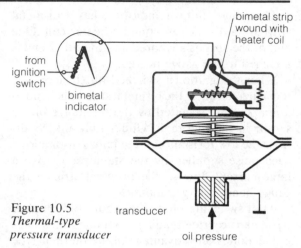

Figure 10.5
*Thermal-type
pressure transducer*

closed until the bi-metal strip has been heated sufficient to bend the strip and open the contacts. The heating effect needed to perform this action matches the effect on the bi-metal strip in the indicator, so the needle of the indicator will register the appropriate pressure.

Piezo resistor The term *piezō* is a Greek word for pressure, so this type of transducer is based on the change in resistance which occurs when pressure is applied to a special semiconductor crystal.

By arranging this resistor in the simple circuit shown in Figure 10.6, a voltage change can be detected when the oil pressure drops below a given value. This signal is passed to the panel where it is processed to operate the warning display.

Figure 10.4 *Spring controlled diaphragm-type oil pressure indication*

Thermal transducer An indication of the actual oil pressure can be given when a thermal type indictor is used (Figure 10.5). The principle of this type is similar to the thermal fuel contents' gauge described previously.

In the construction shown, the pressure on the diaphragm in the transducer holds the contacts

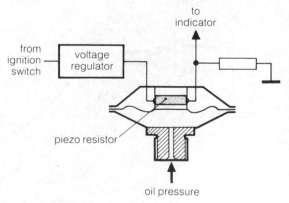

Figure 10.6 *Piezo resistor-type oil pressure transducer*

Tachometer

The common type of tachometer in use today uses the pulses generated by the interruption of the ignition primary current to sense the engine speed.

Figure 10.7 shows a typical circuit for an impulse tachometer. Voltage pulses produced at the negative side of the coil by the circuit breaker allow a transistor array in an I.C. chip in the tachometer to count and then convert the pulses to a steady current; this can be measured by a meter and indicated on a suitably calibrated scale, e.g. when a 4-cylinder, 4-stroke engine is operating at 3000 rev/min, there are 6000 voltage pulses per min or 100 per second: this 100 Hz frequency signals the tachometer to register a reading of 3000 rev/min. Any alteration in the signal frequency produces the appropriate response by the instrument.

Analogue and digital displays are used to indicate speed.

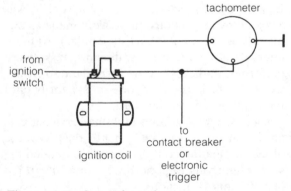

Figure 10.7 *Circuit for tachometer*

Speedometer

Indication of road speed by electrical means overcomes the need for a cumbersome flexible cable drive which is an essential part of the mechanical system used on cars for many years.

Electric speedometers have been in use for a number of years on many p.s.v. and commercial vehicles. These systems use a permanent-magnet a.c. generator to sense the road speed and a voltmeter, scaled to read 'm.p.h.', as the indicating instrument. Being as this type of system senses the speed of rotation of a shaft, an arrangement similar to a tachometer can be used to indicate road speed.

The systems used on modern cars are more compact and less costly than the generator system.

Transistorized pulse generator The speedometer system uses a transducer to generate a series of pulses that correspond to the movement of the output shaft of the gearbox. The pulse frequency transmitted to the speedometer fitted in the panel indicates the vehicle speed; the number of pulses shows the distance covered.

The transducer has a fixed magnet over which passes a rotating plate. This plate has a series of projections that pass across the magnetic pole to generate the pulse. Drive to the plate from the gearbox output shaft is by means of a short length of conventional speedometer cable.

The weak pulse is amplified within the transducer by a small solid-state circuit having two transistors.

The circuit, shown in Figure 10.8 has a single cable joining the transducer to the speedometer. Pulses are passed through this cable to an I.C. chip within the speedometer; this chip counts and converts the pulse to an analogue signal. Needle operation of the speedometer is produced by an action similar to that given by a normal voltmeter.

Maintenance of basic instruments

Most systems can be divided into three sections: sender, wiring loom and instrument unit. When a fault develops, the method of diagnosis is to check-out each section in turn.

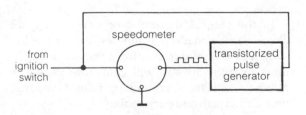

Figure 10.8 *Transistorized pulse generator-type speedometer*

Sensor check   The sensor or transducer is disconnected from the system and its condition is checked. This is carried out by:

- Checking the insulation and circuit resistance with an ohmmeter. This instrument *must not be used* when the sensor contains solid-state devices because the current supplied by an ohmmeter can cause damage.
- Substituting a test instrument which generates a test signal to simulate the signal given by a good sensor (Figure 10.9). If the indicating unit on the panel functions correctly when the test signal is generated, then the fault is most probably in the sensor. Before fitting a new sensor it is recommended that the original sensor is rechecked: in many cases the fault is due to a bad connection.

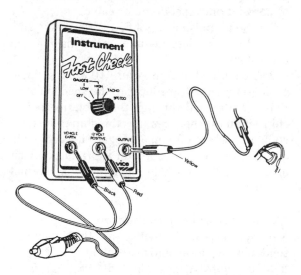

Figure 10.9   *Instrument tester (Austin-Rover)*

In the case of thermal-type gauges, a quick check can be made if a jumper lead is used to short out the sender unit for a few seconds. On most instruments this will cause the needle to move towards the maximum position if the gauge unit and circuit are serviceable.

Cable check   The purpose of a cable is to transmit messages by passing current along the cable. Any unintentional resistance in the cable, or at a connection point, affects the 'message' received by the instrument, so close attention should be given to this vunerable section of the circuit.

Although cable resistance can be measured by a standard meter check, most manufacturers recommend that a signal-generator test should be applied directly to the instrument end of the cable to isolate faults in the cable.

Instrument check   Direct application of the signal generator to the instrument eliminates the remainder of the circuit. If the panel unit fails to respond to this test, then it is concluded that the unit fitted in the panel is faulty. Naturally this conclusion assumes that the test apparatus is serviceable.

10.2 Vehicle condition monitoring (V.C.M.)

In the past, the service state of a vehicle was assessed by inspections; these were made as frequently as every 4800 km (3000 miles). At these intervals, preventative maintenance tasks were performed and checks were carried out to ensure that the vehicle could operate satisfactorily until the next inspection was due.

Against a background of rising servicing cost and inconvenience to the owner by the loss of use of the car, manufacturers gradually extended the service intervals to periods of about 20 000 km (12 000 miles) or more.

During this long period, both in time and distance travelled, fluid levels fall and vital components wear, so to improve safety and avoid breakdown due to system failure, a monitoring system has been developed.

In addition to the monitoring of systems originally checked at a time when the car was being serviced, other areas are now included which hitherto had been the responsibility of the driver, e.g. oil and coolant levels.

The degree of sophistication of the monitoring system depends on the model line. Introduction of V.C.M. generally appears on the 'high-line' models first, but in due course the increased use of the V.C.M. parts reduces the price; as a consequence it allows the system to be used on even the 'low-line' models (Figure 10.10).

Some items such as engine oil pressure and temperature, charging, fuel contents, main beam and directional indicators have been monitored for many years. Nowadays the popular middle-range vehicles have additional facilities to monitor and signal when the:

- Lighting system is defective due to a bulb or circuit failure
- Brake pads need replacing
- Engine oil level is low
- Fluid levels are low in either the cooling or washer systems
- Brake reservoir is low or brake operating system is fauity

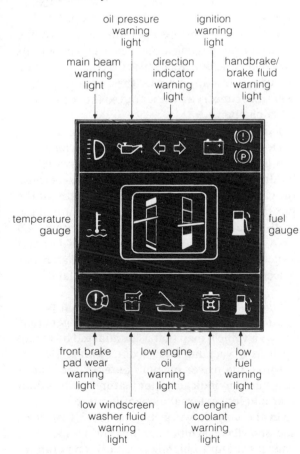

Figure 10.10 *Vehicle condition monitoring*

Warning signals, in various colours and patterns, located in a separate and prominent region in the instrument panel, command the driver's attention by indicating the appropriate warning. On some models the monitoring system is up-rated by the incorporation of a voice synthesizer unit; this uses the radio speaker to communicate operating information or give warnings to support the visual displays.

Most V.C.M. systems use an electronic digital unit to process the information received from the various sensors placed around the vehicle. Many sensors act as a switch. When the warning signal of a defect has to be given, the sensor switch alters its state and either gives, or interrupts, a flow of current in that particular branch of the circuit. The change of current flow triggers an electronic switch in the V.C.M. control unit and a warning display illuminates an instrument panel symbol shaped to conform to the international standard.

If the system is to be trusted by the driver, it is necessary to periodically check the various sensor circuits to ensure that the V.C.M. system is serviceable; this is normally carried out when the ignition is switched-on. A test current, delivered to each sensor and signal lamp for about five seconds, enables the driver to check that the complete system is functioning correctly.

These additional features show why a simple circuit is incapable of meeting modern requirements. The many duties that have to be performed require the fitting of a microprocessor. As the name suggests, the data supplied to this electronic unit is processed and, when necessary, it gives out an electrical signal to activate a warning device.

The principle of a V.C.M. system can be shown by considering some of the sensor arrangements.

Bulb failure
This safety feature is commonly used on vehicles. The arrangement monitors the main lighting system and also includes directional indicators and stop lamps.

Indication of a bulb filament failure, or circuit defect, is generally signified to the driver by lights on a graphical display panel. This shows a plan view, or map, of the vehicle which includes miniature lights to represent the lamps of the vehicle; these display lights become illuminated when a particular lamp is not functioning correctly. On some systems the panel works in the opposite way; it shows the lamps that are in operation instead of illuminating when a fault is present. In most cases an audible warning or extra light on the main panel is used to initially attract the driver's attention; this warning of a defect allows the driver to refer to the vehicle map to identify the faulty lamp.

Many lamp-failure systems, as well as other monitoring arrangements, use a *reed switch* as a sensor.

Reed switch   A reed switch consists of two or more contacts mounted in a glass vial to exclude contaminates. The vial is evacuated of air, or filled with an inert gas, to reduce damage by arcing.

In the type shown in Figure 9.21 the contacts are open and in this position no current will flow between A and B.

Closure of the switch is achieved by using the magnetic flux produced by a permanent or electromagnet. When a given magnet flux acts along the axis of the switch, the contacts close. Opening of the contacts is obtained either by moving a magnet away from the reed or by using another magnet or ferrous-metal vane to divert the magnetic flux away from the reed. When electromagnetic operation is required, a coil is wound around the switch to create the magnetic field.

Reed switches can also be used for sensing:

- The position of a component part (fuel level indicator)
- Speed of movement of a rotating part (speedometer)

Limiting frequency of a reed switch is about 600 Hz, but well before this frequency is reached a contact bounce of duration about 1ms is evident. Life of the reed depends on the application, but in a typical case is of the order of 10^8 cycles. The electrical load must not exceed the manufacturer's recommendations.

Bulb failure module   Figure 10.11(a) shows a module as used on some Austin-Rover cars. Four bulb-failure monitor units are often used: two at the front and two at the rear. Each monitor unit is mounted adjacent to the lamp it is sensing; this eliminates a false signal being sent if a short circuit occurs in the wiring to the lamp.

Each monitor unit consists of a reed switch mounted inside a wire coil (Figure 10.11(b)). The coil is connected in series with the lamp, so when the lamp is operating correctly the magnetic flux produced by the coil closes the reed switch; this passes current to illuminate the appropriate segment of the display panel.

A different construction of reed switch, shown in Figure 10.12 (overleaf), is used on Audi vehicles. This differential relay, a double coil wound around the reed switch, monitors two lamps; each coil provides the current for one lamp.

When the lamps are operating normally, the equal current passing around each coil creates its own magnetic flux. Since the two coils are wound in opposite directions, the opposing flux polarities cancel each other out so the reed switch remains open.

Failure of one lamp filament, or fuse, reduces or stops the current flow in one coil, so a magnetic flux is built up which causes the reed contacts to be drawn together. This action occurs because the closed contacts provide a shorter path for the flux through the centre of the coil. Closure of the reed-switch contacts allows a current, controlled by a load resistor, to pass to the E.C.U.; this switches-on a central warning lamp and activates a speech synthesizer to announce the fault.

Graphic display unit   This unit forms a part of the instrument panel, or console, and generally shows a vehicle map that is illuminated by means such as a vacuum fluorescent display. Besides warning the driver of 'bulb outage', the map is often used to indicate other features such as door ajar and low external air temperature.

In the unit shown in Figure 10.13 (overleaf) the snowflake symbol for low air temperature changes colour to highlight critical temperatures of 4°C and 0°C. This type of display requires 18

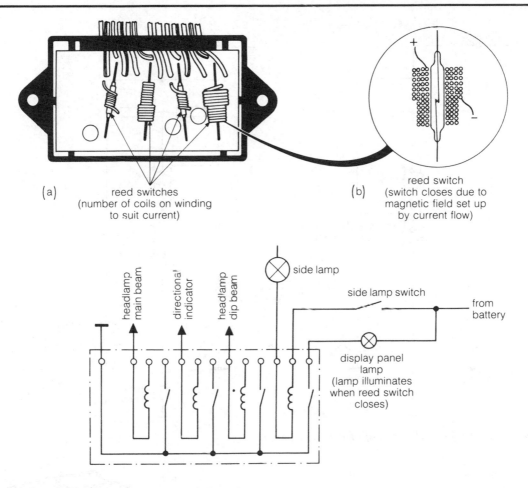

Figure 10.11 *Bulb failure module (Austin-Rover)*

terminals; 3 are used for ignition voltage, earth and sidelamp supply, and 15 are used to supply the logic signals for triggering the map segments.

Brake lining wear

A sensor set into the brake friction material detects when the friction material reaches the end of its life. This is a useful safety feature and it eliminates the need for periodic inspections to assess the degree of wear of the material.

A simple system is shown in Figure 10.14(a) (overleaf). This has an insulated metal contact, buried into each friction surface, which rubs against the drum or disc when the friction material has worn down to its limit. When contact is made with earth a signal lamp illuminates a symbol on the instrument panel.

Although this system is cheap, its reliability is poor. This is because the presence of an open circuit in the system prevents the warning system operating even though the friction material has worn to its limit. This is overcome by using the closed-loop system as shown in Figure 10.14(b).

The loop sensing system has a wire buried in the friction material. This is arranged so that the wire is ground away and cut when the friction material is worn to its limit, e.g. worn to a thickness of less than 2 mm.

The brake sensors are arranged in series with each other and the two cables forming the circuit

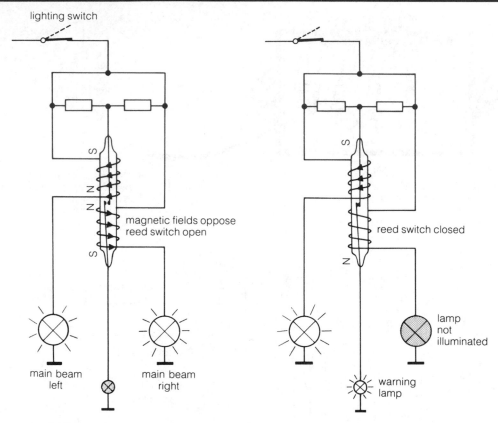

Figure 10.12 *Differential relay bulb failure unit (Audi)*

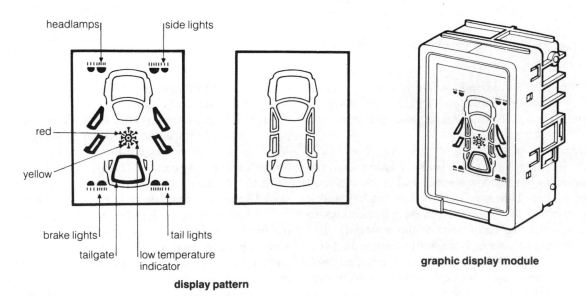

Figure 10.13 *Graphic display unit*

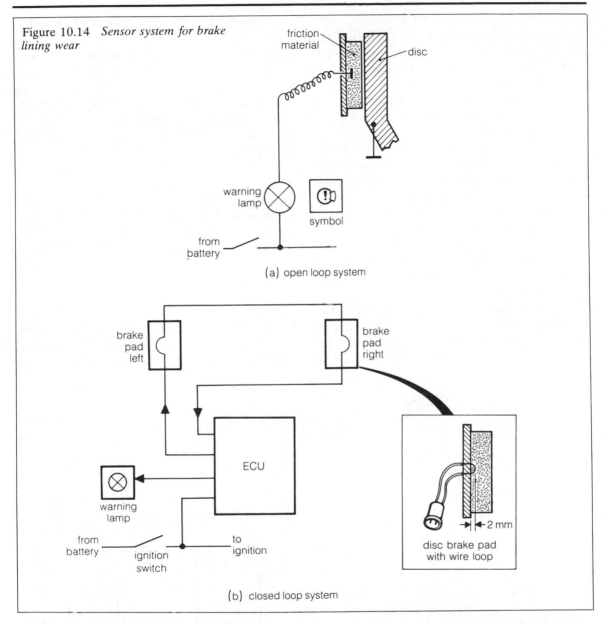

Figure 10.14   *Sensor system for brake lining wear*

friction material

disc

warning lamp

symbol

from battery

(a) open loop system

brake pad left

brake pad right

ECU

warning lamp

from battery

ignition switch

to ignition

2 mm

disc brake pad with wire loop

(b) closed loop system

loop are connected to an E.C.U. This processor unit is programmed to pass a current to the lamp for about five seconds after switching-on the ignition to ensure the system is serviceable.

After the initial check period, the illumination of the warning lamp indicates that a brake inspection is necessary. If a brake check shows that the material is serviceable then tests should be carried out to locate the cause of the open circuit.

When the external circuit is found to be satisfactory, but the lamp still remains 'on', then a fault is present in the E.C.U; this usually requires replacement.

Damage to the E.C.U. will occur if current from an ohmmeter is passed to the E.C.U; this unit should be disconnected when checks are being made to the external circuit.

Engine oil level

One method of sensing the oil level is to use a 'hot wire' dipstick (Figure 10.15). In this arrangement, a resistance wire is set inside a hollow plastics moulding, which carries the normal dipstick oil level marks. Two cables from the dipstick are connected, via the main loom, to the E.C.U.

Operation of the system relies on the increase in resistance of the 'hot wire' when a small current of about 0.2 A is passed through the wire for about 1.5 seconds. This current is supplied only when the ignition is initially switched-on, or when the ignition is switched-on after the engine has been inoperative for over 3 minutes.

When the wire is immersed in oil, the heat from the interrogating current is dissipated to the oil, so the temperature of the 'hot wire' sensor remains constant. But when the oil level drops more than 3 mm below the 'low' mark, the sensor temperature and wire resistance increases. If the resistance during the period of current flow differs from that given when the current is switched-off, the E.C.U. activates the warning lamp.

Faulty operation of this sensitive system will occur if a resistance develops in the circuit, e.g. when a pin of a multi-plug connector becomes dirty. Manufacturers recommend that before carrying out diagnosis checks, all contact pins are cleaned with a typist's eraser pencil or a piece of writing paper.

It is essential to disconnect the E.C.U. from the sensor circuit before attempting to measure the resistance of the 'hot wire'; a typical resistance when the dipstick is dry is 7.5–8.5 Ω. Faults in the circuit can be pin-pointed by comparing the resistance at the sensor with the resistance at the pins of the multi-plug at its connection point with the E.C.U. (Figure 10.16).

Fluid levels

Most V.C.M. systems monitor the levels in the reservoirs of the cooling, windscreen washer, brake system and fuel tank. Float switches are the

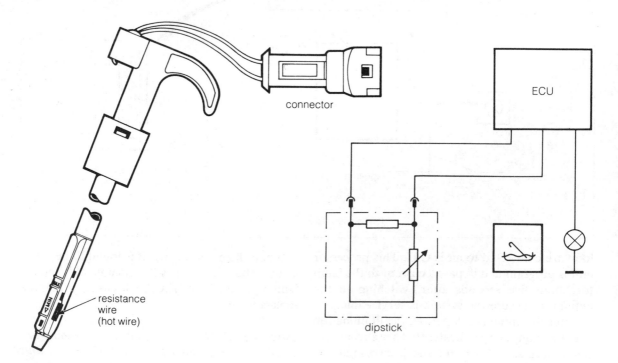

connector

resistance
wire
(hot wire)

dipstick

ECU

Figure 10.15   *Engine oil level sensor*

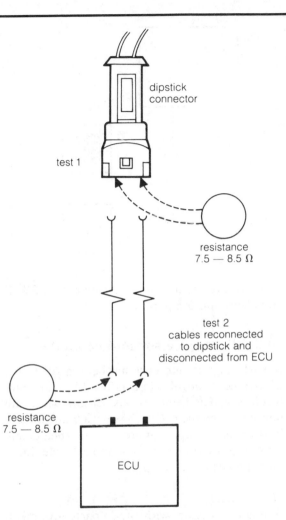

test 1

dipstick
connector

resistance
7.5 — 8.5 Ω

test 2
cables reconnected
to dipstick and
disconnected from ECU

resistance
7.5 — 8.5 Ω

ECU

Figure 10.16 *Resistance test of engine oil level sensor circuit*

mark on the coolant reservoir. The warning is triggered by a sensor magnet as it moves away from the reed switch. At a given point the reed switch opens; this reduces the current flow in the sensor circuit. When the E.C.U. detects this fall in current it passes a signal to the instrument display panel.

To avoid a false signal due to fluid 'slosh', the E.C.U. is programmed to receive a continuous sensor signal for about 8 seconds before it activates the warning display.

Fail-safe sensing is obtained by fitting a fixed resistor in parallel with the reed switch; this provision enables the sensor circuit to be monitored for short- and open-circuit defects.

Figure 10.17(b) (overleaf) shows an alternative construction for a reed float switch in a coolant reservoir. A ring magnet is set in the float and the reed switch is positioned so that the magnet closes the switch when the minimum level is reached. The path to earth provided by the switch signals the E.C.U. to activate the appropriate warning display.

Low coolant level a.c. probe   This a.c. sensor has two flat metal blades that protrude from an insulated boss which is screwed into the coolant reservoir. Two resistors, mounted within the boss, are connected to the blades and to the supply leads which carry a high-frequency a.c. current.

The liquid level is detected by monitoring the impedance of the circuit. When the two blade electrodes are not immersed in liquid, the impedance increases; this raises the a.c. voltage across the sensor terminals and allows the E.C.U. to signal the warning display.

Wiring faults are detected by measuring the resistance across the two resistors; when its value changes, the E.C.U. signals the fault.

A.C. operation overcomes the polarization effects given by d.c. systems and avoids the build up of contaminants on the probe.

Air temperature
Advance warning of the risk of ice on the road allows the driver to adjust speed to suit the conditions.

cheapest and simplest, but other types of probe sensor are used such as those involving a.c. impedance and hot-wire techniques.

Float switch   The fluid level sensor shown in Figure 10.17(a) (overleaf) is fitted on the side of the fluid reservoir. It consists of a reed switch which is activated by a small permanent magnet fitted in the float.

The display warning signal is illuminated when the fluid level falls to the minimum level; typical settings for a small car are 25% of the washer capacity, 7 litres of fuel in the tank or the low

Figure 10.17 *Float*

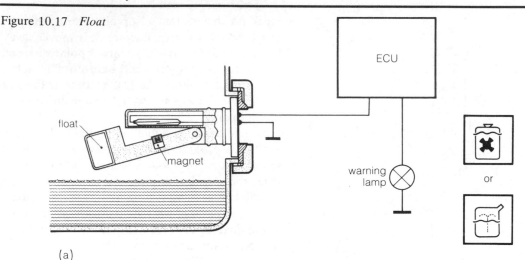

(a)

(b)   switch closes when float drops

temperature resistance coefficient 4.3%/K (K = absolute temperature in kelvin).

10.3 Microprocessor and computer

In past chapters reference has been made to the Electronic Control Unit (E.C.U.). This unit receives signals from various sensors which are interpreted by the E.C.U. against a built-in set of instructions; an appropriate action is then taken. This decision function is a typical role for a microprocessor or computer.

Microprocessor

A microprocessor, with other parts around it, forms a microcomputer. It was introduced by the American semiconductor company Intel when the company was developing a large-scale integrated circuit for a calculator. To meet the specification would have required a very expensive 'large scale integration' (l.s.i.) circuit; this would have had a limited application. Instead the company divided the *hardware* circuit into two: one part processed the data and the other part functioned as a memory unit. These two parts required only an instruction program (software), for storage in the memory, that applied to the particular application; this meant that the hardware could be used for many other purposes. This breakthrough, together with the large volume production that followed, made the microproces-

Sensing of the external air temperature is provided by a special thermistor fitted in a position which is exposed to the air flow (Figure 10.18).

The sensor is mounted in a circuit supplied from the E.C.U. A typical thermistor unit has a resistance of 1 kΩ at 25°C and a negative

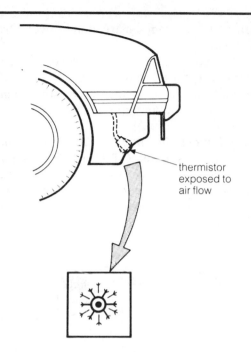

Figure 10.18 *Air temperature sensor*

sor a common part of many machines that required a control device which had to follow either a fixed or variable program.

Because of the way the microprocessor performs this 'thinking' process, it is sometimes called an 'electronic brain'.

System control
The way a microprocessor controls a system can be seen by referring to the V.C.M. system covered in Chapter 9.

Signals supplied to the computer from various sensors placed around the vehicle provide the input information (Figure 10.19). After compar-

ing these input signals with a stored program of instructions in the memory, the microprocessor performs the required duty, e.g. consider a brake monitoring unit. When the brakes are in good condition, a current is supplied from the sensor in the brake pad; this signal is stored in the memory unit when the computer is manufactured. At given intervals when the car is in use, the microprocessor checks the current from the brake sensors and compares it with the data stored in the memory. If the two signals are similar, the microprocessor takes no further action, but if the input signal does not correspond with the stored information, then it issues a command to the instrument panel. Also at this time it may be instructed to send a signal to a voice synthesizer to warn the driver of the detected fault (Figure 10.20, see overleaf).

To enable the microprocessor to perform in a set way, it must be given a description of the tasks it has to carry out. The description is known as the *software* and it is conveyed to the microcomputer by a *program*; this is a list of instructions that is held in a *program store*.

Binary representation
The microprocessor operates on a digital system and uses either high- or low-voltage pulses to transmit messages from one part to another. It is common to use the symbols 1 and 0 to represent high- and low-voltage pulses respectively. Since the microprocessor can recognize only these two levels of signal, the program that dictates its method of operation must be coded in terms of 1's and 0's, i.e. binary notation.

Numbers in everyday use are based on the *denary system*; this notation is based on ten digits,

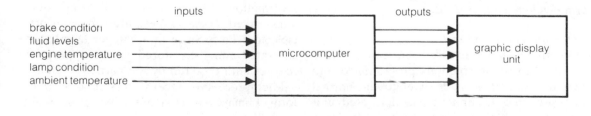

Figure 10.19 *System control*

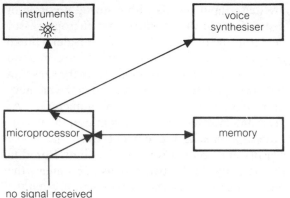

Figure 10.20 *Warning sequence*

0 to 9. Binary notation is based on the number 2 and extra digits represent 'powers of 2'. A binary number of 1 1 1 1 is:

$$1 \times 2^3 + 1 \times 2^2 + 1 \times 2^1 + 1 \times 2^0$$

This value represents a denary number of 8 + 4 + 2 + 1 or 15

Other examples are shown in Table IV on page 50.

Eight binary digits can be used by a microprocessor to represent denary numbers, so by referring to Table IV and inserting 0's to fill the spaces, the binary code 0 0 0 1 1 0 1 1 represents the denary number 27.

Each binary digit is called a *bit* and a group of bits is called a *word*. When a word comprises 8 bits, the word is called a *byte*.

Many microprocessors accept its input data and process instructions at the rate of 8 bits at a time, so programs for these are made-up in this grouping. When instructions are written and supplied to the microprocessor in binary code the term *machine code* is used.

Microcomputer components

Most microcomputers used on cars operate according to set instructions programmed into the unit when it is made. Unlike the normal 'domestic' microcomputer, the dedicated unit fitted on a vehicle generally receives its input signals from various transducers and sensors rather than from instructions given by a keyboard.

The three main components of a vehicle microcomputer are shown in Figure 10.21. This system consists of a microprocessor, memory section and input/output section.

Central processing unit (C.P.U.) The microprocessor acts as the C.P.U. for the computer. This unit is formed on a single silicon chip and its duty is to initiate all actions that take place in the computer; in some case it also controls the action taking place.

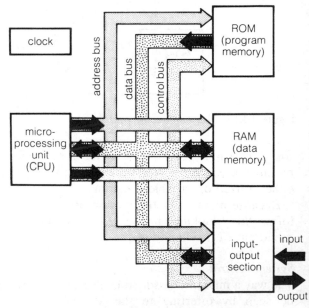

Figure 10.21 *Sections of a microcomputer*

The C.P.U. has the facility to store instructions temporarily while it is processing other data. Information being processed is held in storage 'bins' called *registers*; these are similar in principle to a Parts Store in a garage. Data is first separated and then stored in the appropriate register until required by the particular unit.

Microprocessors receive instructions in the form of binary digits that are made up into words of 8 bits or one byte. When signals such as 00011011 have to be stored, the register is set to hold the 8-bit word in one 'bin', and other 'bins' are used to store other words.

Figure 10.22 shows the registers incorporated in a C.P.U. The *program counter* is the main register and its duty is to record the location, in the memory store, of the instructions that the C.P.U. has to follow. At a given time when the instruction is required, the C.P.U. *fetches* the data from the memory and *executes* the instruction according to the information given to it in the form of a program.

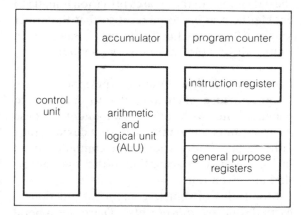

Figure 10.22 *Block diagram of the parts of a microprocessor*

As its name suggests the *arithmetic and logical unit* (A.L.U.) processes information relating to any arithmetic and logical functions that are needed. All data supplied to the C.P.U. requiring addition or subtraction of binary words is directed to the A.L.U. Storage of data that is being used by the A.L.U. during the processing operation is retained in a temporary store called an *accumulator*.

The command unit that directs the processing operation is performed by a *control unit*; this arranges the movement of data between the sections of the computer and provides the appropriate control signals to activate the parts that actually process the data.

Clock signal Movement of data between the various sections is controlled by a time pulse given by a clock; this is given by a quartz crystal that oscillates at a high frequency such as 3.072 MHz. When a voltage pulse of the clock is applied simultaneously to two parts of the computer, then data will pass between the two parts. Other parts of the computer will be inactive unless their contents are 'unlocked' at the same instant by a similar time pulse. This system of data movement allows the various computer sections to be interconnected together in a simple manner by a series of parallel wires called a *bus*. These common wires act as highways for the instructions to pass between the various parts.

Bus These highway interconnections form the communication channels between the various parts of a computer. These highways are multi-lane and in some cases are arranged to give a one-way flow.

A microcomputer bus is divided into three sections which are named according to the signals they carry. These are:

Data bus	Normally eight wires that carry data in both directions between the computer parts.
Address bus	This carries the *address* in the form of a binary code from the C.P.U. to the memory. The signal carried by this bus identifies the actual place in the memory where a given item of information is stored. The bus usually has sixteen wires and signals pass in one direction only.
Control bus	Signals in this bus control the functions of the computer; they select the units required and determine the direction of data movement at a given time. The terms *reading* and *writing* apply to the direction of data movement. Reading means that data is passing to the C.P.U.; writing shows that data is being delivered from the C.P.U.

Memory A memory consists of a number of separate cells which store data bits in a binary form so that they can be read by the microprocessor when it is required at a later time. The capacity of a memory is expressed as the number of cells or the total number of bits it can store at any one time, e.g. 64, 256, 512, 1024, etc. When converted to bytes these capacities become 8, 32, 64, 128.

The memory unit is separated into two parts; one part stores the program for the C.P.U. and the other holds the information data for giving either an input to the C.P.U. or an output from the computer.

A computer *dedicated* to perform only one function, such as the control of a simple fuel system of an engine, only needs fixed instructions that never need altering. A memory for this application is called a *Read Only Memory* (ROM). In this case the C.P.U. will not be able to write to it, so the memory is part of the *firmware* of the computer.

A ROM memory is a non-volatile device; this means that the information remains in the memory the whole time, even when the computer is switched-off. Different types of ROM unit are available; two types are the PROM and the EPROM.

A *Programmed ROM* (PROM) has its data fixed in it when the memory is manufactured whereas an *Erasable Programmable ROM* (EPROM) has a memory that can be altered to suit the application. The latter type is particularly suitable for vehicle-equipment manufacturers because it is cheaper for them to buy a large quantity of standard chips rather than purchase a smaller number of special chips.

A typical EPROM contains a number of cells which are either charged with electrons or left empty to represent the two binary states. When re-programming is necessary, the existing memory is first erased by shining ultra-violet light through a quartz window on the top of the device. After erasing the contents, the device is then connected to a PROM programmer for insertion of the new data.

Some vehicle computers have to process and then temporarily retain a larger quantity of information than that which can be held in the microprocessor. In these cases, an extra memory called a *Random Access Memory* (RAM) is used. This is a volatile device so any data stored in the memory is lost when the power supply is switched-off.

On some vehicles a computer retains information relating to service intervals and any faults it has detected in its own systems or those covered by the V.C.M. layout. This data is held in a RAM unit and to ensure that the information is not lost, the unit has a constant source of power. A separate battery is not always used, so if the battery of the vehicle has to be disconnected, the appropriate instrument readings should be taken before the data is lost. In these cases provision is made to enable the reading to be reset so that the driver knows when a particular service is needed. In future, this feature will be very common.

Peripheral devices Various peripherals are connected to a microcomputer to enable it to communicate with other components. These *interface* units act as input or output devices; they either supply information to the computer or act in a given manner according to the logic signals that are given out.

Monitoring of vehicle-component functions is carried out by *transducers*. These peripheral devices convert physical qualities into electrical voltages. In cases where the transducer emits a voltage of analogue form, the signal must be converted to a digital form before it enters the computer. This is carried out by a small I.C. interface unit called an *analogue-to-digital converter* (A/D). In a similar way, a system requiring a supply of an analogue signal would need a D to A converter placed between the computer and the output system.

Transducers can be divided into two main classes: *passive* and *active*. Passive types have no external power supply so they are generally less sensitive than the powered active types, i.e. the active type gives a larger output signal change.

When an on-board 'calculator-type' computer is fitted for the driver to obtain vehicle performance such as fuel consumption, the input and output peripherals are usually a keyboard and display unit.

Automotive microcomputers are normally fitted with a number of transducers. Figure 10.23

shows a part of an instrumentation system covering a speedometer and fuel-level indicator.

When many sensors are fitted the computer is made to *pole* each one in turn; this allows each signal to be processed and sent to the instrument panel before moving on to work on the information sent from the next sensor in the polling order. An alternative method allows the computer to continue to perform its routine tasks until a transducer signals a change; at this time, the transducer *interrupts* the operation of the computer.

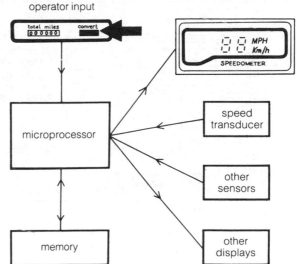

Figure 10.24 *Interrupt applied to speedometer system*

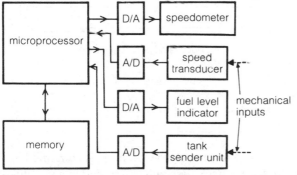

Figure 10.23   *A/D & D/A devices to interface the microprocessor to the transducers*

Figure 10.24 shows how an interrupt is applied to a speedometer system. This layout uses a keyboard to give an input signal to the computer when the driver wishes to change the read-out from mile/h to km/h or vice versa. When this new input is keyed-in, an interrupt senses the change from 0 to 1, or 1 to 0, given by the keyboard; this causes the program to change to suit the new requirement. During the interrupt period the signal data from the transducer are stored in the RAM unit. When the program change is complete the data are supplied to update the panel readout.

10.4 Electronic display

Until about 1980 conventional electro-mechanical analogue instruments were commonly used to give the driver details about the operation of the vehicle.

Technological advances in solid-state display devices and drive circuits, combined with the application of digital electronic systems for the execution of engine and vehicle-control functions, has persuaded many manufacturers to fit a solid-state instrumentation system. This system has the following advantages:

- Faster and more accurate indication
- No moving parts
- Layout is more comprehensive and the display is more attractive
- Improved clarity of display due to the flexibility and format of the display area. Dot matrix, alpha-numeric, ISO symbols and bar graphs may be used
- Greater freedom in display location enables instruments to be sited in the most visible position
- Panel is easier to accommodate due to its compact form.

Figure 10.25 (overleaf) shows an electronic display.

Display technologies fall into two groups: *active* and *passive*. An active display emits light whereas a passive reflects the incident light that falls on it. Current electronic display systems include the following listed overleaf:

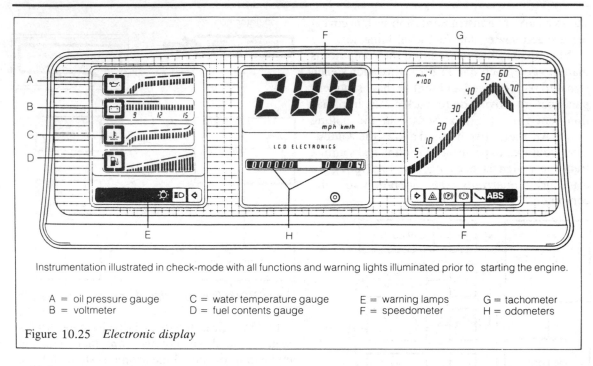

Instrumentation illustrated in check-mode with all functions and warning lights illuminated prior to starting the engine.

A = oil pressure gauge   C = water temperature gauge   E = warning lamps   G = tachometer
B = voltmeter   D = fuel contents gauge   F = speedometer   H = odometers

Figure 10.25   *Electronic display*

- Light emitting diode   LED
- Vacuum fluorescent display   VFD
- Liquid crystal display   LCD
- d.c. electroluminescence   DCEL
- Cathode ray tube   CRT

These systems can be arranged to display data in an analogue or digital form. Systems can be designed as a single unit or can be built up from a series of solid-state modules. Often instrumentation layouts use more than one display system, e.g. LED and VFD.

Light emitting diode (LED)
The widely used LED is one of the simplest forms of display device. In many cases it has displaced the traditional hot-filament lamp used in an instrument panel.

Based on a relatively simple modification of a semiconductor diode, the LED gives a pin-point light of high intensity from the semiconductor junction when a voltage such as 5 V is applied to it (see page 43). Since the forward resistance of an LED is very low, a series resistor must be used to limit the current when there is a risk of the voltage exceeding the normal operating value.

A LED is formed when a diode is combined in a reflective backplate and embedded in a translucent potting material. It is compact, is mechanically robust and has a life in excess of 50 000 hours.

Light radiation patterns are controlled by the escapulation; if it is transparent the LED functions as a point source with the emitted light restricted to a small angle. When the escapulation is translucent the light is diffused over a wide angle up to about 80°. Since the escapulation acts as a lens, the light pattern can be varied to suit the application.

Figure 10.26 shows how a series of LEDs can be arranged to produce a bar-chart pattern for a tachometer. The line formed by the LEDs can be designed to change colour as the engine speed rises. Common colours are red, orange, yellow and green; these colours are obtained by varying the proportions of arsenic and phosphorus in the diode material.

A numeral display based on a simple 'figure of eight' is obtained by using seven LED segments (Figure 10.27). With each segment connected to a common earth, the display of a given digit is achieved by applying a voltage to the appropriate connections numbered 1 to 7 in the diagram. A series of separate LEDs can be used to form letters of the alphabet; this can display simple messages. An alternative display is to use LEDs

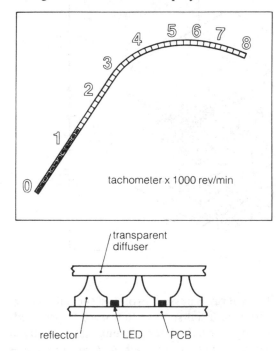

tachometer x 1000 rev/min

section through instrument

Figure 10.26   *LED display as applied to a tachometer*

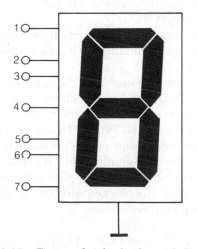

Figure 10.27   *Figure of eight display with 7 segment LEDs*

to illuminate an *annunciator*: a visual signal to indicate a given situation. These signals can represent complete words or be made to display an ISO-recommended symbol.

Vacuum fluorescent display (VFD)

This active display system has a wider colour range than LEDs and even includes a pleasing blue which is difficult to obtain with LEDs. Its general ruggedness and ease of connection to a drive circuit makes the system suitable for the display of numbers, word patterns and bar graphs.

Figure 10.28 (overleaf) shows a VFD display used for a digital speedometer. The anode of this module has 20 small numeral segments, each one coated with a fluorescent substance and connected to a terminal pin.

The segment is activated when the fluorescent surface is bombarded with electrons. When this happens, the surface glows and the yellow-green light that is emitted illuminates the segment.

Electron movement is caused by arranging the construction in a way similar to that used in a triode valve. It consists of a filament that acts as a cathode to emit electrons, and a grid, which controls and evens-out the flow of electrons to the anode. These parts are all sealed in an air-evacuated chamber. This has a flat glass front incorporating a coloured filter to allow the display to be seen in the required colour.

When a current is passed through the thin tungsten filament wires, they heat up to about 600°C and emit negatively-charged electrons. Normally these are attracted to the positively-charged control grid, but when a segment of the anode is also given a positive potential by applying a charge of about 5 V to it, some of the electrons pass through the grid and strike the anode; this causes the anode to glow.

The appropriate segments needed to form the various digits are arranged in a way similar to that used in an LED display.

Liquid crystal display (LCD)

This passive and relatively inexpensive system operates with a low voltage and consumes little power even when a large area is used. Many

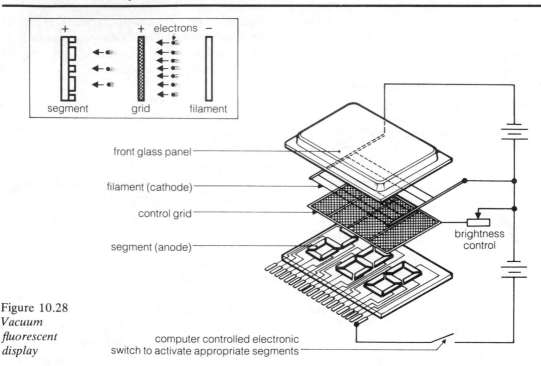

Figure 10.28
*Vacuum
fluorescent
display*

computer controlled electronic
switch to activate appropriate segments

manufacturers select this for their instrumentation system because the display is not washed-out in direct sunlight conditions. It gives a sharp image and by using coloured filters it can be made to present a display in different colours.

The basic system uses reflective light, so when light conditions are poor, back-lighting must be used to make the display visible.

A LCD display utilizes the characteristics of polarized light to produce the image.

Polarized light   A ray of light travels in a wave formation similar to that produced when one end of a length of rope is rapidly moved up and down (Figure 10.29(a)). This action causes a transverse wave, initiated by the 'whip' action, to travel along the length of the rope. In the case of light, the velocity of the wave through space is 300 000 km/s.

Normal light comprises a number of waves which vibrate in many different planes. Figure 10.29(b) shows this effect, but for simplicity, only two waves are shown and represented as AA and BB.

If a light ray is passed through a special polarizing filter material such as that used in polaroid-type sunglasses, the filter prevents the majority of waves from passing through it. Only the waves which vibrate in the same plane as the 'axis' of the filter will pass. A similar action results when a rope is passed through a grid having parallel wires (Figure 10.29(c)).

When vibrations of the light ray are confined to one plane, the resultant light is called *plane polarized light*.

Light rays will not pass through two filters which are 'crossed', i.e. two filters set so that the axis of the second filter is at 90° to the axis of the first filter (Figure 10.29(d)). This can be demonstrated by holding one pair of polaroid-type sunglasses in front of another pair. By rotating one pair it will be seen that a position is reached where the light is blocked out; this is when the polarizing axes are crossed.

Polarized light in an LCD   LCDs are based on a range of strange materials known as liquid crystals; these behave like liquids but have optical

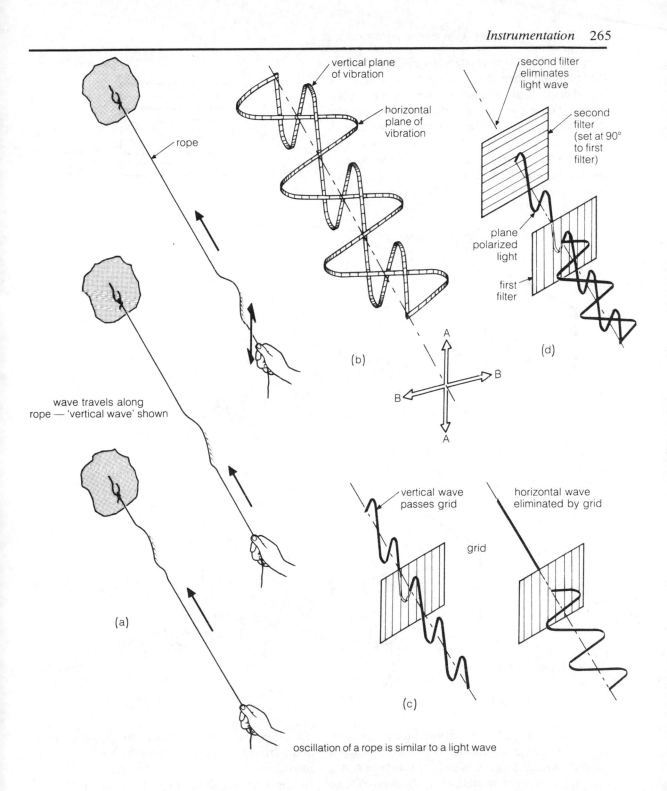

vertical plane
of vibration

horizontal
plane of
vibration

second filter
eliminates
light wave

second
filter
(set at 90°
to first
filter)

plane
polarized
light

first
filter

rope

(b)

(d)

A

B

B

A

wave travels along
rope — 'vertical wave' shown

vertical wave
passes grid

horizontal wave
eliminated by grid

grid

(a)

(c)

oscillation of a rope is similar to a light wave

Figure 10.29 *Polarized light*

properties similar to crystals. In an LCD the long thin nematic (thread-like) molecules forming the liquid crystals are sandwiched between two glass plates. These are placed about 10 μm apart and the crystals are contained in the cell by a perimeter seal. The inner surface of each glass plate is coated with a transparent conductor and polarizing filters, attached to the front and rear of the cell, are set 'crossed' so that their axes are at 90° (Figure 10.30).

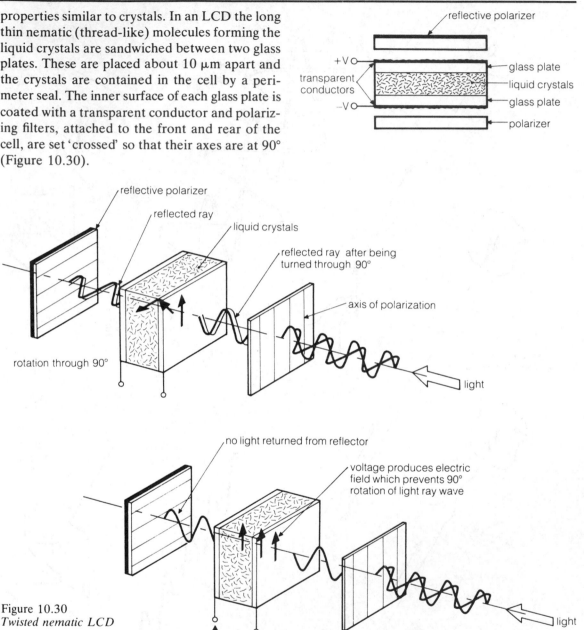

Figure 10.30
Twisted nematic LCD

A common LCD displays a black character against a light background; this type is called a *twisted nematic* LCD. Crystals of this type rotate the plane of light polarization through 90°, so when a reflective surface is formed on the second polarizer, the light is returned by the same path and a bright image similar to the background is provided.

Applying a voltage of 3–10 V at 50 Hz between the two conductor surfaces causes the electric field to rearrange the molecules; this sets them in a position which prevents the 90° rotation of the polarization plane. When the crystals are set in this condition, light cannot pass to the reflector because it cannot get through the second filter. Since no light is returned, a black image is formed and the activated cell forms one part of the character.

Bar graphs or characters are formed by dot or bar segments in a manner similar to a LED unit; each cell requires its individual electrical connections.

Although temperature affects the operation of an LCD, the normal range of −20°C to 80°C of a standard twisted nematic type is satisfactory for automotive use.

Coloured backgrounds can be obtained by backlighting and using colour filters. An alternative method is to use the 'guest-host' arrangement in which a dye (guest) is dissolved in the liquid crystal material (host).

D.C. electroluminescence (DCEL)

An electroluminescent (EL) panel is a solid-state device similar to a LCD cell with the liquid crystal layer replaced by a zinc sulphide-based compound.

This system has many of the advantages of the LCD; in addition it provides a display by emitting light instead of relying on reflected light.

Many colours can be produced by an EL panel. Operation of the system is by d.c. or a.c; often the d.c. electroluminescent panel is used to provide back-lighting for a LCD display.

Cathode ray tube (CRT)

The provision of a single visual display unit (VDU) in the instrument panel enables graphics and other items of driver information, including gauge functions, to be displayed at a comparatively low cost.

The CRT needs a high-voltage supply for its electron beam and it is more fragile and bulky than other displays. Experience gained from the aerospace industry shows that the CRT's brightness, speed of operation, high resolution and simpler driving function makes it very suitable for an instrumentation system.

Driving the display

Electronic display units alphanumeric prompts, patterns and bar graphs; various operating con specific faults.

Irrespective of the type of display, trical connections are needed to supply the various rectangular bars or dot images that form the display.

Standard 7-segment displays (Figure 10.27) require seven electrical connections to form one digit so for the display of vehicle speed, a three-digit stick would require 21 connections plus a number of auxiliary lines. This expensive bulky layout can be reduced by multiplexing, whereby all digits in the stick effectively share the same electrical connections (Figure 10.31). Rapid cycling the driving current between the digits illuminates only one digit at any one time; this cycling action is achieved by making the driving circuit earth each connection A, B and C independently. Being as the human eye retains an image for a short time, the impression is gained that each digit is lit continuously. Flicker is avoided by switching each display segment on and off many thousands of times per second.

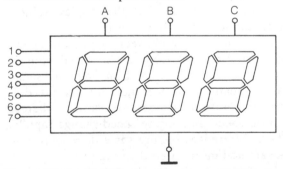

Figure 10.31 *Multiplexing a 7-segment display*

Data sampling

A computer can deal with only one item of information at any one time, so to handle the numerous items of data being fed to it, the multiplexing technique is often used to separate the various signals. Figure 10.32 (overleaf) shows the principle of this system of sampling. In this diagram the *multiplexer* is shown as a switch (M); this selects

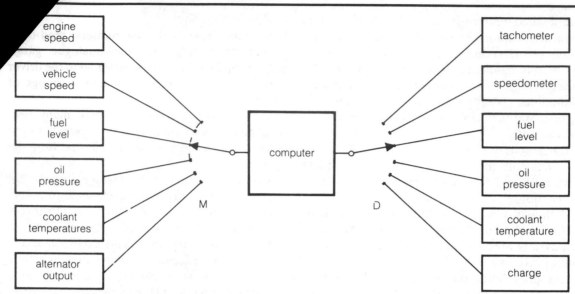

Figure 10.32 *Data sampling by multiplexing*

the signal source and conveys the data to the computer for processing.

After processing, the signal must be transmitted to the correct display area at the right time. This is achieved by fitting a similar switching device, called a *demultiplexer* (D) to the computer output. Naturally the two 'switches' must be timed to ensure that the display subject matches the appropriate sensor.

The switching time between subsequent signals that are received from given signal sources depends on the rate at which an individual signal varies. Quantities such as fuel contents and coolant temperature change very slowly, whereas vehicle speed and engine speed change rapidly. These differences require some data sources to be sampled more often than others.

In addition to the variable sampling periods, a system must also allow for the longer time needed by the computer to process some items of data which are numerous and lengthy in operation; these control functions are programmed into the computer.

System configuration

A typical instrumentation system incorporates analogue displays, on–off warning lamps and digital 7-segment displays. Figure 10.33 shows a configuration of six instruments involving three types of display that are operated by a computer system. Sensor signals, converted into 8-bit digital codes by the A/D unit, are supplied to the C.P.U. by the multiplexer. After processing, the demultiplexer outputs the signals in an 8-bit or on–off form to drive the appropriate area of the display unit.

Electronic speedometer

This system can be actuated by a transistorized pulse-generator type of sender unit driven by a short length of flexible cable from the gearbox output shaft. The sensor shown in Figure 10.34(a) consists of an electromagnetic coil, a four-bladed rotor driven from the gearbox, and a solid-state circuit incorporating two transistors. Two cables are connected to the sensor, one gives a 12 V supply from the battery and the other provides an oscillatory voltage of 1.2–6.5 V at a frequency of 3.5 MHz (Figure 10.34(b), overleaf).

When the rotor pole passes the electromagnet, the oscillation is quenched and this produces a signal similar to that shown. This is passed to the logic circuit in the instrument panel where it is converted into a signal of square wave form.

The speed of the rotor controls the number of pulses generated per second. A typical application produces 5968.8 pulses per mile which the logic board converts to give an output to the speedometer display unit of 1.6588 Hz per 1 mile/hour and 6151 pulses per mile for odometer operation. A stepper motor is often used to drive the odometer.

Speech synthesizer
In addition to the main instrumentation display, some vehicle manufacturers provide a speech synthesizer to convey warning to the driver of critical vehicle operating conditions. This information supplements the normal visual displays and ensures that the driver is made aware of operating problems.

When it is triggered by the onset of a critical operating condition by the main instrumentation mother-board (logic board), the speech synthesizer unit uses the vehicle's audio equipment to announce the warning statement.

Speaker units of audio equipment translate electrical waves into sounds, so by using a computer to generate waves of the required form, it is possible to reproduce human speech. The *phoneme* synthesis technique is one method used. This constructs words from the basic units of sound of a given language. By storing signals that form these sounds in a memory chip, the computer can build up any word, or collection of words, in a manner similar to that used by a human being.

Assuming the memory capacity is adequate, warnings can be stored in several languages, so if the vehicle is supplied to a multi-lingual market, the dealer can select the appropriate language.

Speech processors can be obtained in a single chip form to provide the following basic functions (listed overleaf):

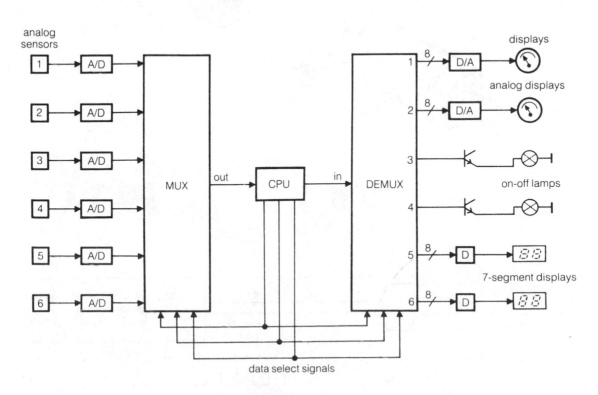

Figure 10.33 *Multiplex system for various displays*

Software-programmable digital filter	Models a human voice, male or female
16K ROM	Stores the data and program
Microprocessor	Controls the:
	(a) flow of data from the ROM to the digital filter;
	(b) assembly of word strings for linking the speech elements together;
	(c) pitch and amplitude information for the control of the digital filter.
Pulse width modulation	Creates a digital signal that can be converted to analogue signal by an external circuit

Figure 10.34 *Electronic speedometer*

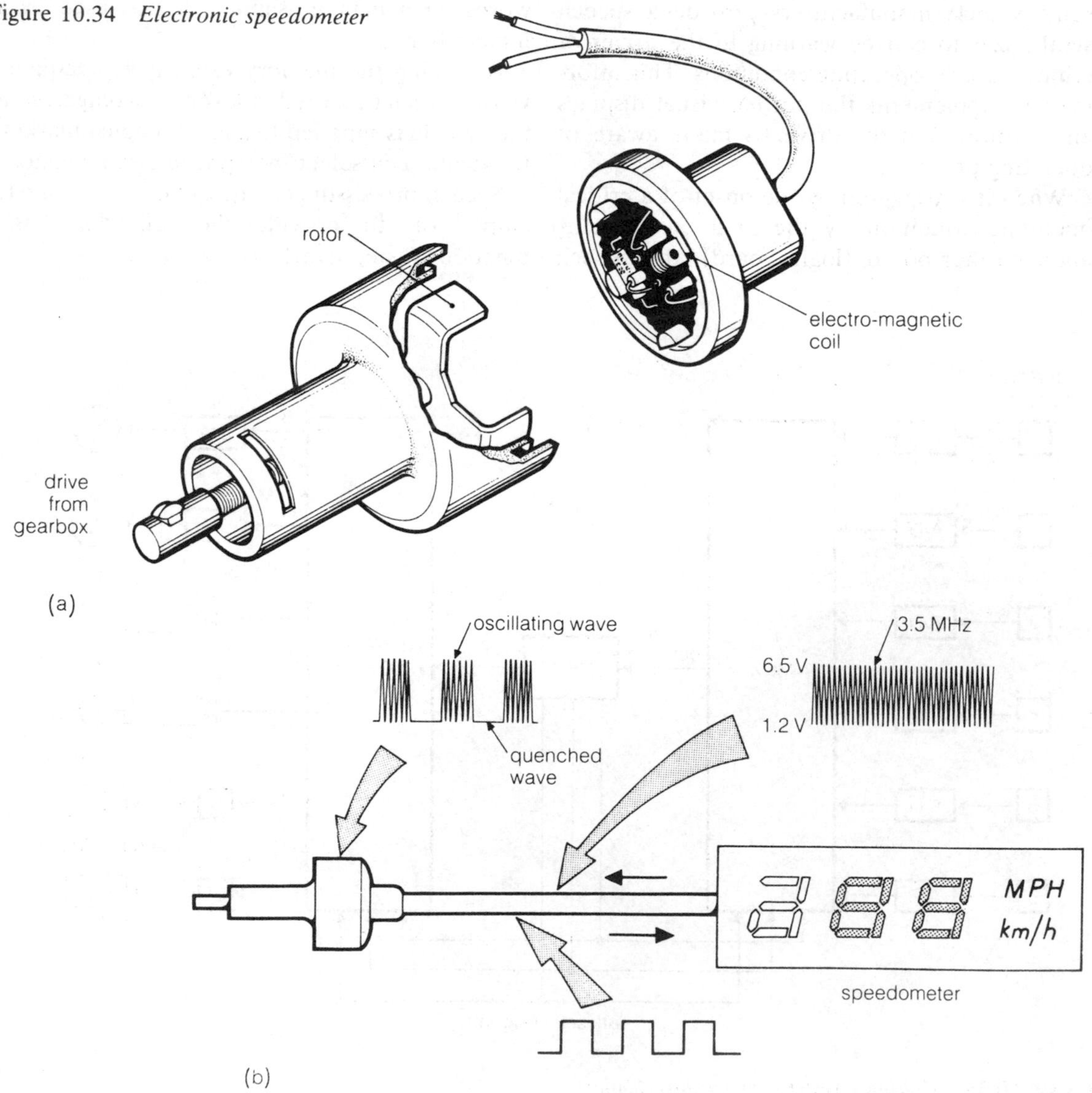

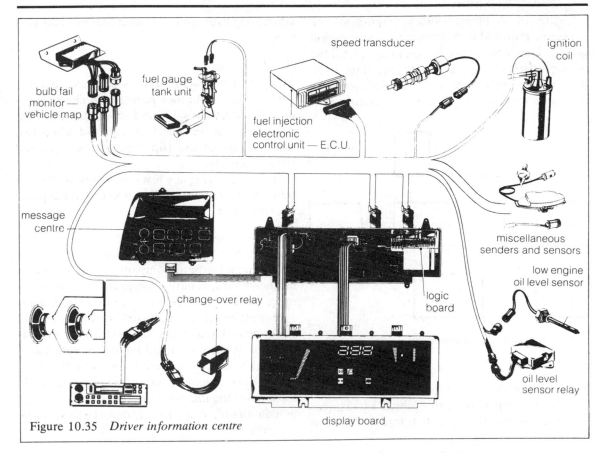

Figure 10.35 *Driver information centre*

Figure 10.35 shows a speech synthesizer module similar to that used on some Austin-Rover cars. This unit also incorporates a LCD message centre and a trip computer.

The V.C.M. feature overrides the trip computer and initiates warnings from the following inputs: engine temperature, engine oil pressure, battery charging, brake pads worn, brake fluid low, parking brake on, lamp failure, doors not shut, outside air temperature, vehicle servicing due, low engine oil, low engine coolant, low fuel level and low screen-wash fluid level.

With this model, warnings are given as a text on the message display panel and audibly by the speech synthesizer. Messages are graded into five groups according to their priority; high-priority messages such as 'low oil pressure' override any other message of lower priority being displayed at that time.

Trip computer

Trip computers have been available for several years; they were first introduced as an after-market accessory.

The use of electronic instrumentation systems and their associated sensor signals make it comparatively easy and cheap to use these signals to provide an input for a trip information computer.

In addition to giving the time and date, a typical module can compute:

- Average speed
- Estimated time of arrival
- Fuel used
- Instantaneous fuel consumption
- Average fuel consumption
- Distance to empty fuel tank
- Average fuel cost per mile

Figure 10.36 shows a block diagram of the hardware required to provide a trip information system. Two sensors, in addition to signals for the odometer and clock, are already part of the basic instrumentation system, so a fuel-flow sensor is the only extra part required by the computer.

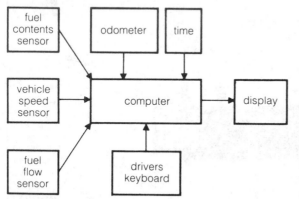

Figure 10.36   *Trip computer system*

Fuel-flow sensor   The signal representing fuel flow can be derived from the vehicle's fuel injection system or from a separate transducer fitted in the fuel line.

One type of fuel line transducer has a turbine and a light interrupter. When fuel flows through the unit, the turbine rotates and this interrupts a beam of light emitted from a LED directed on to a phototransistor (see page 46).

This causes the unit to produce a square-wave output at a frequency proportional to the rate of fuel flow. One model in use produces 10 404 pulses per litre of fuel flowing through the transducer; an accuracy of ±3% is claimed.

10.5 Maintenance and fault diagnosis of electronic displays

At first sight an electronic instrumentation system looks very complicated, but when the layout is considered as a series of separate units, the system becomes easier to understand and simpler to repair. Initially it is necessary to understand the method used to operate the system before attempting to diagnose apparent faults in the system.

Panel self-tests
Many systems incorporate a self-test function whereby the computer performs a check sequence of its main display and speech features. On some systems this check is initiated when two specific keys of the trip computer are pressed simultaneously.

Before applying any test equipment to the system it is recommended that a full test of the panel is carried out.

Need for care
Compared with normal electrical equipment, a display panel with its companion logic board, or mother-board, is easily damaged. With replacement costs in the region of £500–£1000, great care must be taken when working on this type of equipment. Fault-diagnosis tests should be carried out as recommended by the manufacturer, and unless otherwise stated, *full-battery voltage should not be applied to any of the panel inputs*. In most cases the incorrect use of a test meter such as a ohmmeter, can seriously damage the computer circuits.

Checking the driver information centre
The following description applies to an Austin-Rover type of electronic display panel. It is intended that this outline will introduce the reader to some of the main checks. Further study of the manufacturer's literature is necessary before attempting any work on this type of equipment.

Pin connectors   Many models use a number of connectors to link the harness to the panel. The connectors are generally coloured for identification purposes and locking tabs are often used to hold the connectors secure. Pins, and their mating sockets, must not be damaged during the testing operation, especially when meter connections have to be made to the harness. Often damage can be avoided by using a spare connector plug when meter connections have to be made.

Failure of individual segments   Setting the panel to its normal static display mode enables various sections of the panel to be checked. Failure of one or two segments to illuminate indicates that the logic board is transmitting the correct pulse signal via the multiplexed circuit, but the symptom shows that the display segments are not functioning correctly. In this case the display panel normally has to be replaced.

Failure of one instrument unit   If a unit, such as the temperature gauge, fails to operate, the first check is to examine visually the wiring harness between the sensor and the instrument panel. Assuming this is satisfactory, the next step is to use the appropriate test equipment to locate the fault.

Any fault in the system must be situated in one of three sections, namely: sensor, wiring harness, instrument panel or connectors joining these three sections. To locate this fault, the Austin-Rover Company produces 'Fast-Check' testers to simulate the signals produced by the sensors. By substituting these signal generators for the sensors, the remainder of the circuits can be checked (Figure 10.37).

If the display shows the correct reading when the test signal is applied to the system, then the sensor is assumed to be faulty.

When 'no display' is obtained, then the Fast-Check tester is applied direct to the appropriate input socket of the panel. Failure to obtain the correct reading with the tester giving an input at this point shows that the panel is faulty. Conversely, if the tester produces a display, then a fault is present in the harness or connectors.

Computer Fast-Check   This tester reproduces the signals of the fuel flow and speed sensors. The operation of the computer and display unit can be verified by using a method similar to that described previously.

LCD instrument tester   This tester, which is applied direct to the instrument panel, provides reference input signals for the main panel and message centre, so the operation of the complete driver information centre can be ascertained. This test is used generally to confirm the diagnosis when previous tests have suggested that the panel is faulty.

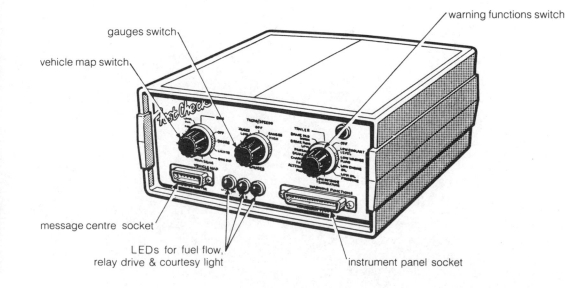

Figure 10.37   *Fast-Check tester*

Dismantling a mother-board

When a part of the instrument panel is found to be faulty and a new part has to be fitted, the complete assembly normally has to be removed. As in the case of other electrical components, the battery should be disconnected before work is commenced.

Separation of the mother-board from the display panel should be carried-out in a clean area. Also, precautions should be taken to prevent static electricity from the human body damaging the I.C. chips; body static can be discharged by periodically touching a known earth point.

Panels should be handled by touching the edges only, and surfaces such as the non-reflective display window and face of the display board should not be fingered.

11 Engine fuelling

11.1 Carburettor systems

For many years the carburettor has been used to provide the engine with petrol and air suitably proportioned, mixed and atomized to form a combustible mixture.

Metering requirements
Under ideal conditions the air–fuel ratio is 15:1 (by weight). This proportion is called the *chemically correct ratio* because when this air–fuel mixture is ignited, it burns completely to form carbon dioxide (CO_2) and water (H_2O).

When a richer mixture such as 12:1 is supplied to an engine, the fuel consumption is increased. Also undesirable gases such as carbon monoxide (CO) and nitrogen oxides (NO_x), are exhausted from the engine.

Less pollution of the atmosphere occurs when the engine is operated on a weaker mixture, e.g. 17:1. Although slightly weaker mixtures give the best economy, the power output is not so good. In addition, a weak or lean mixture is more difficult to ignite, is prone to detonate (i.e. cause combustion knock), and since it burns slower, is more likely to overheat the engine.

With the many problems associated with weak mixtures, the need for its use might well be questioned. However, the overriding reason for its use is the need for the engine to conform to the stringent statutory regulations that exist in many parts of the world in respect of exhaust pollution and fuel economy.

Special lean-burn engines are now used to provide a cleaner and more efficient power unit, but in order to satisfy statutory regulations, which demand that the engine shall remain in tune for a given period, a close control of the air–fuel mixture is needed. It is in this control area that electronics out-performs other methods.

Electronic carburettor
Electronic actuation is used to control various carburettor functions; these include:

Cold-starting mixture	Automatic 'choke' control
Speed control	Slow-running speed adjustment to allow for temperature variation of the engine and ambient (outside) conditions
Fuel cut-off	Improves economy and emissions by cutting-off the fuel both on overrun and when the engine is switched-off (without this cut-off many lean-burn engines would continue to run due to dieseling)
Mixture control	Varies the air–fuel ratio to suit the conditions

Cold-starting mixture control
In the past, many mechanically-operated automatic chokes have lacked sensitivity because of the limitations of the sensor system. This has resulted in unreliable operation, high emissions and high fuel consumption, especially when the vehicle is operated on short journeys.

Figure 11.1 (overleaf) shows an S.U. type of carburettor in which the cold-starting mixture is supplied by an auxiliary carburettor. Petrol–air metering and mixing takes place in a separate chamber which is activated by a cylindrical valve called a 'rotary choke'. This valve is rotated via a 9:1 reduction gearbox by a stepper motor that is energized by an E.C.U.

Figure 11.1 *Carburettor with auxiliary cold-starting system (simplified)*

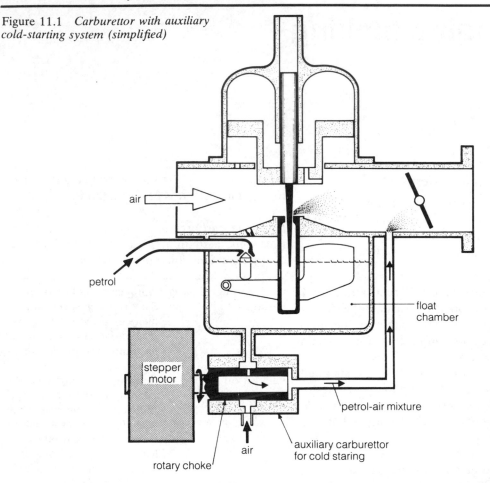

air

petrol

float chamber

stepper motor

petrol-air mixture

auxiliary carburettor for cold staring

air

rotary choke

The E.C.U. receives signals that sense ambient air temperature, engine coolant temperature and engine speed (Figure 11.2). Temperatures are measured by thermistors and engine speed is sensed from pulses generated by the ignition primary circuit. When these signals indicate the need for 'choke' operation, then pulse signals are sent from the E.C.U. to the stepper motor. This provides mixture enrichment to suit the initial start conditions. In addition, this feature also weakens-off the mixture as the engine warms up.

Stepper-motor control of the 'choke' works in conjunction with a vacuum valve that enriches the mixture when a cold engine is accelerated (Figure 11.3, overleaf). This valve controls the quantity of air that is allowed to enter the 'choke' system. With this feature the mixture supplied by the 'choke' can be set closer to the ideal; in consequence better economy is achieved.

A permanent-magnet-type stepper motor is used; this has a step angle of 7.5° and works through a range of three revolutions.

Pulse signals to the motor are supplied from the E.C.U. by five cables: one supply and four returns. This circuit arrangement enables the direction of flow through the motor windings to be controlled by earthing out the appropriate return.

Speed control

The throttle opening required to sustain a satisfactory idling speed is critical especially when the engine is cold and the oil is 'thick'. Slight alteration of the throttle, or variation of the engine

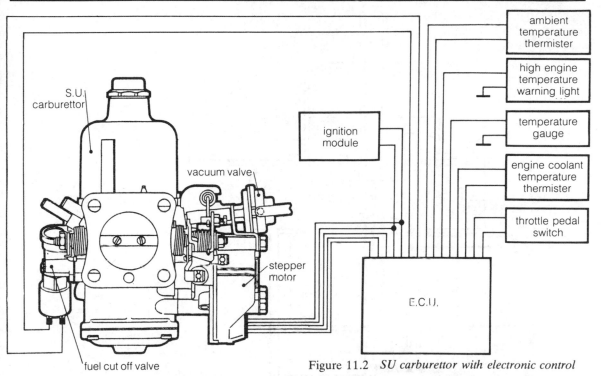

S.U. carburettor

vacuum valve

stepper motor

fuel cut off valve

ignition module

ambient temperature thermister

high engine temperature warning light

temperature gauge

engine coolant temperature thermister

throttle pedal switch

E.C.U.

Figure 11.2 *SU carburettor with electronic control*

load, causes the engine to either race or stall, so it is difficult to set a fixed throttle stop to suit all operating conditions.

Mechanically-operated chokes, manual and automatic, normally include a throttle jack to slightly open the throttle when the engine is cold. These systems give a generous throttle opening to prevent stalling during the warm-up period.

Electronically-controlled carburettors, such as the S.U. type, include a provision for adjusting the idling speed to suit the engine conditions. Being as the speed is continually monitored, the throttle stop can be reset at 1-minute intervals. This feature allows a lower idling speed, so a lower fuel consumption is obtained.

Idling-speed control uses the stepper motor that operates the automatic choke. The first revolution of the motor controls the throttle stop setting for fast idle; the remainder of the movement controls the mixture enrichment.

When the engine is warm and the E.C.U. detects that a stall is likely, a pulse signal is sent to the stepper motor; this causes the engine speed to be increased by 100–200 rev/min. By polling the engine speed every 60 seconds, the E.C.U. is able to keep the engine running at a low speed.

(Stepper motor control is also used on some Weber carburettors for idle-speed control.)

Fuel cut-off valve

Improvements in fuel consumption and exhaust emission levels are achieved by cutting-off the fuel supply when the engine speed exceeds about 1200 rev/min with the throttle closed. The S.U. carburettor achieves this by reducing the air pressure in the float chamber. A solenoid valve, controlled by an E.C.U., is fitted in an air passage connecting the top of the float chamber with a 'vacuum' region between the venturi and the throttle valve (Figure 11.4, overleaf).

When the solenoid is energized and the vacuum valve is opened, a depression is formed above the fuel in the float chamber; this reduction in pressure stops the flow of fuel to the venturi.

Solenoid energization occurs at half-second intervals to prevent stalling and extends for a maximum duration of 9 seconds. The E.C.U. is

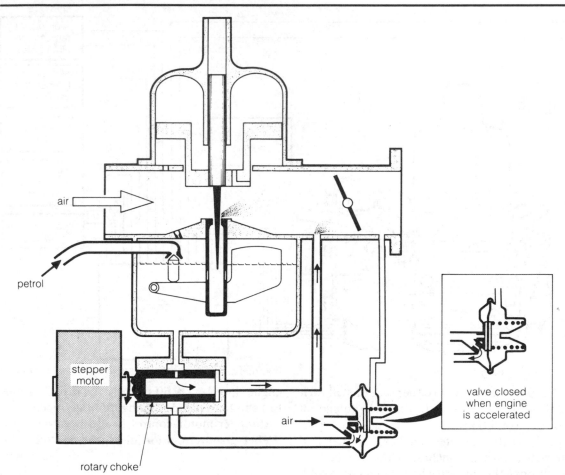

air

petrol

stepper
motor

rotary choke

air→

valve closed
when engine
is accelerated

Figure 11.3   *Vacuum valve to vary mixture when engine is accelerated*

programmed to energize the solenoid only when the air and coolant temperatures are above 6°C and 80°C respectively.

An alternative arrangement used on a Bosch–Pierburg Ecotronic carburettor closes the throttle valve completely during over-run at speeds above 1100–1400 rev/min. This design has a slow-running fuel outlet on the air-intake side of the throttle, so all fuel supplies are cut-off when the throttle is closed. Throttle actuation is by a diaphragm which is controlled electronically from an E.C.U. (Figure 11.5).

Anti-run-on valve

Nowadays many carburettors are fitted with a solenoid operated valve to cut-off the idling mix-

ture when the engine is switched-off. This prevents the engine running-on or 'dieseling' due to the residual heat in the combustion chamber. Lean-burn engines are prone to this problem.

Carburettors that close the throttle to cut-off the fuel during overrun often use this throttle-closing provision to prevent run-on.

Mixture control

Constant-choke carburettors require a compensation system to prevent the air–fuel mixture becoming too rich as the engine load is increased. Although these systems have met past requirements, the stricter standards demanded by current emission and economy regulations, together with the need for extra carburettor refinements,

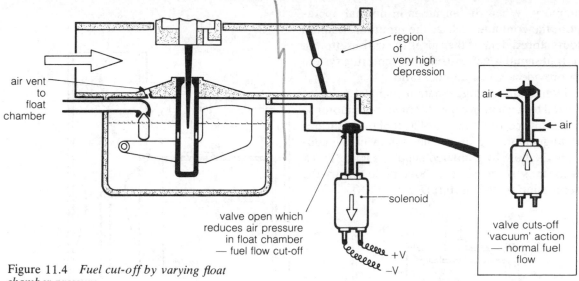

Figure 11.4 *Fuel cut-off by varying float chamber pressure*

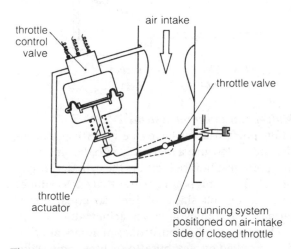

Figure 11.5 *Fuel cut-off by throttle actuation*

make electronic control an attractive solution. The alternative is to use petrol injection but a p.i. system is generally more expensive.

In most electronic carburettors, the basic layout resembles a simple carburettor with the final adjustment of the air–fuel ratio dictated by an E.C.U. The output from the E.C.U. controls a separate metering system; this supplements the fuel provided by the basic system and gives an air–fuel ratio to suit the conditions as sensed by the various transducers. The main sensing system signals the engine speed and load. In addition electronic carburettors have extra transducers to sense other important variables that affect the air–fuel ratio requirements (Figure 11.6).

Generally the E.C.U. is a computer which is pre-programmed with the mixture requirements for a particular engine. Engine tests during development of the engine determine the air–fuel ratio needed to suit the various operating conditions; this data is then entered and locked into its

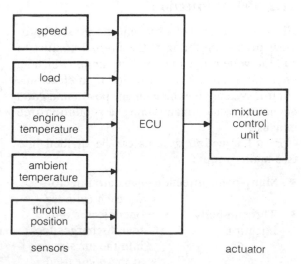

Figure 11.6 *Mixture control system*

memory. When the engine is in normal operation, the computer relates the input information to its stored data; it then gives the carburettor an output signal which instructs the metering system to provide a given air–fuel ratio.

Digital or analogue output signals are converted by a vacuum regulator or stepper motor to physically control the rate of flow of fuel. Various arrangements are used; one system in use controls the flow by a tapered needle (Figure 11.7). Another system uses the choke plate to vary the depression in the venturi (Figure 11.8).

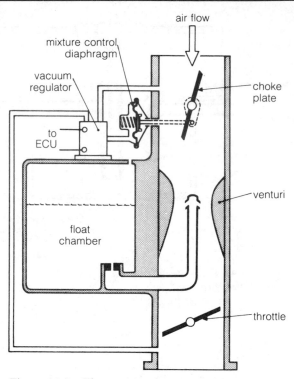

Figure 11.8   *Electronic carburettor – choke plate control*

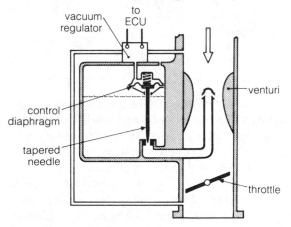

Figure 11.7   *Electronic carburettor – tapered needle control*

11.2 Petrol injection

Electronic control of a fuel injection system provides precise metering of the air–fuel mixture to suit the wide range of conditions under which an engine operates. The sensitivity of an electronic control system gives high engine power and good economy while maintaining a pollution-free exhaust.

Petrol injection systems can be divided into two main groups:

- Multi-point injection – separate injectors for each cylinder
- Throttle body injection – one injector only that discharges fuel into the air stream at the point used by a carburettor.

Multi-point injection systems

Multi-point systems have one injector per cylinder situated to give a fuel spray into the air stream at a point just before it enters the cylinder. Injection of fuel at this point ensures that each cylinder receives its full share of fuel, so equal power output from the cylinders is achieved.

Overlap of induction strokes of adjacent cylinders supplied by a carburettor causes some cylinders to rob other cylinders of their full charge. This induction robbery drawback is minimized by a multi-point fuel-injection system so an improved all-round engine performance is obtained. Unfortunately, the multi-point system is more expensive than a system which uses either a carburettor or throttle body injection.

Injection of the petrol takes place in the induction manifold. Normally the fuel spray is directed towards the inlet valve as shown in Figure 11.9. This *downstream* spray is produced by a pressure of about 2 bar (30 lbf/in²) which is either 'timed'

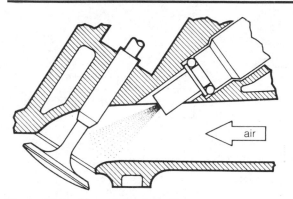

Figure 11.9 *Downsteam injection*

or 'continuous'. The former method gives an intermittent spray from injectors which open at least once every cycle whereas the continuous method delivers a constant stream of fuel at a rate proportional to the quantity of air that is entering the engine.

Multi-point systems are controlled by either mechanical or electronic means. The former uses a mechanical system to measure the air and fuel, whereas the latter measures, meters and injects the fuel by electronic means.

Most modern mechanical systems need some form of electronic control to make the system sensitive to changes in temperature and pressure, so this need, together with electrical operation of the pump, makes the system very different from early types that were completely mechanical.

Electronic systems   The first fully-electronic system was introduced by Bendix in the USA in 1950. In 1967 a similar unit was developed by Bosch and fitted to a Volkswagen car. Since that time, electronic injection has become the common system for many luxury and sport-type cars. With the greater emphasis placed on cleaner exhaust products, the electronic system is now used in the popular car market.

A number of different types of fully-electronic system is in use. The main difference between them is the way in which the air flow is measured; the two main systems are:

● Indirect or pressure-sensed airflow
● Direct air-flow measurement.

Pressure-sensed airflow   This system uses a manifold absolute pressure (MAP) sensor to measure the manifold depression. Signals from the MAP sensor are passed to the E.C.U. and, after taking into account the data received from other sensors, the E.C.U. signals the injector to open for a set time; this is proportional to the quantity of air that the engine is receiving.

Bosch D-Jetronic   This is a good example of a pressure-sensed system (the 'D' stands for *Druck* which is the German word for 'pressure'). Figure 11.10 (overleaf) shows a typical layout of this system.

In this layout the quantity of air induced into the engine depends on the manifold pressure and the throttle opening. These two variables are measured by an MAP sensor and a throttle position switch respectively.

The electrical control system performs two duties: it signals the start of injection and determines how long the injector is to stay open. The latter controls the quantity of fuel to be mixed with the air charge, so the duration increases as engine speed, or load, is increased.

Commencement of injection is triggered by either a switch in the ignition distributor or a sensor situated adjacent to the flywheel. For 6-cylinder engines, the injectors are operated in sets of 3, i.e. 3 injectors spray at the same time.

Direct air-flow systems   Systems using this principle normally use solenoid-operated injector valves; these have a variable opening time to suit engine speed and load conditions.

Bosch L-Jetronic   This was one of the first designs to use electronic sensing of the air flow. In this case the 'L' stands for *Luft* which is the German word for 'air'.

Measurement of air flow is obtained by using a:
● vane or flap
● hot wire
The principles of these two sensors are described on page 230.

Vane or flap metering   Figure 11.11 (overleaf) shows the layout of a system similar to a Bosch

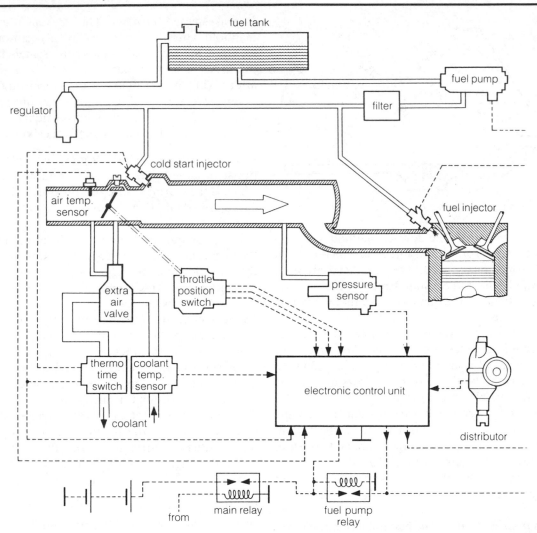

Figure 11.10   *Bosch D-Jetronic electronic system for fuel injection*

L-Jetronic. Fuel pressure, produced by an electrically-driven fuel pump, is maintained constant at 2 bar by a pressure regulator. Injector valves are actuated once per crankshaft revolution and the length of the opening pulse is computed by an E.C.U. from the signals supplied from the air-flow meter, throttle position switch and engine block temperature sensor.

During cold-start and warm-up conditions, the cold lubricating oil produces a greater drag, so to compensate for this an auxiliary air valve allows a small quantity of air to by-pass the throttle. This action is similar to the fast-idle feature provided in a carburettor.

Air flow to an engine is irregular, so this causes pulsations which make it difficult for the air-flow sensor to measure the flow accurately. To minimize this problem the induction manifolds of fuel-injected systems incorporate a plenum (air) chamber. A volume of about 0.8–1.2 of the engine capacity is normally sufficient to damp the pulsations and smooth the air flow.

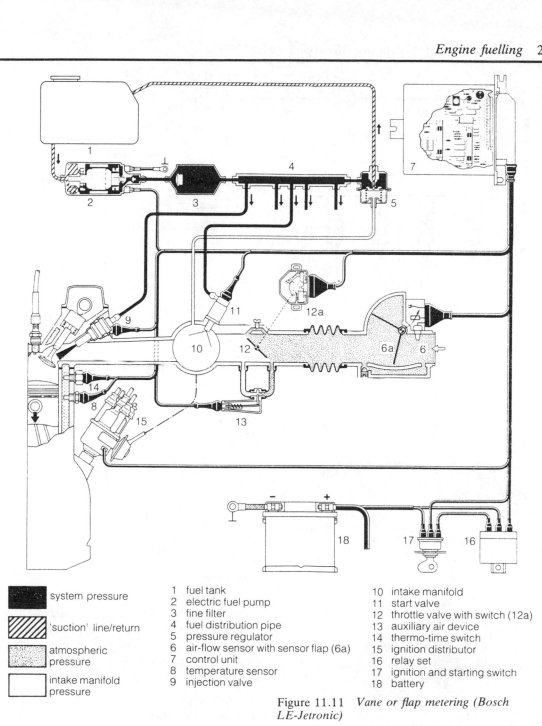

■ system pressure	
▨ 'suction' line/return	
░ atmospheric pressure	
□ intake manifold pressure	

1 fuel tank
2 electric fuel pump
3 fine filter
4 fuel distribution pipe
5 pressure regulator
6 air-flow sensor with sensor flap (6a)
7 control unit
8 temperature sensor
9 injection valve

10 intake manifold
11 start valve
12 throttle valve with switch (12a)
13 auxiliary air device
14 thermo-time switch
15 ignition distributor
16 relay set
17 ignition and starting switch
18 battery

Figure 11.11   *Vane or flap metering (Bosch LE-Jetronic)*

Hot-wire metering The Lucas EFI and the Bosch LH-Jetronic use this principle. Both systems have a basic layout similar to many of the previous systems described, but extra refinements are incorporated to improve the engine's performance. A hot-wire sensor measures the inducted air mass directly and the indicated value takes into account the density and temperature of the air. Consequently, the closely-controlled air–fuel ratio limits exhaust

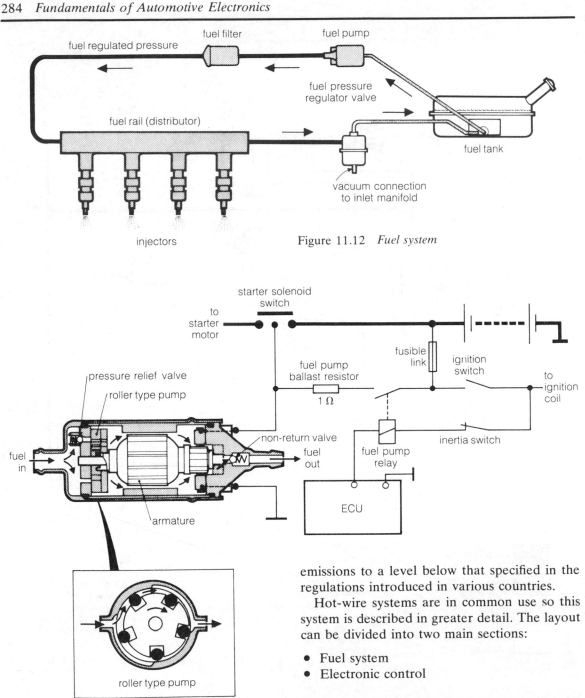

fuel filter

fuel pump

fuel regulated pressure

fuel pressure
regulator valve

fuel rail (distributor)

fuel tank

vacuum connection
to inlet manifold

injectors

Figure 11.12   *Fuel system*

starter solenoid
switch

to
starter
motor

fusible
link

ignition
switch

pressure relief valve

roller type pump

fuel pump
ballast resistor

1 Ω

to
ignition
coil

non-return valve

fuel
in

fuel
out

fuel pump
relay

inertia switch

armature

ECU

roller type pump

emissions to a level below that specified in the regulations introduced in various countries.

Hot-wire systems are in common use so this system is described in greater detail. The layout can be divided into two main sections:

- Fuel system
- Electronic control

Figure 11.13   *Fuel pump and electrical supply circuit*

Fuel system   The purpose of a fuel-supply system is to provide the injector with adequate fuel at a pressure sufficient to allow the injector to give good atomization (this means that the fuel is mechanically broken up into fine particles).

The layout shown in Figure 11.12 includes the following parts:

Pump This is normally a roller-type pump driven by a permanent-magnet electric motor (Figure 11.13). Rotation of the pump moves the rollers outwards and seals the spaces between the rotor and casing. As the fuel is carried around with the rotor, the combination of the rotor movement and the decrease in volume causes an increase in pressure.

Fuel from the pump passes through the motor; this aids cooling. Although sparks are generated in the motor, combustion does not occur because insufficient oxygen is present.

The pump is designed to supply more fuel than is needed; excess fuel is recirculated back to the tank. This feature reduces the risk of vapour-lock problems.

Two ball-valves are fitted in the pump: a non-return valve at the outlet and a pressure-relief valve to limit the maximum pressure.

The pump is controlled by the E.C.U. via a fuel pump relay. Supply to the pump is taken through a ballast resistor which drops the voltage to 7 V. This resistor is shorted-out when the engine is being cranked to compensate for the lower battery voltage.

On switching-on the ignition, the pump motor runs for a short time to fully pressurize the system. After this initial period the pump is stopped until the engine is cranked.

For safety reasons an inertia switch is fitted in the supply line to the pump relay. This switch opens if it is jolted so, in the event of a collision, the pump ceases to operate. The switch can be reset by pushing down a protruding plunger.

Fuel pressure regulator This controls the operating pressure of the system and is set to maintain a constant pressure difference (e.g. 2.5 bar or 36 lbf/in²) above the manifold pressure irrespective of the throttle opening. It consists of a spring-loaded diaphragm and ball-valve (Figure 11.14).

Manifold depression depends on throttle opening, i.e. engine load, so when the opening is small

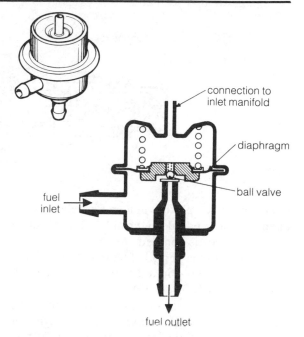

connection to inlet manifold

diaphragm

ball valve

fuel inlet

fuel outlet

Figure 11.14 *Pressure regulator*

the high depression encourages more fuel to leave the injector. To compensate for this, the fuel system operating pressure is lowered when the manifold depression is high. This is achieved by connecting one side of the regulator to the induction manifold. At times when the engine is operated under a light load, the regulator valve is slightly opened and the pressure is reduced.

The pressure controlled by the regulator is as follows:

Engine condition	Manifold depression	Typical operating pressures
Idling	Very high	1.8 bar (26 lbf/in²)
Full throttle	Very low	2.5 bar (36 lbf/in²)

Injectors The duty of an injector is to deliver a finely-atomized spray into the throat of the inlet port. In addition, the injector must vary the quantity of fuel to suit the engine operating conditions; this is achieved by varying the time that the injector is open.

The required conical spray pattern is obtained by pumping the fuel through a pintle-type nozzle.

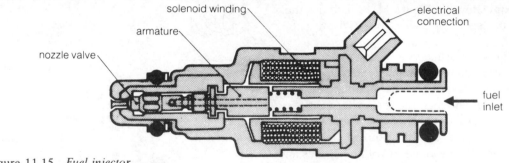

Figure 11.15   *Fuel injector*

Fuel flow takes place when the nozzle valve is opened by a solenoid (Figure 11.15). Movement of the valve is limited to about 0.15 mm (0.006 in) and the period of time that the valve is open varies from about 1.5 to 10 milliseconds (0.0015 to 0.0100 second).

This variation in opening time alters the amount of fuel that is supplied to each cylinder per cycle. Open time depends mainly on the rate of air flow and engine speed, but engine temperature and fuel temperature also have a bearing on the amount that needs to be delivered.

All injectors are electrically connected in parallel, so for 4- and 6-cylinder engines they all open and close at the same time. When the engine is warm, this opening normally occurs once every other revolution.

When the engine is cold-started extra fuel must be injected; this is provided by increasing the frequency of injections.

Injector opening is obtained by using a solenoid winding of resistance about 16 Ω. This solenoid overcomes a return spring which is fitted to hold the nozzle against its seat when it is closed.

Electronic control   The control system of a fuel-injection system must arrange for the correct quantity of fuel to be injected at the right time. To meet this requirement, the quantity of air entering the system, together with the engine crankshaft position must be measured accurately. The quantity of air entering the engine dictates the amount of fuel needed and the crankshaft position, acting on pulses from the ignition coil, signals when injection should commence.

The brain of the control system shown in Figure 11.16 is the E.C.U.

Electronic control unit   During recent years, advances made in E.C.U. technology have made it possible for the E.C.U. microprocessor to accurately meet the requirements of the engine. This means that the operating performance of a modern system depends largely on the quality of the peripheries, namely the accuracy and efficiency of the various sensors and actuators, particularly the former.

Normally the E.C.U. consists of a number of integrated circuits and many hybrid modules containing various semiconductors which are all mounted on one or two printed circuit boards. Input and output signals are communicated by a wiring harness that is connected to the E.C.U. by a multi-pin connector.

The E.C.U. computer unit interprets the data received from the sensors and, after calculating the duration of the injection time, it signals this message, together with the time of opening, to the injectors.

Many designs of electronic fuel-injection system work on an analogue computing principle, using electrical signals which are variable so the memory capacity of such a system is limited. In consequence, it is suitable only for engines with a relatively simple fuelling requirement.

Digital control   More recent developments use digital control units. These allow quite complex fuelling needs to be stored in a micro-sized silicon chip. One system uses a 5 mm square chip con-

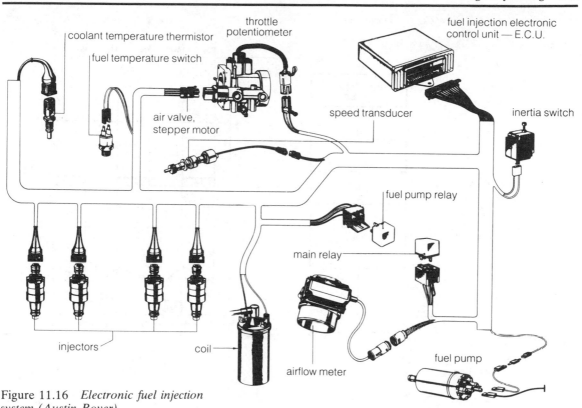

Figure 11.16 *Electronic fuel injection system (Austin-Rover)*

taining about 5000 transistors as a part of its control circuitry.

The digital units function by sensing the presence or absence of a voltage pulse, so greater stability is achieved with less drift especially when a change in temperature occurs.

Most digital E.C.U. units store the fuelling requirements in a digital memory pictorially represented by a 3-dimensional map. This gives a standard injection pulse length for 16 different engine speeds and 8 different engine loads. Once the standard pulse has been determined, it is then modified to take into account the conditions that exist at that time, namely coolant and air temperatures. Correction must also be made for battery voltage, because the action of the injector depends on the voltage applied to it.

Many different control units are in use; the following description covers the basic principles of a Bosch E.C.U. as fitted to L-Jetronic systems.

E.C.U. operation The block diagram of the E.C.U. shown in Figure 11.17 (overleaf) has five sections for processing the pulse so as to make it suitable for operating the injector.

Pulse-shaping This circuit converts the signal sensed from the ignition coil's l.t. to a rectangular form. The circuit incorporates a monostable multivibrator. This 'switch' changes its state from 0 to 1 on receipt of a signal, but reverts back to its original state 0 after a fixed time.

Frequency divider Located on the same chip as the pulse shaper, the frequency divider converts the number of pulses given by the input signal to one per crankshaft revolution. A 4-cylinder engine has 4 ignition pulses per 2 revolutions, so the frequency divider halves the input frequency. In a similar way 6-cylinder and 8-cylinder engines have to be divided by 3 and 4 respectively.

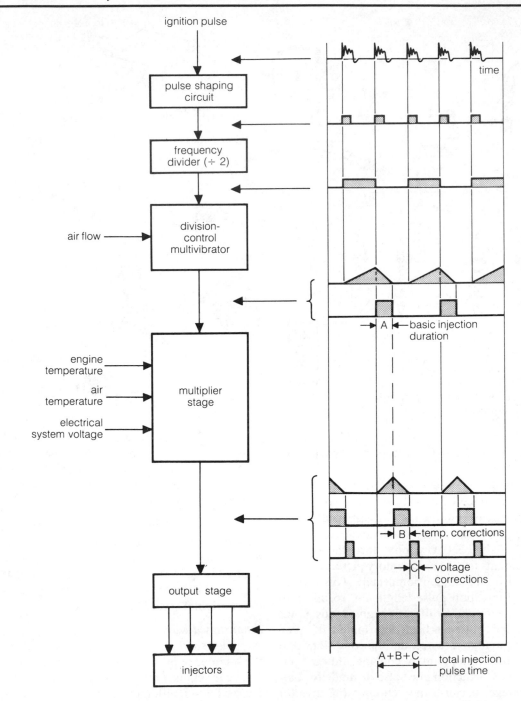

Figure 11.17 *Electronic control functions*

The frequency divider uses a bi-stable multi-vibrator or flip-flop; this circuit rests in one or other of its two states, 0 or 1 and requires a pulse to switch it from one state to the other.

Division-control multivibrator This circuit determines the standard injection pulse by using input signals of:

- Engine speed, from the frequency divider
- Air flow, from the air-flow sensor

Since this pulse duration is proportional to the air flow during the induction stroke, the air flow must be divided by the engine speed. The charge–discharge action of a capacitor is used to perform this division.

Multiplier stage Standard pulses, appropriate to the engine speed and air flow are modified in duration by this stage to take into account the actual engine conditions. Signals from the temperature and throttle sensors allow the circuit to calculate a correction factor which is used to multiply and lengthen the standard pulse; this gives a pulse duration suitable for conditions such as warm-up and full load.

Output stage Further modification of the pulse duration is needed to compensate for changes in battery voltage, since any change affects the injector opening time. The pulse voltage is then amplified to enable a current of about 1.5 A to be produced for energization of the injector solenoid. This final stage is performed by a Darlington driver circuit.

The actual injection period differs slightly from the pulse signal duration produced at the output stage; this variation is due to the time needed to operate the valve. The time delay that occurs on opening and closing is called *response delay* and *drop-out delay* respectively.

Other features In addition to these basic functions, an E.C.U. fitted to a modern injection system incorporates extra control circuits to include:

- Cranking enrichment
- After-start enrichment

- Hot-start enrichment
- Acceleration enrichment/deceleration weakening
- Full-load enrichment
- Over-run fuel cut-off
- Idle-speed control

Cranking enrichment When the engine is rotated at cranking speed, the E.C.U. provides double the number of injection pulses to satisfy cold-start conditions.

After-start enrichment The E.C.U. provides extra fuel by extending the injection pulse duration for a given time after the engine has been started. This feature is provided for all engine temperatures but the enrichment period is much shortened when the engine is hot.

Hot-start enrichment When the fuel in the fuel rail (gallery) is very hot, the E.C.U. lengthens the pulse duration to compensate for the change in fuel density.

Acceleration enrichment/deceleration weakening Signals from the throttle potentiometer and air-flow sensor indicate when the engine is accelerating or decelerating. Under these conditions, the E.C.U. enrichens or weakens the mixture accordingly by increasing or decreasing the pulse respectively.

Full-load enrichment Slight enrichment is required when the engine is put under full load. When the sensors indicate this condition, the E.C.U. lengthens the pulse duration.

Over-run fuel cut-off The E.C.U. cuts-off the fuel so as to improve economy and exhaust emissions when the following conditions are sensed:

- Engine speed is above 1500 rev/min
- Throttle is closed
- Coolant temperature is above 25–30°C

Idle-speed control Idle speed is controlled by a stepper motor energized by pulses supplied from the E.C.U. A valve moved by the motor varies the quantity of air that is allowed to by-pass the throttle valve.

The main engine sensors provide the main signals for this feature but extra data is required when the vehicle is fitted with an automatic gearbox and/or air conditioning. Idle speed must be increased if the gearbox is set in 'drive' or if the air-conditioning unit is 'switched-on'.

Throttle body injection

A single-point injection system provides a comparatively simple replacement for a carburettor.

Replacing a carburettor with this system gives the advantages:

- More accurate metering of the fuel and air since electronic control is used
- Better atomization of the fuel over the speed range, especially at part-load

These advantages result in good economy and less pollution from the exhaust. Furthermore the control system can easily be programmed to include fuel cut-off and enrichment require-

ments. Also, it can take into account signals from other sources which have some influence on the air–fuel requirements.

Figure 11.18 shows a layout similar to a Bosch Mono-Jetronic system. This shows a single solenoid-operated injector situated centrally in the air intake. It is supplied by a pressurized fuel system similar to that used in a multi-point layout.

Injected fuel is directed into a venturi-shaped region around the throttle, so the increased air speed at this point is used to further break-up the fuel.

Air flow is measured by a flap or hot-wire sensor. On this housing is mounted the E.C.U. so a compact layout is obtained. Since the E.C.U. contains a microcomputer it can easily process data from additional sensors other than the throttle and temperature units shown.

Working in conjunction with these sensors the E.C.U. varies the fuel flow to give deceleration cut-off and enrichment during cold-start, warm-up, acceleration and full-load operation.

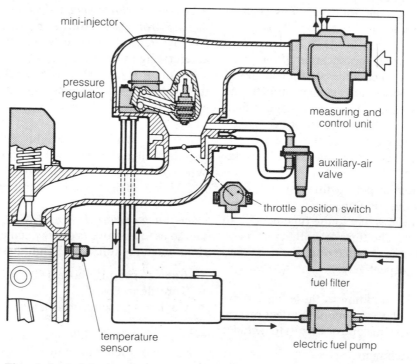

Figure 11.18   *Throttle body injection*

When the engine is cold, fast idle of the engine is provided by an auxiliary air valve.

It is claimed that this single-point injection system is capable of giving the same metering accuracy and long-term stability as a multi-point injection system.

11.3 Maintenance and fault diagnosis of fuel systems

In general the periodic maintenance of a fuel system involves checking the system for security of electrical connections and fuel lines. At the time when this check is carried out, an inspection should be made for fuel leaks.

Fuel-injection systems incorporate a fuel filter so this must be changed at the appropriate time, e.g. every 48 months or 80 000 km. Direction of fuel flow through the filter is important so it should be fitted as indicated by the arrow on the filter casing.

Carburettor systems
Manufacturers provide a test sequence for checking the electronic operation of the carburettor. Often these tests involve the use of a multimeter.

The following summary illustrates the method of checking a unit such as an electronically-controlled carburettor.

1. Initial check After completing a check of the basic engine systems, the coolant thermistor is then disconnected; in this state the panel should register a hot engine.

A cold engine is then simulated by connecting the thermistor cable to earth. This action should cause the stepper motor to operate through its full travel; the throttle jack should also move.

2. Stepper motor Five cables are connected to the stepper motor: one common feed and five returns. Using a multimeter between the common feed, and each of the other cables in turn, should show a given resistance, e.g. 12–15 Ω.

3. Coolant temperature thermistor The bulb is placed in water and the temperature is raised. Using an ohmmeter, the resistance is checked at given temperatures and the results are compared with the specification. Typical values are:

Temperature (°C)	Resistance (Ω)
20	2350–2650
40	1050–1250
60	550–650
80	300–360

4. Ambient air-temperature sensor This n.t.c. thermistor is checked in a way similar to the coolant thermistor. Resistance values are compared against a table of values.

5. Fuel cut-off solenoid This normally has a resistance of 20–25 Ω and is operated at battery voltage. After checking that the resistance is correct, the harness is disconnected and voltage is applied to the solenoid; a click should be heard as it energizes.

Fuel-injection systems
Before carrying out specific tests on a fuel-injection system, basic engine checks, such as plugs, timing, etc., should be made to ensure that these items are serviceable.

For safety reasons, it should be noted that any test involving the disconnection of a fuel line must take into account that the line is pressurized to the extent of about 2 bar.

As in the case of many other electronic components, items such as E.C.Us can easily be damaged by workshop test equipment (and personnel), so care must be exercised when testing any part connected to an E.C.U. Three cases are often specified; these are:

Boost starting	Using a high-speed charger as a starting aid.
Electric welding	Induced voltage from an electric welding plant. In some cases it is recommended that the battery earth is disconnected; in some situations it may be necessary to isolate the E.C.U.
Steam cleaning	Heat, together with steam, creates a damaging environmental hazard for the E.C.U.

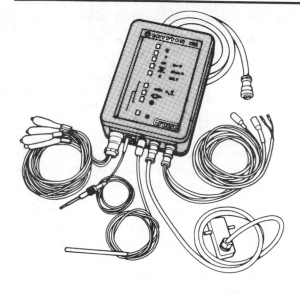
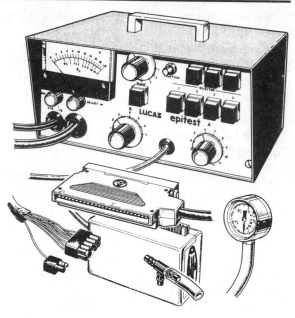

Figure 11.19   *Fuel injection test equipment*

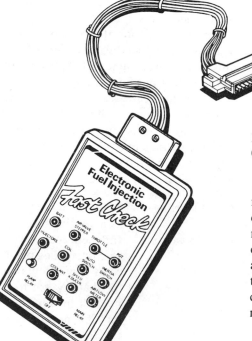

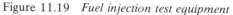

Figure 11.20   *Electronic fuel injection tester (Austin-Rover)*

Full diagnostic tests of a fuel-injection system normally require the use of specialized test equipment (Figure 11.19). Without this equipment, tests are limited to items which can be checked with a multimeter. By incorporating the appropriate meters and switches in a dedicated test unit, it is possible to cover the numerous checks in a shorter time. This leads to quicker and more accurate fault diagnosis.

Austin-Rover have developed a comparatively cheap tester that uses a series of LEDs to signal to the operator the results of checks made on the main parts of a hot-wire E.F.I. system (Figure 11.20).

The Fast-Check tester is connected to the 25-pin main harness plug that is normally joined to the E.C.U. The test sequence covers the following:

- Operation of the main and pump relays by supplying each relay, in turn, with a current
- Indication by LED of battery voltage, air valve stepper motor circuit resistance, throttle POT circuit, supply from ignition coil, automatic transmission inhibitor switch, inertia switch, coolant sensor, road-speed transducer and air flow meter resistance

- Injector operation by supplying current to each injector in turn
- Pump operation

Results given by this Fast-Check pin-point the area of the system that is faulty. More detailed tests can then be made.

In addition to the electronic checks, other tests are made to the hydraulic system to determine the fuel pressure and the correct functioning of the injectors.

12 Engine management

12.1 Engine systems

A system is a collection of interacting parts. When these parts are linked together they perform a particular function. This function may be complete in itself or it may form just a part of a larger system. Conversely, a system may be divided into a number of smaller systems, i.e. a number of sub-sections.

In the case of an engine, ignition arrangements and fuel supply layouts are two examples of systems. Along with other parts they form a larger system, namely the engine or power system. A carburettor and a distributor are examples of sub-systems. In many ways the division of a major unit into systems is similar to a family structure. Human beings are grouped into families which, in turn, are assembled or divided into nations or individuals respectively.

The behaviour of a large family of mechanical components is difficult to investigate when the whole assembly is complete, because each part responds in a different way to a given change. To simplify this study the assembly is split up into smaller sections; this allows the function of each part to be analysed. When this study has been completed, it is possible to see the effect, or interaction, of individual parts on the complete system.

System function and performance An examination of a system often involves two separate studies; an analysis to determine its *function* and a study of its behaviour to ascertain its *performance*.

A functional study is known as a *qualitative analysis* because it establishes the basic qualities of a system and indicates the role that each part of the system fulfils. The manner in which the system performs its given duty, and the effectiveness of a given system in respect to its performance, is called a *quantitative analysis*.

Whereas a qualitative study gives the basic information about the fundamental duties performed by each main part of a system, the quantitative analysis is concerned with the 'filling-in' of the operational details and the measurement of the effectiveness of its operation.

A qualitative analysis of an electronic ignition system is shown in Figure 12.1. Representation of the main parts by blocks in this diagram allows the function performed by each part to be shown. This block diagram layout is a simple method commonly used in electronics to show the arrangement of the various sub-sections and the paths taken by the control signals as they pass through the system.

A quantitative analysis involving a component's performance is often expressed in the form of a mathematical equation. By comparing the

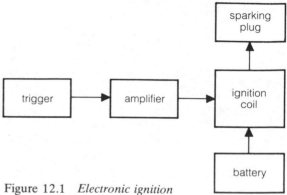

Figure 12.1 *Electronic ignition system (qualitative analysis)*

equations of alternative parts, the designer is able to make an accurate, non-subjective judgement which helps him to select the most suitable system or part for a particular application.

System modelling   The idea of using mathematical equations to represent the performance of individual parts can be extended to the complete system. When mathematics is used in this way, the equation covering the particular system or sub-section is called a *mathematical model.*

Combining the models and supplying the mathematical data of individual models to a computer enables the designer to observe the overall performance. In addition to this important feature, the system's response can be determined when any part of the system is altered.

A modern electronic system has to operate over a wide range of conditions, so with the aid of this modelling facility, it is possible to vary the signal from each part and observe the result of the change on the complete system. These modern design aids enable the designer to predict accurately the performance of a major system well before it is put into production. This allows numerous modifications to be tried to see if it is possible to improve the performance or establish a simpler and cheaper system. Prior to the use of the computer for this purpose, the introduction of a new system required a program of practical tests on the actual components; this took many hours.

Control systems   A control system is an arrangement which directs the operation of a main system.

The commands given by the control system should ensure that the main system performs according to a given program. This program will have been devised to achieve a given performance; in the case of a power unit, this may be the production of a given power, the achievement of a set economy or the limitation of a given exhaust product.

To achieve a set program within the parameters of the control system's scope, the system must be able to respond quickly and accurately to

changes in the operating conditions, maintain a stable control and be able to separate valid input signals from others that are induced into the sensing lines by electrical disturbances, i.e. the system should have 'noise immunity'.

The two main systems of control are:

- Open-loop
- Closed-loop

Open-loop control   This type of control sends commands to the main system but does not have the ability to check or monitor the actual output of the main system (Figure 12.2).

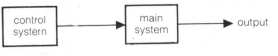
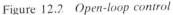

Figure 12.2   *Open-loop control*

Assuming the main system is an engine, then once a control signal has been delivered, the engine will produce its output, but this output will not always be the same if the engine-operating conditions alter, e.g. if the engine-control system does not take into account an ambient condition such as air temperature, then any variation in this condition will alter the power output and will not be corrected by the control system.

Although an open-loop control system is suitable for many applications, it cannot be used where the engine has to operate within narrow limits such as the case where exhaust emissions must be controlled closely to meet environmental legislation requirements.

Open-loop controls are still commonly used for fuel supply and ignition systems. In these applications the fuel mixture and ignition timing setting are arranged so that they follow independent programs. Ideally any change in the air–fuel ratio should be accompanied by an alteration in the timing, but in many cases this is not so. Consequently, the engine performance is lower than expected, economy is poor and a high exhaust emission results. To overcome this problem, many new vehicles now use closed-loop control systems.

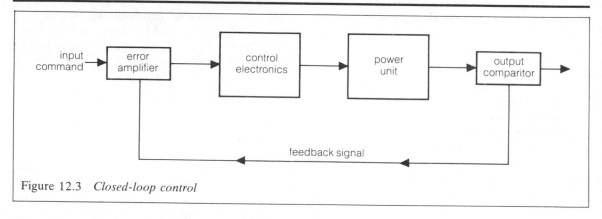

Figure 12.3 *Closed-loop control*

Closed-loop control   This is similar to an open-loop system but has one very important addition; it has a means for measuring the output and feeding back a signal to allow a comparison to be made between the command signal and the system's output.

Figure 12.3 shows the principle of a closed-loop system. In this diagram the feedback signal from the output sensor is passed back to the input where it is compared by use of an error amplifier. This intensifies and processes the signal to allow it to be compared with the input command. If the output differs from that commanded by the input, the command signal is altered until the required output is obtained.

When this control system is applied to an engine, the feedback facility allows any variation in the output to be corrected. This provides a more accurate and stable output than is possible with an open-loop system. Furthermore the system can be made to respond quickly and correct for any changes in the operation conditions. If these conditions are not compensated, then the output would be much different from that intended; in some cases this uncorrect condition can result in extensive damage to the engine.

Two forms of closed-loop control are:

- Proportional control
- Limit cycle control

A *closed-loop proportional control system* uses a sensor in the output to generate a signal proportional to the output, therefore the magnitude of the feedback signal indicates the system's output.

A *closed-loop limit cycle control system* uses the feedback to signal when a given limit is exceeded. The output sensor is inoperative during the normal operating range, but when a preset limit is exceeded, the feedback circuit passes a signal back to the input; this allows an alteration to be made to the command input.

This type of control has a number of automotive applications, e.g. combustion knock and fuel control.

Engine mapping

Open-loop digital systems store ignition-timing and fuel-mixture data in a memory unit of an E.C.U. The data stored in the individual cells of the computer's memory can be represented graphically by a *characteristic map*. Information for this 'graph' is obtained by carrying out a series of tests on the engine; the program for these tests is called *engine mapping*.

These tests measure the performance and investigate the effects of each variable that has some bearing on the output of the engine. When the effects are known, the settings that give the best performance can be determined and noted.

A dynamometer is an essential item for these tests since it can be programmed to simulate road conditions. The engine is loaded by means of this 'brake' and the torque, power output, economy, and emissions are measured against speed and other factors that have some effect on the engine output (Figure 12.4).

Performance curves plotted by hand or by computer show graphically the behaviour of the

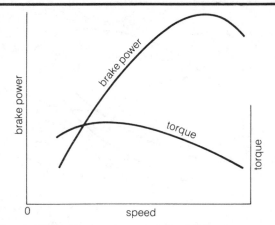

Figure 12.4 *Engine characteristic, full load test (throttle full open)*

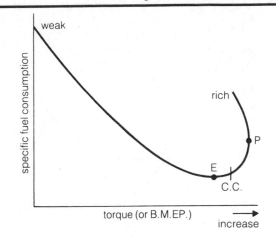

Figure 12.5 *Torque/consumption loop*

engine when it is subjected to changes in the following:

- Speed
- Load (throttle opening)
- Ignition timing
- Air–fuel ratio
- Engine and ambient temperatures

The performance curves derived from the tests are called engine *maps*. Some of the more important maps are included here.

Torque/consumption loop   This map is obtained by varying the air–fuel ratio and measuring the fuel consumption and torque output for each setting. Speed is kept constant during each test so a series of tests is needed to cover the engine operating range.

Figure 12.5 shows a characteristic 'fish-hook' shaped map which is obtained when the engine is operated under full-load. On the *y*-axis (vertical axis) the specific fuel consumption (SFC) is plotted; these values are obtained from

$$SFC = \frac{fuel\ consumption\ (kg/h)}{brake\ power\ (kW)}$$

The specific fuel consumption indicates the quantity of fuel that is needed to produce one unit of power.

The map shows that when the engine is run on a weak mixture, the SFC is high and the torque

output is low. As the mixture is enriched, the consumption falls to a point E where maximum economy is achieved. Enriching the mixture past this point gives an increase in torque but at the expense of fuel; maximum torque and power occurs at point P. It will be seen that the chemically correct (c.c.) ratio of 15:1 gives neither maximum torque nor maximum economy; to achieve these outputs, the mixture must be slightly enriched and slightly weakened respectively.

Exhaust emission/air–fuel ratio   Before exhaust emission regulations were introduced, the mixture supplied was based on the air–fuel ratios required to give either maximum power or maximum economy. Unfortunately the 12–15% enrichment from the c.c. ratio to give maximum power also gives a high emission of health-damaging exhaust gases such as carbon monoxide (CO), hydrocarbons (HC) and nitrogen oxides (NO_x).

Figure 12.6 shows the relationship between the formation of undesirable gases and the air–fuel ratio. This map shows the need to avoid operating the engine on an enriched mixture if exhaust pollution is to be kept to a minimum. Comparing the results shown in Figure 12.6 (overleaf) with Figure 12.5 indicates that 'lean-burn' engines designed to operate with minimum exhaust pollution suffer a considerable increase in consump-

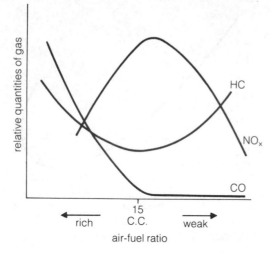

Figure 12.6   *Exhaust emissions*

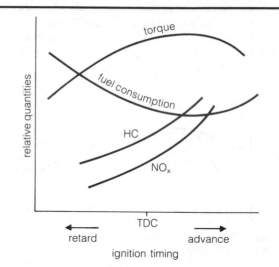

Figure 12.7   *Effect of varying ignition timing*

tion and decrease in power if the air–fuel ratio is weakened beyond the economy point E. The tolerance is very small, so close control of fuel metering is needed if satisfactory output, combined with freedom from engine damage, is to be achieved.

Spark timing and engine performance   The general effects of varying the spark timing are well known. Maximum power over the speed range is achieved when the spark is timed so that maximum gas pressure occurs at 12° after t.d.c. Reducing the spark advance for a set engine speed reduces the power, increases the consumption and, as a result of the slower burning mixture, overheats the engine. An over-advanced spark also gives poor performance but in addition is likely to cause combustion knock (detonation) which quickly damages pistons and causes a high level of noise from 'pinking'.

In addition to these effects, spark timing also alters the exhaust products. Figure 12.7 shows a map obtained from running an engine at one set speed and load, and altering the ignition timing. This map allows the best setting of the spark to be determined.

Factors affecting spark timing   Spark timing to achieve a set power depends on three main factors:

- *Speed*: As the speed increases, the crank moves through a larger angle in the time taken for the gas to burn.
- *Load*: As the load is increased, the throttle has to be opened a larger amount to maintain a set speed. This increases the gas filling of the cylinder and as a result of the higher compression pressure, the flame rate is increased and the gas burns quicker.
- *Air–fuel ratio*: A weaker mixture takes longer to burn than the c.c. ratio.

By using a series of maps the effects of these three factors can be determined; as a result the optimum timing can be established. The following is a summary of the timing requirements. The spark timing advance is increased when the:

- engine speed is increased
- air–fuel ratio is weakened

The spark timing advance is decreased when the:

- engine load is increased
- exhaust emission of HC and/or NO_x is too high.

Fuel mixture requirements are allied to engine load because when the engine is under light-load, or if the vehicle is 'cruising', a weaker mixture is supplied for economy purposes. Conversely, a full-load condition indicates that high engine power is needed, so a less-weak mixture has to be provided by the fuel system.

Being as the air–fuel ratio is dictated by the load on the engine, the spark timing need only be responsive to load and speed. For this reason most timing maps are based on these two variables.

Three-dimensional maps After performing a series of engine tests at different loads to determine the optimum angle of advance with respect to speed, a large number of maps is obtained. The number can be reduced to one by using the three-dimensional form as shown in Figure 12.8. The three axes of the map x, y and z represent speed, spark advance and load respectively. Accuracy of the timing requirement is dependent on the number of tests used to construct the map. In the simple map shown, a total of 60 timing settings is used.

To determine the spark advance for a speed of 32 rev/second (1920 rev/min) and a half-load condition, the 32 point on the x-axis is located and the line from this point is followed until it

intersects the half-load line; at this intersection the height of the map indicates the advance. In this case the angle is 52°.

Fuel mixture map A three-dimensional map is used to show the fuel requirements of an engine. In this case the three factors are: speed (x), air–fuel ratio (y) and load (z). Plotting these on the appropriate axis indicates the air–fuel ratio that is needed to suit the conditions of speed and load (Figure 12.9, overleaf).

This map is often called a 'lambda' map. The term *lambda* is the name of the Greek letter 'L'. The symbol λ is used when the mixture is chemically correct, or to use a more technical term, the stoichiometric ratio. At this ratio:

$$\lambda = \frac{\text{Supplied quantity of air}}{\text{Theoretical air requirement}} = 1$$

When the 'excess air factor' represented by the symbol λ is less than 1, there is insufficient air for combustion, i.e. the mixture is rich. Conversely when λ is more than 1, there is excess air; the mixture is weak.

To summarize:

λ = 0.95 mixture rich
λ = 1.00 mixture correct
λ = 1.05 mixture weak.

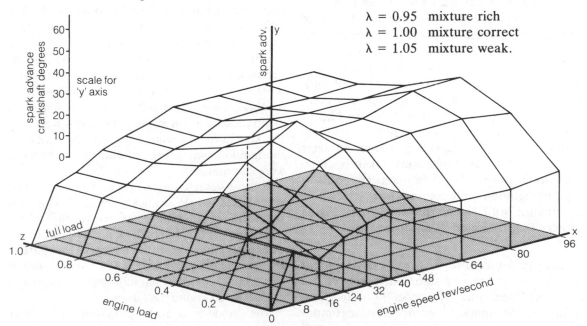

Figure 12.8 *Typical spark advance map (simplified)*

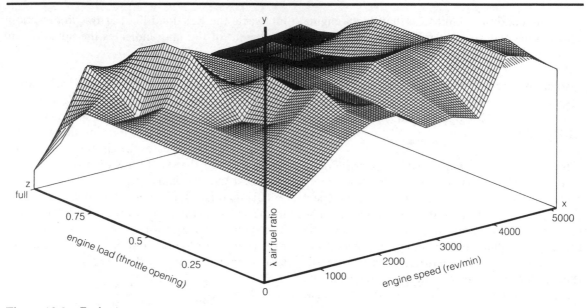

Figure 12.9   *Fuel mixture map*

12.2 Open-loop engine control systems

Control arrangements based on the open-loop principle have been in common used since the engine was first introduced many years ago.

Prior to about 1935 the driver had the job of acting as a 'feedback' system, because in addition to manipulating the main controls, he also had to adjust the ignition timing and set the air–fuel mixture. This demanded considerable skill since it involved the setting of the ignition to a point where the engine was just knock-free and keeping the mixture at the point where it was just rich enough to develop maximum power. In later years, these driver duties were taken over by open-loop systems. These automatically controlled the ignition and fuel-mixture settings in accordance with signals received from engine sources that indicated its speed and load.

Ignition timing control   An early development was the introduction of automatic timing by use of a centrifugal timer. This speed-sensitive unit advanced the spark to ensure that maximum cylinder pressure was maintained at about 12° after t.d.c. over the full speed range.

This type of automatic control assumes that the burn-time between spark and maximum pressure remains constant in time. Weakening the air–fuel mixture makes this burn-time longer, so when economy carburettors came into use in the late 1930s, the weaker mixture delivered by this type of carburettor during part-load operation required a larger advance than that given by the centrifugal timer. This was provided by a load-sensitive vacuum control unit. The spring-loaded diaphragm used the manifold depression to sense engine load and since the carburettor also used this source for the same purpose, the actions of the ignition timing control and the carburettor were harmonized.

In later years greater precision of the timing was needed, so electronics were used. This took the form of an E.C.U. in which the timing requirements are stored in a memory unit. The timing data must relate the timing of the spark to the two factors previously taken into account by the original systems, namely engine speed and engine load (Figure 12.10).

The modern open-loop system is a large improvement over earlier designs, but one drawback is that it has to assume that the engine is in

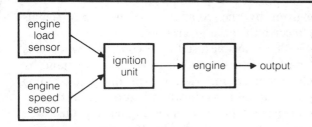

Figure 12.10   *Ignition timing system*

the same condition as that used when the memory was programmed. When this is not so, the timing given by the look-up table in the E.C.U. will be unsuitable for the engine; as a result, emission, economy and power will all suffer.

Fuel-mixture control   Development of the constant-choke carburettor brought about mixture-compensation systems to correct for enrichment of mixture with increase in load/speed; this was followed by economy systems which weakened the mixture during part-load operation.

Methods for improving mixture distribution and poor atomization at low speeds received considerable attention from carburettor manufacturers, but the benefits of fuel injection with respect to these problems did not become attractive until stricter emission controls were introduced.

Both carburettor and fuel-injection systems must be capable of sensing engine load and engine speed, so these factors must be measured by the main sensing system. Mechanical and elec-

tronic sensing systems suffer the same drawback as the ignition unit; they can only follow the program introduced when the engine is made. Variable factors such as air leaks past the pistons, valve guides, throttle spindles, etc. are not taken into account, so output will suffer. The introduction of more sensors improves the situation and in this area, electronic systems can take into account many more variables than those used with mechanical systems.

Combined ignition and fuel-supply systems
Previous studies show that both the ignition timing and fuel-supply systems require sensors to measure engine speed and engine load. Duplication of the basic sensors and control electronics is uneconomic, so in many cases the two systems are combined to form a single Engine Management System.

Combination of the two systems allows other sensing signals to be used jointly by ignition and fuel systems so greater precision of control is possible. These additional peripheral devices include the measurement of ambient and engine temperatures and other factors that affect the operation of the engine (Figure 12.11).

Use of engine maps   To achieve precise control of an engine that has no means for feeding-back output data, the control system must be programmed with very accurate information relating to the setting that is required for each condition under which the engine is expected to operate. This program should ensure that the input command produces the expected engine response.

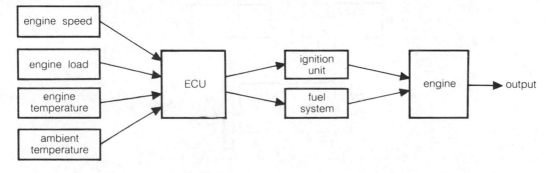

Figure 12.11   *Combined ignition and fuel system*

Control maps compiled during the engine development stage show the settings of the main systems in relation to the variables which affect the particular systems. The control data indicated by the maps is programmed permanently into the computer's memory unit.

The computer uses the stored control data in various ways; one way is to operate the computer so that it calculates the ignition and fuel mixture settings each time the speed or load conditions change. This computation may take as long as 100 milliseconds to perform and since this operation may have to be repeated at 50 millisecond intervals, a quicker method is needed. This is achieved by using *look-up tables*.

Look-up tables   A look-up table is a list of related values stored in the memory unit of a computer. This table relates the output settings given by the computer to the input signals received from a given sensor.

The principle used by a computer when it is performing this duty is easier to understand by considering the example of the spark-timing look-up table as shown in Figure 12.12; this has been constructed from the engine map illustrated in Figure 12.8. To simplify the example, only the speed factor is considered at this stage.

Assuming the engine is running at 32 rev/second (1920 rev/min), the appropriate sensor signals this speed to the E.C.U. After entering the computer the signal initially is converted by an encoder into the 8-bit digital form, 00100000; this is the binary code for the number 32. The code is then stored in one of the registers of the CPU and the memory is searched until a similar code is recognized. When the search has been completed and the result verified, the memory

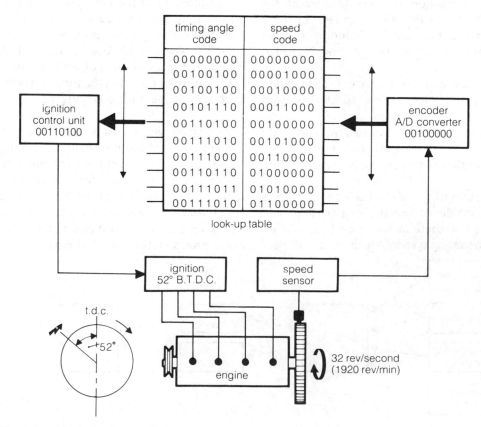

Figure 12.12   *Spark timing look-up table*

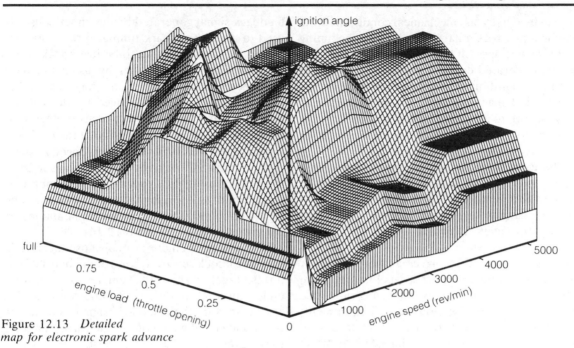

ignition angle

full
0.75
0.5
engine load (throttle opening)
0.25
0

1000
2000
3000
4000
5000
engine speed (rev/min)

Figure 12.13 *Detailed map for electronic spark advance*

unit then issues another binary code such as 00110100. This is the code that the computer has been programmed to write whenever it reads the code 00100000, i.e. the matching value in its look-up table. The 8-bit code is the spark timing instruction to the ignition control unit, so, after it has been deciphered, the control unit sets the spark to occur at 52° before t.d.c.

Although this example is limited to a table of 10 values, a modern computer has a much larger table.

In addition to the spark advance/speed table, extra tables covering other variables are stored in the memory unit. Spark advance depends mainly on load and speed, so the advance given in the 'load look-up table' must be added to the value given by the 'speed look-up table'. This calculation is performed by the computer very quickly because it involves only the addition of the values given by the look-up tables.

To obtain this method of control, development must progress through the following steps:

1. Mapping of the prototype engine to determine the best settings.
2. Construction of maps to show graphically the

required settings to suit the varying operating conditions.
3. Programming the computer's look-up tables with the data contained in the maps.
4. Testing the engine to verify that the computer is giving out control instructions in accordance with the requirements.

Detailed maps To achieve the best possible engine performance, the various maps should closely follow the requirements of the engine. This involves the use of maps with many more reference points than those used in Figure 12.8 and the fitting of an engine management computer that has a memory unit of sufficient capacity to store the detailed look-up tables. An example of a detailed map is shown in Figure 12.13.

12.3 Closed-loop engine control systems

The accuracy of setting the ignition timing and metering the fuel in an open-loop system is as good as the mechanical condition of the engine and the program for the control function. Systems using the open-loop principle can only follow a program set-up after testing a prototype engine in a first-class mechanical condition. If the engine is not in a similar condition then the timing and fuel settings will be incorrect. This problem can be illustrated by highlighting the problem of setting the ignition timing of an older type engine with the aid of a strobe light. Although the timing may be set to the angle recommended by the manufacturer, it may not be the ideal setting for the actual engine being tuned.

These problems can be minimized by using a closed-loop control system. This system can be applied independently to manage ignition and fuel metering or can be combined to give a *Full Engine Management System*.

Ignition control   The correct ignition timing for an engine is where combustion is just free from knock. On modern engines this setting cannot be obtained accurately without special equipment; this is because knock occurs before it can be detected by the human ear. Many factors affect the maximum spark advance that can be used before the onset of knock; these include:

- Mechanical condition of engine
- Compression ratio
- Octane rating of fuel
- Volumetric efficiency
- Throttle opening
- Shape of combustion chamber
- Air–fuel ratio
- Engine temperature
- Carbon deposits

To program for all these variables would require a very sophisticated open-loop system having many sensors together with an elaborate map containing a very large number of ignition angles. This is not possible so a simpler map is used for normal open-loop systems; this map gives suitable ignition angles to meet general needs but maintains a margin of safety to keep the engine free from knock and damage.

This safety margin can be reduced with the result that better power and improved emissions can be obtained if a sensitive knock-control system is used. When this closed-loop control system is used in conjunction with a lambda map, the feedback signal generated by the knock sensor is used to adjust the spark timing so that combustion in the chamber is set to be just knock-free.

Initially the timing is set by using the data contained in the characteristic map. This basic timing setting is then advanced by the E.C.U. until the knock sensor detects a given degree of knock. When the sensor signals that the engine is knocking, the timing advance is reduced in steps of 1.5 crankshaft degrees until the sensor indicates that the engine is knock-free. The continual repetition of this sequence ensures that the timing for each individual cylinder is maintained at the optimum angle (Figure 12.14).

Turbocharged engines can also use the signals from the knock sensor to control the boost of the turbocharger. When the engine is operating under full-load conditions and knock is detected by the sensor, the feedback signal activates a command signal which opens the turbocharger's waste-gate and reduces the boost.

The knock-control system incorporates a safety circuit which recognizes malfunctions of the system. At times when a fault is detected, a warning light on the instrument panel is activated and the ignition timing advance is reduced sufficiently to prevent damage to the engine.

Further details of the knock-control system is covered on pages 144 and 235.

Closed-loop dwell-angle control   This feature is used to ensure that the correct primary current is achieved under conditions of differing battery voltage, engine speed and temperature.

This is particularly important when the engine is cold-started, because at this time the voltage applied to the coil is much lower than normal. To allow for this, a longer time must be allowed for the primary current to build up, i.e. the dwell must be lengthened. Conversely, if the dwell angle is set electronically at too large an angle, then power loss and heating of the ignition system is experienced.

Like other closed-loop systems, the dwell-angle control circuit uses a feedback circuit to signal when the primary current exceeds a predetermined value. When the signal is passed to

Figure 12.14 *Knock control system*

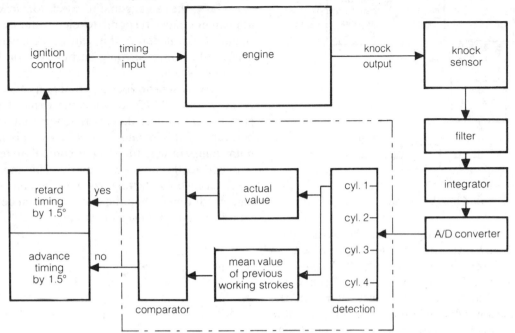

the dwell-angle control section, the dwell angle is reduced accordingly.

Fuel control The need for a closed-loop fuel control system is highlighted in countries such as the USA that have severe emission regulations. In the early 1970s, the Environmental Protection Agency in the USA introduced the Clean Air Act. This stipulated that vehicles produced in 1975 had to reduce emission limits by about 90%.

Initially, open-loop systems that controlled fuel supply and ignition settings were used, but these control methods did not satisfy the regulations that came into force in later years. To meet the very tight controls over emissions, it was necessary to use a catalytic converter in the exhaust system to convert pollutants by chemical means into safe products. These three-way catalytic converters effectively oxidized CO, CH and NO_x, but they suffered the drawbacks of high production costs and poor economy. Furthermore, the chemical converter can be used effectively only when the engine is operated in

the region of the chemically correct mixture ratio; hence the need for an efficient feedback control system that is capable of maintaining the air–fuel ratio in this region.

Lambda sensor The need for a mixture control system that meets most emission regulations is filled by using a Lambda closed-loop system. This arrangement uses a sensor in the exhaust to detect the mixture being burnt in the engine cylinders. Signals fed back from this sensor enable the fuel-control system to vary the fuel quantity whenever the sensor detects that the mixture is either too rich or too weak.

The Lambda sensor is an oxygen sensor which produces a voltage pulse when no oxygen is present in the exhaust gas; this occurs only when the air–fuel ratio is on the rich side of stoichiometry (chemically correct ratio).

The system layout is shown in Figure 12.15 (overleaf). This diagram shows the feedback circuit through which the digital pulse signal is conveyed when no oxygen is present in the exhaust gas. The principle of the Lambda Exhaust Gas Oxygen (ESO) sensor is described on page 234.

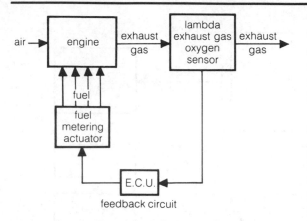

Figure 12.15 *Closed-loop fuel control system*

When the sensor signals that the oxygen content is below a certain limit, a voltage pulse from the EGO sensor commands the fuel system to decrease the fuel supply so as to weaken the mixture. Shortly after this fuel alteration the EGO sensor will detect that oxygen is present in the gas; this will cause its voltage output to fall and as a result the output from the sensor will change to zero. When the controller discovers that the feedback signal has ceased, it instructs the fuel system to enrich the mixture. Mixture control is obtained by oscillating the mixture between the rich and weak limits; in this way the controller is able to keep the air–fuel ratio within a mixture range that is near-correct. Since modern sensors are capable of working to a very small tolerance with respect to the air–fuel ratio, the risk of severe exhaust emission is considerably reduced; a limit of 0.05 from the required value of 15:1 is typical for many sensors.

Figure 12.16 shows the voltage variation when the air–fuel ratio is varied. The abrupt change in voltage between the point where the excess air factor (λ) is greater than 1.0 to the point where λ is less than 1.0 is the region where the voltage pulse is produced for use as an output signal. This signal gives a limit-control type of cycle so the EGO sensor may be considered as a switch which opens and closes to signify weak and rich mixtures respectively, i.e. the EGO gives a digital output with the two states, 0 and 1 used to identify the weak and rich mixtures.

Since this control system is expected to operate very fast, the arrangement used for mixture adjustment must respond to electronic control signals without delay; this can be achieved by using either an electronic carburettor or fuel-injection system.

The EGO sensor does not function efficiently below about 300°C, so when the exhaust temperature transducer detects that the mean temperature of the exhaust gas is below this minimum temperature, the engine-control system is switched to the open-loop mode. This mode is used during cold-starting, when the engine is idling, and at times when the control system detects a fault in the EGO sensor circuit.

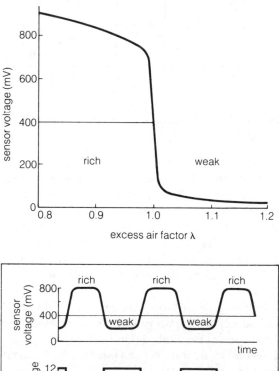

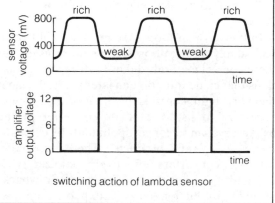

Figure 12.16 *Operation of lambda sensors*

Occasions arise when the engine has to be operated outside the range of the EGO sensor; these are during acceleration and deceleration. At these times the other sensors signal the engine condition and, in response to these signals, the E.C.U. selects the open-loop mode until normal steady conditions are re-established. The EGO sensor will operate in conjunction with a lead-free fuel only.

12.4 Maintenance and fault diagnosis of engine management systems

Modern engine-control systems need many sensors to supply the E.C.U. with operational data to enable the system to control the engine effectively. Cables from these sensors and associated circuitry require many multi-pin connectors.

Connector problems

Under ideal conditions the connectors used on motor vehicles are normally adequate for their purpose, but when they are subjected to a hostile under-bonnet environment and exposure to water, salt, oil and dirt, the connectors are likely to cause a number of problems.

Manufacturers attempt to minimize these problems by using some form of flexible cover to prevent the ingress of contaminates, but since the initial cost of the connector must be kept low, the chance of partial failure is present, especially after the connector plug has aged.

During routine maintenance, and at times when a fault is being diagnosed, attention should be paid to the condition of the connectors, particularly those plugs situated in exposed positions. Security and assessment of the condition of the connector plug covers are two important routine checks that should be made. Where an intermittent fault is experienced, it is often impossible to locate a cable connector fault by normal meter tests; in these cases attention should be directed to cleaning the contact pins of all connectors in the circuit that have the problem.

Some manufacturers recommend the actual method for cleaning the contact surfaces; these methods range from the use of an ink eraser to spraying the surfaces with a special cleaning fluid. EMERY CLOTH SHOULD NOT BE USED for two reasons; it removes the contact surface and is likely to create a short-circuit due to the electrical conductivity of the emery dust.

Self-diagnosing systems

It is anticipated that in the near future most management systems will incorporate their own fault-diagnosis circuit. Already many systems in use have a monitoring circuit which either signals to the driver when a fault is present, or controls the system in such a way that the fault does not seriously damage the engine. In these cases the E.C.U. resets the control system to enable the vehicle to 'limp home' and be driven to the garage for repair.

Some engine management computers have a built-in self-diagnosing feature that displays, when instructed by a technician, the area in which the fault is present. Systems of more advanced design have a facility that allows the transmission of information relating to a malfunction to a larger computer installed in the workshop.

Self-diagnosis by light signal The Toyota Computer Controlled System (TCCS) uses a coded light signal to indicate the cause of a malfunction in the system.

When a fault develops, the E.C.U. registers the sub-system in which the fault is present, into its memory. This information remains in the memory even after the engine is switched-off.

Ten possible faults are monitored by the system. Five of these faults are capable of producing an engine stall, so in these cases a warning light on the instrument panel advises the driver to 'check engine'.

When the malfunction corrects itself, the warning light goes out, but the E.C.U. still holds the information in its memory; this is particularly helpful to the diagnostician when an intermittent fault recurs.

Access to the memory data is obtained by short-circuiting a test terminal; this causes the panel lamp to flash at a rate which allows the particular fault to be identified by reference to the code shown in the repair manual. Figure 12.17 shows the lamp behaviour and the faults associated with each one of the three examples.

In addition to this memory feature, this computer also incorporates a fail-safe function that avoids an engine stall due to faulty operation of MAP, coolant temperature and intake air-temperature sensors. Malfunctions in one or more of these areas cause the computer to make the following adjustments (see opposite):

MAP	Sets the ignition timing to 10°; b.t.d.c. and maintains a constant injection duration
Coolant temperature	Sets injection duration by assuming a temperature of 80°C
Air temperature	Sets injection duration by assuming a temperature of 20°C

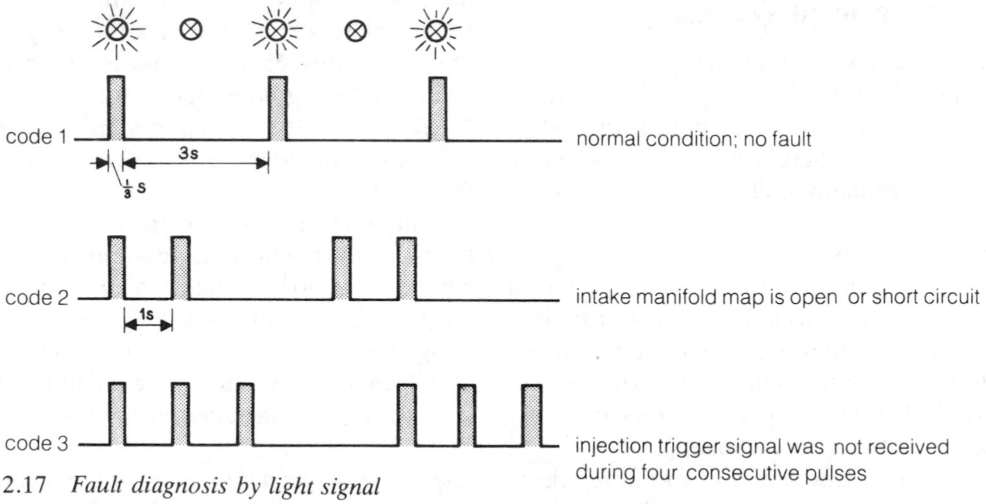

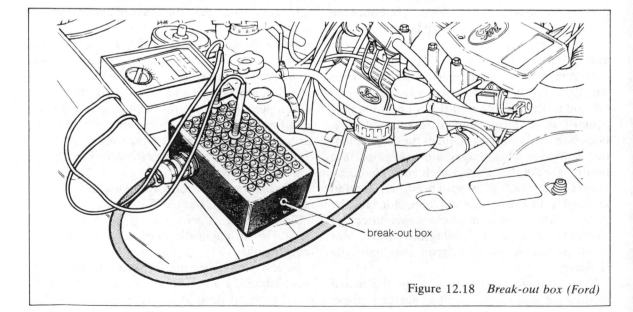

code 1 — normal condition; no fault

3s   ⅓ s

code 2 — intake manifold map is open or short circuit

1s

code 3 — injection trigger signal was not received during four consecutive pulses

Figure 12.17   *Fault diagnosis by light signal*

break-out box

Figure 12.18   *Break-out box (Ford)*

Breakout box tests A breakout box provides a series of contact plug sockets that enables test meters to be connected into the various circuits to allow fault diagnosis to be carried out.

Figure 12.18 shows a breakout box used on Ford vehicles. This box has 60 sockets and a provision for the connection of these sockets to the multi-plug which normally fits into the E.C.U.

Tests conducted with the aid of this box cover many sub-sections. Its use minimizes the prob-lems of connecting test equipment to the wrong pin and making ineffective connections to the test meters.

Each socket is numbered, so, when used in conjunction with a fault-diagnosis chart, the test-ing procedure is simplified. As with many other tests of electronic equipment, great care must be taken when using a multimeter to measure re-sistance. This is because the current supplied by the meter can damage many of the components in an electronic control unit.

13 Other electronic applications

13.1 Vehicle speed control

Holding the vehicle at a near-constant speed for a long period while cruising on a road such as a motorway is very tiring for the driver. This task can be made easier and less fatiguing if an electronic system is used to control the throttle opening to keep the vehicle speed constant even though the resistance acting against the vehicle may vary due to changes in the wind or road gradient.

Nowadays a *cruise control system* is fitted as original or optional equipment to a number of vehicles. Although various layouts are used, the functional role of the main components in each system is similar.

Basic system

The principle of a typical closed-loop cruise control system is shown by the block diagram in Figure 13.1. The brain of the system is the control unit; this has two inputs: the command speed signal, set by the driver to indicate the desired speed of the vehicle, and the feedback signal, which signifies the actual speed of the vehicle.

When the feedback signal shows that the vehicle speed is different to the command speed, the control unit transmits a signal to the actuator. This adjusts the throttle of the fuel system in a direction which quickly restores the vehicle's speed to that specified by the driver.

On all vehicles there is a delay between the opening of the throttle and the alteration in vehicle speed. A cruise control system incorporates in its circuitry a feature to take this into account. This reduces the severe oscillation in speed which would otherwise occur when the system attempts to equalize the actual speed with the command speed.

Special attention to safety is needed to ensure that the driver can quickly and automatically over-ride the cruise system when the vehicle has to be braked. This is achieved by fitting a switch on the brake pedal that cuts-off the electrical supply to the control system.

A similar switch is fitted to the clutch pedal of vehicles fitted with manual transmission systems to prevent over-revving of the engine when a gear is changed. This would occur due to the drop in

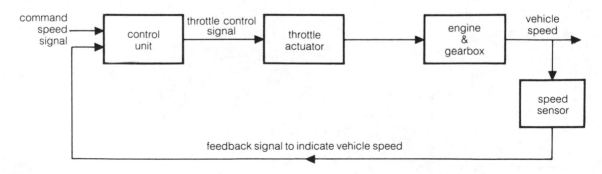

Figure 13.1 *Cruise control system*

road speed and the response of the cruise control system during the time that a gear change is taking place.

Main components
The main parts of a basic system are shown in Figure 13.2.

Driver's command switch Normally the command switch is a stalk-type switch mounted on the steering column and positioned so that it is in easy reach of the driver's hand. Most switches have three positions such as; 'activate/set', 'off' and 'reactivate'.

In the first position the vehicle can be accelerated as long as the button is depressed. When the desired speed is reached, the button is released; this 'sets' the cruise control unit so that it holds the vehicle to the speed that existed at the time the button was released.

The 'reactivate' or 'resume' position is used to reset the system after it has been switched-off by brake or clutch pedal operation.

Actuator Movement of the throttle valve is obtained by means of an electric or pneumatic servo motor. This unit is capable of moving the throttle smoothly and gradually in both directions.

The electric-type actuator is often a d.c. permanent-magnet motor so in this case the return motion is obtained by reversing the direction of current. This current is supplied in the form of pulses of short duration of the order of a few tenths of a second so this enables the throttle to be moved through small angles in a smooth precise manner.

Pneumatic actuators are similar in construction to the vacuum unit used in a distributor to control ignition timing. 'Vacuum' for the operation of the actuator can be provided by the engine-induction manifold but greater precision is obtained when the 'vacuum' is developed by an electrically-driven pump; this ensures that the pressure is constant. Return motion of the servo diaphragm is given by a spring.

Brake and clutch switches In addition to the electrical cut-out switches, the pneumatic system also incorporates brake and clutch vent valves. These valves allow the 'vacuum' in the servo actuator to be dumped to make the system inoperative when either the brake or clutch pedal is moved.

Electrical servo-motor systems often have a disconnect-relay to perform a duty similar to the pneumatic dump valves. This relay switches-off the complete system and cuts-off the fuel pump

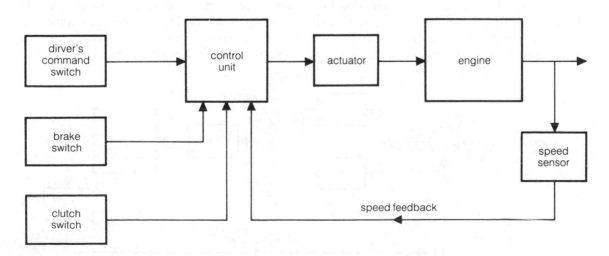

Figure 13.2 *Main components of cruise control system*

to overcome the problem that would arise if the servo motor failed to return the throttle due to seizure of the motor. In the event of this type of malfunction, the accelerator pedal is fitted with an extra switch, called a *drag switch*; this switch reactivates the pump when the driver depresses the throttle.

Speed sensor   The speed sensor is normally connected to the speedometer drive. In cases where the speedometer is electronic, the signal from the speedometer transducer is also used to provide the feedback signal to the E.C.U.

An alternative type of sensor uses permanent magnets glued to the drive shaft to generate an a.c. pulse in a pick-up coil situated adjacent to the drive shaft.

Whereas some speed transducers generate an analogue signal voltage which rises proportional to speed, other systems use a digital signal. In the latter case an increase in speed produces a larger number of pulses in a given time.

Control unit   This unit is the brain of the system; it provides an output signal to the actuator to adjust the throttle whenever it detects that the actual speed of the vehicle is above or below that set by the driver.

The closed-loop controller fitted in the control unit compares the system's output measured by the sensor to that indicated as an input command. Difference between the two signals, obtained by simple subtraction, is called an error signal; this is used to adjust the control signal so that it is proportional to the error signal.

A *proportional control system* by itself lacks sensitivity since a comparatively large error signal is required to make final adjustments to the speed. To overcome this problem, vehicle speed controllers generally use a *proportional-integral system* as shown in Figure 13.3. In this diagram it will be seen that the command signal is the sum of the outputs from the two blocks A and B. The circuit of the proportional block A gives an output that is in proportion to the error signal, whereas the integral block B produces a variable-rate output designed to reduce the error to zero in a short time without oscillation.

The cruise-control system can be designed for analogue or digital operation. The analogue system uses a capacitor charge set by the driver by the switch operation to represent vehicle speed. Detection of the charge on the capacitor is achieved by a very high input impedance amplifier; this outputs a voltage to an error amplifier to represent the command speed.

A digital circuit stores the input speed command and the vehicle speed as numbers. Feedback pulses from the sensor are counted against a clock signal and these are subtracted from the command speed pulse to obtain the error number. This is then passed through the two paths previously outlined. Alternatively, the error signal can be fed to a microcomputer for processing, so with this method of control a more sophisticated programme can be used.

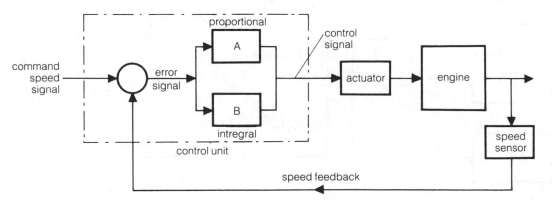

Figure 13.3   *Proportional-integral cruise control system*

Data in a digital system is stored in the form of digital codes so when this is compared with an analogue controller, the digital system is less affected by changes in temperature and humidity.

Fault diagnosis

Electronic operation of this system is similar to many other sensor–E.C.U.–actuator systems so fault diagnosis is carried out in a similar manner. A dedicated test analyser is supplied for some makes of cruise-control system; this test set enables the main units of the system to be accurately checked in short time.

13.2 Anti-skid brake system

The majority of drivers cannot judge the point at which a wheel starts to skid because far too many factors are involved. This problem, combined with the driver's over-reaction to an emergency, normally results in skidding of the vehicle. When the wheels are skidding the accident risk is high; this is because the driver experiences loss of directional control due to yawing of the vehicle, ineffective steering and a longer stopping distance. Statistics show that over 10% of accidents are due to the brakes locking-up.

An anti-lock or anti-skid braking system, often abbreviated to A.B.S., overcomes these problems; this makes the vehicle much safer to drive especially when tyre-to-road adhesion is poor.

Basic principles

Anti-skid systems are used on cars and heavy vehicles. Articulated-type heavy vehicles benefit considerably from A.B.S. because these vehicles are prone to jack-knifing (rear wheels of tractor locking) and trailer swing (trailer wheels locking). Cars that tow caravans are also subject to jack-knifing.

Both the car and heavy vehicle requires a similar electronic control layout (Figure 13.4). This has a sensing system to measure the wheel speed so that a control unit can determine when the wheel is at the point of skidding. When this critical point is reached, the control unit operates a valve in the brake system to release the pressure on the brakes. Having averted the initial skid, the E.C.U. then opens a brake valve to increase the pressure until the skidding point is reached once again. This cycle, involving a pressure release–apply–release–apply, recurs at all times that the brakes are fully applied. By repeating the cycle between 4 and 10 times per second, maximum

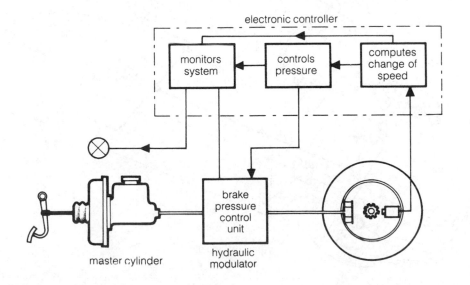

Figure 13.4 *ABS control system (closed loop)*

braking can be achieved even though the road conditions may vary.

A main difference between the light and heavy vehicles is the type of brake system to which the A.B.S. layout is added. Heavy vehicles use compressed air to operate the brake system, whereas a light vehicle has a hydraulic brake system. This means that although the electronic control system is similar, the method of control of the actual brake system is different.

Hydraulic brake A.B.S. components

The layout of a typical A.B.S. unit for a rear-wheel-drive car is shown in Figure 13.5.

Wheel-speed sensors   Normally three wheel sensors are fitted: one to each non-driving wheel and one to the propeller shaft flange on the axle. An inductive-type sensor is used which gives a voltage impulse-type signal as the tooth on the rotating reluctor ring passes the sensing head.

Electronic controller   This E.C.U. evaluates the signals transmitted from the sensor. After processing, it determines whether the brake pressure needs to be decreased, maintained constant or increased. The computer in this module is programmed to recognize the rate of change of wheel speed which immediately precedes the lock-up point. At this instant, it gives an output signal to lower the brake pressure slightly.

Hydraulic modulator   This unit is the actuator part of the control circuit and its duty is to vary the pressure in the fluid lines in accordance with the electrical signals it receives from the E.C.U.

Fluid passage to, or from, the brake lines is controlled by solenoid-operated valves; one valve is needed for each brake circuit. Fluid released from an individual line flows to an accumulator. This is a chamber fitted with a spring-loaded piston which temporarily stores hydraulic pressure so that it can be re-supplied to

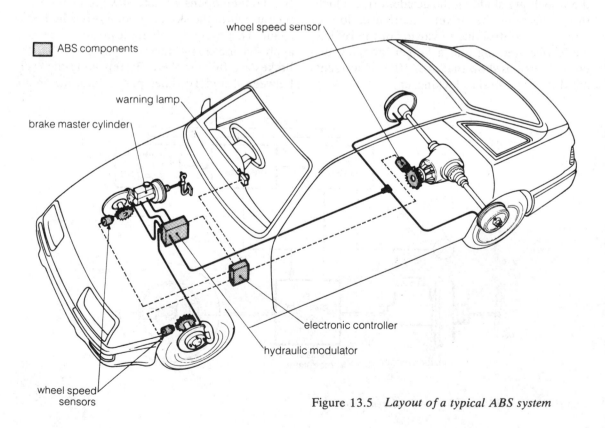

Figure 13.5   *Layout of a typical ABS system*

the brake line if the E.C.U. signals that extra fluid pressure is needed. Return of fluid to the lines via the accumulator is achieved by a pump.

The solenoid valves have three working positions:

1. Open to the reservoir to release pressure.
2. Open to the pump to increase pressure.
3. Closed to allow either pressure to be maintained during one phase of the anti-lock cycle or normal operation of the brakes by the master cylinder.

E.C.U. operation

The digital controller consists of an input amplifier, computing unit, power stage and monitoring unit.

Input amplifier   This unit filters and reshapes the a.c. waves from the sensors into rectangular pulses suitable for operating the computer.

Computing unit   A computer with its microprocessor and memory circuit reads the signals sent by the sensors. After comparing the signal with a reference signal obtained from the wheel diagonally opposite the sensor, it is able to detect the approach of a skid by computing the rate of change in the wheel speed (Figure 13.6).

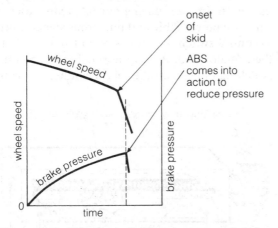

Figure 13.6   *Wheel speed change at onset of skid*

Power stage   This provides an output signal to operate the solenoid valves. Digital pulses from the computer allow the power stage to deliver a regulated current to the appropriate solenoid; this moves the valve in the desired direction.

Monitoring unit   Every time the vehicle is started or stopped the circuit is monitored. If a fault is present, a warning lamp on the instrument panel is illuminated. A defect discovered by the monitoring circuit when the vehicle is in use causes the unit to activate the warning lamp; for safety reasons it also switches-off the A.B.S. system. In this state, the vehicle can still be driven but the driver loses the anti-lock feature.

13.3 Other electronic developments

The automatic electronic field is rapidly changing and many exciting applications have either recently been introduced or are in the process of development. A few of these applications are surveyed in this chapter.

Electrical seat adjustment

A number of vehicles in the upper-price range are fitted with an electronic control system for adjustment of the seat position. In addition to the motorized drives to alter the longitudinal, height and backrest positions, many systems have an electronic control unit. This incorporates a memory unit to allow individual seating positions to be recalled so that the seat and driving mirrors can be readjusted at a press of a button.

Figure 13.7 (overleaf) shows the layout of a typical system. This uses four electric motors to adjust the settings of the seat and a separate memory unit to store four seating positions.

Rotary motion given by the motor is transmitted by shafts, flexible and rigid, to gearboxes sited in suitable positions to give the required seat movements. These gearboxes often have a worm and wheel drive, which acts directly on to the seat frame to control front and rear height adjustment, or uses the worm wheel to drive a rack and pinion gear to provide the longitudinal adjustment.

The memory feature uses four potentiometers to sense the seat position; these signal the settings of the four adjusters. Figure 13.8 shows one of

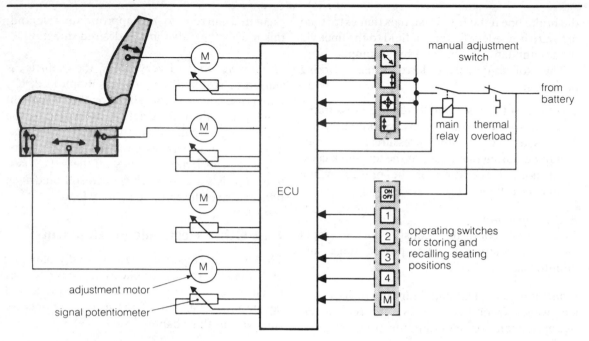

Figure 13.7　*Electrical seat adjustment*

these potentiometers. It consists of a threaded spindle that drives a slider along a resistive surface; its operation is similar to a conventional potentiometer (see page 33). The voltage signal transmitted to the E.C.U. depends on the position of the slider, so when the seat position is set, the driver operates the memory button; this commands the E.C.U. to store the voltage signal for future reference when it has to reset the seat position.

Multiplexing
The number of electrical units and the need for monitoring systems have increased, so this has meant that the number of cables has increased to carry these signals and feeds. Although these cables are tied together to form a compact harness, the bulk, and weight of the looms and connectors makes accommodation difficult, especially in the driver's control area. In the past, areas around the instrument panel have caused problems, but this has been relieved partially by the use of printed circuit boards. In addition to the bulk problem, the grouping together of sup-

ply cables often causes the cables at the centre of the loom to overheat; as a result of the increase in resistance, the efficiency of the system is lowered.

One method of overcoming the problem is to use a *remote switching* system. This system, which has been in use during recent years, uses a common power cable and numerous signal cables to control switching devices mounted adjacent to the actuator unit. A relay normally performs the switching duty so all the relays can use the same

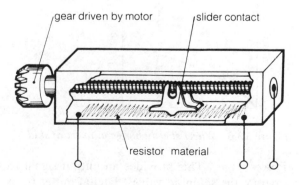

Figure 13.8　*Potentiometer for sensing seat position*

power supply. The signal cable to each relay only carries a small current so the cables can be smaller than those originally used. This type of system is often used for electric windows, heated rear windows and other systems that consume high power.

A modern alternative is a *multiplex control system*. This uses a 'ring main' around the vehicle to which are connected the lamps, actuators and motors that make up the electrical equipment of the vehicle. Switching of these units is obtained by transmitting coded electrical or optical signals along a second cable, called a *bus*. This system is similar to that used in a computer (see page 259).

Adjacent to each consumer unit is fitted a *decoder*; this device is able to recognize when its own special digital code is transmitted along the bus. On receipt of the message that follows the initial call-up code, the decoder operates a relay which activates the unit according to the command.

A single data bus is used to carry messages for a number of units, so a time allocation system is utilized to ensure that each unit has exclusive use of the bus during its time slot. A separate time slot is provided for each unit; the process for dividing the time is called *time division multiplexing* (MUX).

Figure 13.9 shows the principle of multiplexing as applied to some of the electrical units fitted at the rear of a vehicle.

When the driver closes the switch to operate a particular unit, an *encoder unit* controlled by a microprocessor, sends a series of binary voltage pulses along the bus. The code transmitted consists of two signals: an initial code in the form of an address to enable the decoder to recognize its 'call sign' and a command to instruct the unit to perform a set function.

Suppose the driver wishes to heat the rear window. After closing the switch on the control panel, the microprocessor transmits a coded signal through the bus during its time slot and the decoder responds by energizing the relay controlling the current flow to the heater element. The coded signal is repeated many times a second until the driver switches-off the heater. This causes a change in the command signal with the

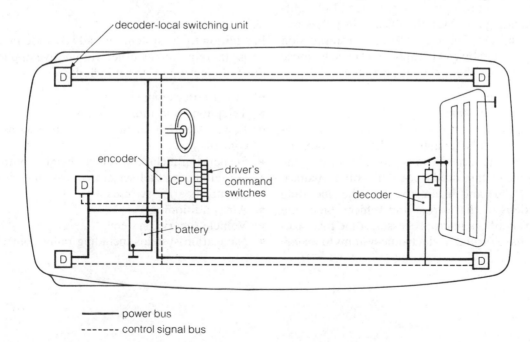

Figure 13.9 *Multiplex control system*

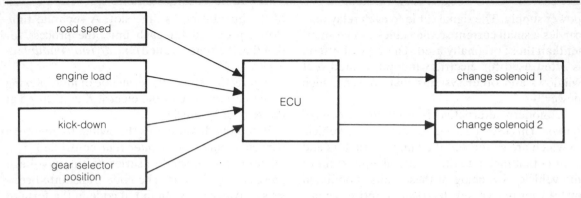

Figure 13.10   *Transmission control system*

result that the decoder de-energizes the relay and opens the heater circuit.

The signal current needed to operate this system is very low, so good circuit connections are needed. Furthermore, precautions must be taken to reduce 'noise' in the line, since this can alter the digital pulse signals in a manner that affects the operation of the system.

Multiplex systems that use fibre optics to carry the signal data are being developed for automotive use. A fibre optic is a thin strand of light-conducting glass through which light pulses can be transmitted. This method of data transmission is not affected by electrical fields especially those radiated from the h.t. ignition system.

Transmission control

Most automatic transmission systems use hydraulic controls to activate the clutches/brake bands to obtain the various gears. In the past the hydraulic system used a collection of mechanical cables; rods and levers to sense the operating conditions of the engine and vehicle. Since the invention of the microprocessor, some manufacturers have turned to electronic systems to under-

take the sensing role because electronics are more sensitive in this area.

A basic control system is shown in Figure 13.10. Data from the input sensors is passed to the computer which is programmed to change the gear when the input signals dictate that a change is necessary.

Often the actuators are solenoid-controlled ball-valves which govern the oil pressure acting on the change valves.

Other developments

In addition to the systems already covered in this book, there are many other automotive applications of electronics; these include:

- In-car entertainment
- Telephone and radio systems
- Power steering including speed-responsive controls
- Suspension including vehicle height control
- Diesel engine fuel-system control and cylinder pre-heating systems
- Air conditioning
- Vehicle security systems
- Navigation systems including radar control.

Index